The Oxford Book of Sea Stories

The Oxford Book of
Sea Stories

~~~

Selected by
TONY TANNER

Oxford New York
OXFORD UNIVERSITY PRESS
1994

Oxford University Press, Walton Street, Oxford OX2 6DP

Oxford New York Toronto
Delhi Bombay Calcutta Madras Karachi
Kuala Lumpur Singapore Hong Kong Tokyo
Nairobi Dar es Salaam Cape Town
Melbourne Auckland Madrid
and associated companies in
Berlin Ibadan

Oxford is a trade mark of Oxford University Press

British Library Cataloguing in Publication Data
Data available

Library of Congress Cataloging-in-Publication Data
The Oxford book of sea stories / selected by Tony Tanner.
p. cm.
1. Sea stories, English. 2. Sea stories, American.
I. Tanner, Tony.
PR1309.S4095 1994 823'.010832162—dc20 93–23093
ISBN 0–19–214210–0

10 9 8 7 6 5 4 3 2 1

Printed on acid-free paper in
the United States of America

I would like to dedicate this anthology to my travelling step-daughter, Barbara: 'daylight and champain discover not more.'

Contents

Acknowledgements

I would like to acknowledge the generous help afforded me by Professor Robert Foulke in the preparation of this anthology. Professor Foulke knows more about fiction concerning the sea than anyone I have encountered in person or print, and he magnanimously allowed me access to a number of his bibliographies which duly mobilized me in a number of rewarding directions. I would also like to thank my editor at OUP, Michael Cox, for invaluable help selflessly provided.

Introduction

Navigare necesse est, vivere non necesse.

(It is necessary to sail the seas,
it is not necessary to live.)
Motto of the Hanseatic League

I MAKE no apologies for starting this introduction with a long quotation from a little-known work by John Ruskin—*The Harbours of England.*

Of all things, living or lifeless, upon this strange earth, there is but one which, having reached the mid-term of appointed human endurance on it, I still regard with unmitigated amazement . . . one object there is still, which I never pass without the renewed wonder of childhood, and that is the bow of a Boat . . . the boat's bow is naïvely perfect: complete without an effort. The man who made it knew not that he was making anything beautiful, as he bent its planks into those mysterious, ever-changing curves. It grows under his hand into the image of a sea-shell; the seal, as it were, of the flowing of the great tides and streams of ocean stamped on its delicate rounding. He leaves it when all is done, without a boast. It is simple work, but it will keep out water. And every plank thenceforward is a Fate, and has men's lives wreathed in the knots of it, as the cloth-yard shaft had their deaths in its plumes.

Then, also, it is wonderful on account of the greatness of the thing accomplished . . . in that bow of the boat is the gift of another world. Without it, what prison wall would be so strong as that 'white and wailing fringe' of sea. What maimed creatures were we all, chained to our rocks, Andromeda-like, or wandering by the endless shores, wasting out incommunicable strength, and pining in hopeless watch of unconquerable wave? The nails that fasten together the planks of the boat's bow are the rivets of the fellowship of the world. Their iron does more than draw lightning out of heaven, it leads love round the earth.

Then also, it is wonderful on account of the greatness of the enemy that it does battle with. To lift dead weight; to overcome length of languid space; to multiply or systematise a given force; this we may see done by the bar, or beam, or wheel, without wonder. But to war with that living fury of waters, to bare its breast, moment after moment, against the unwearied enmity of ocean—the subtle, fitful, implacable smiting of the black waves, provoking each other on, endlessly, all the infinite march of the Atlantic rolling on behind them to their help—and still to strike them back into a wreath of smoke and futile foam, and win its way against them, and keep its charge of life from them; does any other soulless thing do as much as this?

No one has described better the miracle of a boat at sea than Ruskin. As he says, 'that rude simplicity of bent plank, that can breast its way through the death that is in the deep sea, has in it the soul of shipping. Beyond this, we may have more work, more men, more money; we cannot have more miracle'. Fiction about the sea necessarily has much to do with the infinitude of the ocean, the 'living fury of waters', the 'unwearied enmity' of the endlessly, endlessly, rolling black waves—and 'the death that is in the deep sea'. (Most of the stories I have included involve death, or a perilously and invariably hair-raisingly close avoiding of it.) But this fiction also brings out the curiously unique 'naïve perfection' of a ship—epitomized for Ruskin in the beauty, mystery, and wonder of the bow. Quite simply, it released men from land-imprisonment—without it, he makes you feel, world history would hardly have begun. It keeps out water, it cuts through and strikes back the remorselessly threatening waves, it sustains men's lives as they, impossibly, 'walk' on the unsustaining waters. 'Does any soulless thing do as much as this?' 'The glory of a boat is, first, its steadiness of poise—its assured standing on the clear softness of the abyss; and, after that, so much capacity of progress by oar or sail [we would have to add—or steam] as shall be consistent with this defiance of the treachery of the sea.' We cannot have more miracle.

But also, the ship creates a special 'fellowship' among and between all who sail in her. On land, people have an endless variety of commitments, relationships, obligations, loyalties, and so on; at sea, everyone—it has entered into common parlance—is 'in the same boat'. And that boat is a lifeboat at all times breasting and seeking to avoid the death-by-water. The intensity of the 'fellowship' which thus ensues, or is thereby created (and the possible strains within it), is thus considerably heightened, and sharpened, and *tested*, as these stories variously reveal and explore. Once a writer has set his ship to sea, he could indeed say with Ruskin—'every plank is thenceforward a fate'.

In discussing his principles of selection for his admirable *Oxford Book of the Sea*, Jonathan Raban says he wanted to compile 'a book of the *sea*—not a book of voyages, naval battles, shipboard life, fishing, or any of the other activities that take place in, or on, or at the edge of, the sea . . . the focus is squarely on the water itself'. This nicely brings out the difference between his collection and mine (I hope it makes for a complementarity), since my stories are preoccupied with precisely such activities—in, on, or at the edge of, the water itself. My selection effectively starts with an elderly seaman exclaiming: 'ships are all right; it's the men in 'em . . .' and the stories that follow are, exactly, about the men in the ships.

The *men* in them. This points to a glaringly obvious fact about sea stories. All the ones in this collection are by men and, to all intents and purposes, exclusively concern men (there are a few exceptions in the stories by Henry James, Bret Harte, F. Scott Fitzgerald, E. M. Forster, and Malcolm Lowry). There are plenty of emotional states and temperaments and reactions on display—bravery, cowardice, panic, terror, incredulity, boredom, aggression, humour, resolve, endurance, madness—but there is, affectively, effectually, no sexual desire. (No trace of homosexuality seems ever to be found in this genre—male-bonding, comradeship, crewmanship of course; but nothing sexual. Though it has to be said that there are some distinctly homoerotic passages in *Moby Dick*.) In the absence of women, desire seems some-how to be forgotten, repressed, occluded, or otherwise left behind or put aside. For men grappling with the sea and with boats to manage, this is doubtless as it should be. And there certainly are other things in life besides sex. But this absence, omission, excision—call it what we will—has implications when it comes to telling stories. For there is a strong case to be made for saying that narration is intimately involved with desire. Homer's epics effectively start with the abduction of Helen. Let me make my point somewhat digressively by quoting from a story by Henry James called, indeed, 'The Story in it'. Basically it is simply a conversation between three people, a man and two women—one adulterous, one chaste (or as the story has it—'honest'). The subject under discussion is whether you can have a narrative—a romance, a drama—without a 'bad' woman. The man—Voyt—main-tains that you cannot, and I will run together some of his arguments.

The honest lady hasn't—can't possibly have adventures. . . . Behind these words we use—the adventure, the novel, the drama, the romance, the situation, in short as we comprehensively say—behind them all stands the same sharp fact that they all, in their different ways, represent. . . . The fact of a relation. The adventure's a relation; the relation's an adventure. The romance, the novel, the drama are the picture of one. . . . If a relation stops, where's the story? If it doesn't stop, where's the innocence? It seems to me you must choose. It would be very pretty if it were otherwise, but that's how we flounder. Art is our flounderings shown.

These, of course, are metaphorical 'flounderings'—the confusions we can get into on account of the ill-controlled circuitings of desire. But it is precisely this, says James (or his character), that makes for narrative. There must be a 'relationship'. We flounder, and there is something to tell. (It is interesting that James's first story—here included—is very involved with the sea. Of course, he is hardly to be considered as a writer of sea stories, but it is worth noting that he uses sea imagery—sinking, drowning, floundering, shipwreck, lost moorings, faulty navigation, insecure anchorage, etc.—more than any other major novelist. From *this* point of view he is indeed one of the great writers of 'sea' stories, but that is not our concern here.) Now where does that leave sea stories? They, of course, are habitually about *literal* flounderings—one way or another. Ships encounter storms, or find themselves becalmed; icebergs are hit, propellers fall off; *men go wrong*, mutinies occur, pirates may board—perhaps, even, monsters may rise from the deep. There *are* relationships—of fear, combat, survival (and in 'The Secret Sharer' Conrad charts one of the most extraordinary relationships between men at sea ever imagined and analysed). But there is, as it were, no occasion for desire. Art is our flounderings shown—certainly. But these men are in real water, and not a woman in sight—or indeed, perhaps, in mind.

No women, no desire—this lack puts a constraint on sea narrative. But there certainly still *is* a 'story in it', as I trust this collection sufficiently shows. I have to say that I was surprised during my researches to discover how much writing about the sea is either autobiographical or historical—how we finally sailed to there, or the voyage of the Something. It almost seems as if fiction is pre-empted, unnecessary. What you often find instead is a kind of moody, metaphysical brooding on the obvious analogue of voyage and life. Here, for instance, is Hilaire Beloc in his immensely popular *The Cruise of the Nona*.

Indeed, the cruising of a boat here and there is very much what happens to the soul of man in a larger way. We set out for places which we do not reach, or reach too late; and, on the way, there befall us all manner of things which we could never have awaited. We are granted great visions, we suffer intolerable tediums, we come to no end of the business, we are lonely out of sight of England, we make astonishing landfalls—and the whole rigmarole leads us along no whither, and yet is alive with discovery, emotion, adventure, peril, and repose.

This is unexceptionable and fair enough, but it is not narrative. Still, I am insisting, there *are* very remarkable sea *stories*.

'And then went down to the ship.' Thus, with a nod to Homer, Ezra Pound begins his own epic odyssey through world literature and the modern world. Literature and voyaging are arguably coeval, since there is no literature without a departure, a setting out, a setting forth; and very often that has meant a forsaking of land for the more uncertain element of the sea. A voyage is invariably a quest—whether the object in view is a golden fleece, a white whale, or simply the safe delivery of a cargo. Any voyage leads almost inevitably to adventures since the natural elements are so imponderable, and people sequestered and thrown together in a boat prove to be unusually volatile and unpredictable. 'Trust the high seas to bring out the irrational in man', says Conrad's Marlow, and it is a matter of constant surprise just what the sea can bring out in those who, for whatever reason, go down to it in ships.

Something should be said about the principle of selection for these stories. I open the volume with the chapter called 'Initiation' from Conrad's *The Mirror of the Sea*. It seemed an ideal way to 'initiate' the reader into the volume, as it were, and it manifests that sort of autobiographical narrative which is so characteristic of much writing about the sea. Apart from that, all the stories are self-contained and by English and American writers—with one exception. An anthology of sea stories without Melville would be *Hamlet* without the prince, and since he has long novellas but no short stories about the sea, I have included his account of the final three days of the chase of Moby Dick which concludes that novel, and which comprises one of the high points in American literature. (I have included the earlier, simpler story, 'Mocha Dick', since it was an important seed for Melville's rich harvesting—and it also reminds us what a vital industry whaling was for America in the first half of the nineteenth century.) Because I wanted to include self-contained short stories, there are some notable names omitted. In particular, those of Marryat and James Fenimore Cooper, who only wrote (to my knowledge) full-length novels about men and the sea. In a way, these two writers represent two major and distinct traditions in sea fiction. Marryat is in an older English tradition (with

Smollett as a predecessor); a retired naval officer, he settled back and
wrote simple, picaresque, and episodic stories about adventures at
sea—with sailors in general being inevitably sentimentalized. Cooper
was something different and, arguably, was the first novelist to raise
sea fiction to the level of high art. We tend to associate Cooper with
the Leatherstocking novels, almost mythically concerned with America's western frontier, with its forests, prairies, diminishing wilderness,
and vanishing Indians. But in his own time, he was often considered
as primarily a novelist of the sea. Indeed, one American critic, Thomas
Pilbrick, makes the claim that 'the sea novel as we know it owes its
inception to the meeting of maritime nationalism and romanticism in
the work of James Fenimore Cooper'. Pilbrick reminds us that before
1850 the American frontier was first and foremost maritime, and that
America itself was regarded as primarily a maritime power. As he
says, during the first half of the nineteenth century American ships
and seamen challenged or even displaced British hegemony of the
most important areas of maritime activity, and by the late 1830s the
United States dominated the passenger trade between New York and
England. (See Thomas Pilbrick, *James Fenimore Cooper and the De-
velopment of American Sea Fiction* (Harvard UP, 1961).) Cooper was
the novelist of maritime America and produced a new kind of sea
fiction. Conrad referred to Cooper as 'one of my masters' and 'my
constant companion'. It seems very appropriate that he wrote about
both Marryat and Cooper in an essay entitled, indeed, 'Tales of the
Sea'. He is impeccably courteous about both writers, but it is clear
that he is registering a crucial difference between them. Marryat is

the enslaver of youth, not by the literary artifices of presentation, but by the
natural glamour of his own temperament. . . . To the artist his work is interesting as a completely successful expression of an artistic nature. . . . To this
writer of the sea the sea was not an element. It was a stage, where was displayed
an exhibition of valour, and of such achievement as the world had never seen
before. . . . His novels, like amphibious creatures, live on the sea and frequent
the shore, where they flounder deplorably. . . . His naïveties are perpetrated in
a lurid light. . . . His adventures are enthralling; the rapidity of his action
fascinates; his method is crude, his sentimentality, obviously incidental, is
often factitious. His greatness is undeniable.

That last tip of the cap to Marryat is a generosity on Conrad's part,
perhaps a gesture of solidarity from a fellow seaman. But he has made
it clear that he cannot regard Marryat as an artist.

But

at the same time, on the other side of the Atlantic, another man wrote of the sea with true artistic instinct. . . . For James Fenimore Cooper nature was not the framework, it was an essential part of existence . . . he loved the sea and looked at it with consummate understanding. In his sea tales the sea interpenetrates with life; it is in a subtle way a factor in the problem of existence, and, for all its greatness, it is always in touch with the men, who, bound on errands of war or gain, traverse its immense solitudes. . . . He knows the men and he knows the sea. His method may be often faulty, but his art is genuine. The truth is within him. The road to legitimate realism is through poetical feeling, and he possesses that . . .

Conrad's discrimination and insight stand—sea fiction became an art with Cooper.

In view of this, it is no accident that—after Conrad's 'Initiation'—both the first and next to last stories in this collection are by American writers. I do not think it is too much to say that it was American writers who effectively invented, and developed, the sea story as a separate and recognizable genre. Given the importance of the sea and navigation to island Britain, it is rather surprising how few novelists of the sea have emerged here. Pilbrick maintains that Sir Walter Scott in *The Pirate* (1822) was the only British novelist before Conrad to attempt an extensive use of nautical materials. Conrad is, without question, the greatest of all writers of sea fiction—and, of course, he is not, by origin, British. Shortly after *The Pirate*, Cooper wrote *The Pilot* (1823)—a work in which, unprecedently, the ocean is the main setting and seamen the principal characters—and sea fiction is set to take off in a way it never did in Britain.

I have discussed Marryat and Cooper in this way because, although they are not actually represented in this volume, they are the progenitors. Of course, subsequent writers wrote very differently of the sea and the men on it—more darkly, more frighteningly, more metaphysically, more psychologically, more pessimistically, more fantastically—as the stories by Poe, Hopkinson, Melville, Harte, Crane, London, Kipling, Wells, and Forster included here reveal. But it was Marryat and, above all, Cooper who launched the genre. And because I do think that the short story about the sea is, initially, a distinctively American genre, I want to conclude by saying a few words about the second story in the collection—Washington Irving's 'The Voyage'. It might hardly seem to qualify as a short story—more, perhaps, as a slight autobiographical sketch. Yet, published in 1819, there had been nothing like it, and it was to act as an influence on

the American sea story for many years to come. There is first of all
the sense that for an American to sail to Europe is, in itself, an
adventure. But more, there is a realization of the effect the sea voyage
can have on the emotions and imagination. The story is *there*—

a wide sea voyage severs us at once—It makes us conscious of being cast loose
from the secure anchorage of settled life and sent adrift upon a doubtful world.
It interposes a gulph, not merely imaginary, but real, between us and our
homes—a gulph subject to tempest and fear and uncertainty, rendering distance palpable and return precarious.

The sea is not incidental—a 'stage' in Conrad's word; it actively works
on the mind of the voyager.

To one given to day dreaming and fond of losing himself in reveries, a sea
voyage is full of subjects for meditation: but then they are the wonders of the
deep and of the air, and rather tend to abstract the mind from worldly themes.
I delighted to loll over the quarter railing or climb to the main top of a calm
day, and muse for hours together, on the tranquil bosom of a summer's sea. . . .
There was a delicious sensation of mingled security and awe with which I
looked down from my giddy height on the monsters of the deep at their
uncouth gambols. . . . My imagination would conjure up all that I had heard
or read of the watery world beneath me.

We are already no very great distance here from Ishmael's famous
mast-head reveries on the *Pequod*—this is a kind of imaginative in-
teraction with the awesome totality of the sea which we do not find
in British writing before Conrad.

   And as well as leaving, there is arriving—which can, in itself, be a
desolating experience. 'I alone was solitary and idle. I had no friend
to meet, no cheering to receive. I stepped upon the land of my
forefathers—but felt I was a stranger in the land.' Irving writes as a
tourist, but sea stories, of course, can as often involve passengers as
seamen. And in this seemingly slight piece, Irving has unerringly
traced a basic parabola of sea writing. In Conrad's words (from *The
Mirror of the Sea*)—'Landfall and Departure mark the rhythmical
swing of a seaman's life and of a ship's career. From land to land is
the most concise definition of a ship's earthly fate.' And on that ship,
'there is a Fate in every plank'. It is not necessary to live: it is necessary
to sail the seas.

Initiation

〜〜〜

JOSEPH CONRAD

'SHIPS!' exclaimed an elderly seaman in clean shore togs. 'Ships'—and his keen glance, turning away from my face, ran along the vista of magnificent figure-heads that in the late seventies used to overhang in a serried rank the muddy pavement by the side of the New South Dock—'ships are all right; it's the men in 'em. . . .'

Fifty hulls, at least, moulded on lines of beauty and speed—hulls of wood, of iron, expressing in their forms the highest achievement of modern ship-building—lay moored all in a row, stem to quay, as if assembled there for an exhibition, not of a great industry, but of a great art. Their colours were grey, black, dark green, with a narrow strip of yellow moulding defining their sheer, or with a row of painted ports decking in warlike decoration their robust flanks of cargo-carriers that would know no triumph but of speed in carrying a burden, no glory other than of a long service, no victory but that of an endless, obscure contest with the sea. The great empty hulls with swept holds, just out of dry-dock with their paint glistening freshly, sat high-sided with ponderous dignity alongside the wooden jetties, looking more like unmovable buildings than things meant to go afloat; others, half loaded, far on the way to recover the true sea-physiognomy of a ship brought down to her load-line, looked more accessible. Their less steeply slanting gangways seemed to invite the strolling sailors in search of a berth to walk on board and try 'for a chance' with the chief mate, the guardian of a ship's efficiency. As if anxious to remain unperceived amongst their overtopping sisters, two or three 'finished' ships floated low, with an air of straining at the leash of their level headfasts, exposing to view their cleared decks and covered hatches, prepared to drop stern first out of the labouring ranks, displaying the true comeliness of form which only her proper sea-trim gives to a ship. And for a good quarter of a mile, from the dockyard-gate to the furthest corner, where the old housed-in hulk, the *President* (drill-ship, then, of the Naval Reserve), used to lie with her frigate side rubbing against the stone of the quay, above all these hulls, ready and unready, a hundred and fifty lofty masts, more or less, held out the web of their rigging like an immense net, in whose close mesh, black against the sky, the heavy yards seemed to be entangled and suspended.

It was a sight. The humblest craft that floats makes its appeal to a seaman by the faithfulness of her life; and this was the place where one beheld the aristocracy of ships. It was a noble gathering of the fairest and swiftest, each bearing at the bow the carved emblem of her name as in a gallery of plaster-casts, figures of women with mural crowns, women with flowing robes, with gold fillets on their hair or blue scarves round their waists, stretching out rounded arms as if to point the way; heads of men helmeted or bare; full lengths of warriors, of kings, of statesmen, of lords and princesses, all white from top to toe; with here and there a dusky turbaned figure, bedizened in many colours of some Eastern sultan or hero, all inclined forward under the slant of mighty bowsprits as if eager to begin another run of 11,000 miles in their leaning attitudes. These were the fine figure-heads of the finest ships afloat. But why, unless for the love of the life those effigies shared with us in their wandering impassivity, should one try to reproduce in words an impression of whose fidelity there can be no critic and no judge, since such an exhibition of the art of ship-building and the art of figure-head carving as was seen from year's end to year's end in the open-air gallery of the New South Dock no man's eye shall behold again? All that patient pale company of queens and princesses, of kings and warriors, of allegorical women, of hero-ines and statesmen and heathen gods, crowned, helmeted, bare-headed, has run for good off the sea stretching to the last above the tumbling foam their fair, rounded arms; holding out their spears, swords, shields, tridents in the same unwearied, striving forward pose. And nothing remains but lingering perhaps in the memory of a few men, the sound of their names, vanished a long time ago from the first page of the great London dailies; from big posters in railway stations and the doors of shipping offices; from the minds of sailors, dock-masters, pilots, and tugmen; from the hail of gruff voices and the flutter of signal flags exchanged between ships closing upon each other and drawing apart in the open immensity of the sea.

The elderly, respectable seaman, withdrawing his gaze from that multitude of spars, gave me a glance to make sure of our fellowship in the craft and mystery of the sea. We had met casually, and had got into contact as I had stopped near him, my attention being caught by the same peculiarity he was looking at in the rigging of an obviously new ship, a ship with her reputation all to make yet in the talk of the seamen who were to share their life with her. Her name was already on their lips. I had heard it uttered between two thick, red-necked fellows of the semi-nautical type at the Fenchurch Street railway station, where, in those days, the everyday male crowd was attired in jerseys and pilot-cloth mostly, and had the air of being more conversant with the times of high-water than with the times of the trains. I had noticed that new ship's name on the first page of my morning paper. I had stared at the unfamiliar grouping of its letters, blue on white ground, on the advertisement-boards, whenever the train came to a standstill alongside one of the shabby, wooden, wharf-like platforms of the dock railway line. She had been named, with proper observances, on the day she came off the stocks, no doubt, but she was very far yet from 'having a name'. Untried, ignorant of the ways of the sea, she had been thrust amongst that renowned company of ships to load for her maiden voyage. There was nothing to vouch for her soundness and the worth of her character, but the reputation of the building yard whence she was launched headlong into the world of waters. She looked modest to me. I imagined her diffident, lying very quiet, with her side nestling shyly against the wharf to which she was made fast with very new lines, intimidated by the company of her tried and experienced sisters already familiar with all the violences of the ocean and the exacting love of men. They had had more long voyages to make their names in than she had known weeks of carefully tended life, for a new ship receives as much attention as if she were a young bride. Even crabbed old dock-masters look at her with benevolent eyes. In her shyness at the threshold of a laborious and uncertain life, where so much is expected of a ship, she could not have been better heartened and comforted, had she only been able to hear and understand, than by the tone of deep conviction in which my elderly, respectable seaman repeated the first part of his saying, 'Ships are all right . . .'

His civility prevented him from repeating the other, the bitter part. It had occurred to him that it was perhaps indelicate to insist. He had recognized in me a ship's officer, very possibly looking for a berth like himself, and so far a comrade, but still a man belonging to that sparsely peopled after-end of a ship, where a great part of her reputation as a 'good ship', in seaman's parlance, is made or marred.

'Can you say that of all ships without exception?' I asked, being in an idle mood, because, if an obvious ship's officer, I was not, as a matter of fact, down at the docks to 'look for a berth', an occupation as engrossing as gambling, and as little favourable to the free exchange of ideas, besides being destructive of the kindly temper needed for casual intercourse with one's fellow creatures.

'You can always put up with 'em,' opined the respectable seaman, judicially.

He was not averse from talking, either. If he had come down to the dock to look for a berth, he did not seem oppressed by anxiety as to his chances. He had the serenity of a man whose estimable character is fortunately expressed by his personal appearance in an unobtrusive, yet convincing, manner which no chief officer in want of hands could resist. And, true enough, I learned presently that the mate of the *Hyperion* had 'taken down' his name for quartermaster. 'We sign on Friday, and join next day for the morning tide', he remarked, in a deliberate, careless tone, which contrasted strongly with his evident readiness to stand there yarning for an hour or so with an utter stranger.

'*Hyperion*,' I said. 'I don't remember ever seeing that ship anywhere. What sort of a name has she got?'

It appeared from his discursive answer that she had not much of a name one way or another. She was not very fast. It took no fool, though, to steer her straight, he believed. Some years ago he had seen her in Calcutta, and he remembered being told by somebody then that on her passage up the river she had carried away both her hawse-pipes. But that might have been the pilot's fault. Just now, yarning with the apprentices on board, he had heard that this very voyage, brought up in the Downs, outward bound, she broke her sheer, struck adrift, and lost an anchor and chain. But that might have occurred through want of careful tending in a tideway. All the same, this looked as though she were pretty hard on her ground-tackle. Didn't it? She seemed a heavy ship to handle, anyway. For the rest, as she had a new captain and a new mate this voyage, he understood, one couldn't say how she would turn out. . . .

In such marine shore-talk as this is the name of a ship slowly established, her fame made for her, the tale of her qualities and of her defects kept, her idiosyncrasies commented upon with the zest of personal gossip, her achievements made much of, her faults glossed over as things that, being without remedy in our imperfect world, should not be dwelt upon too much by men who, with the help of ships, wrest out a bitter living from the rough grasp of the sea. All that talk makes up her 'name', which is handed over from one crew to another without bitterness, without animosity, with the indulgence of mutual dependence, and with the feeling of close association in the exercise of her perfections and in the danger of her defects.

This feeling explains men's pride in ships. 'Ships are all right,' as my middle-aged, respectable quartermaster said with much conviction and some irony; but they are not exactly what men make them. They have their own nature; they can of themselves minister to our self-esteem by the demand their qualities make upon our skill and their shortcomings upon our hardiness and endurance. Which is the more flattering exaction it is hard to say; but there is the fact that in listening for upwards of twenty years to the sea-talk that goes on afloat and ashore I have never detected the true note of animosity. I won't deny that at sea, sometimes, the note of profanity was audible enough in those chiding interpellations a wet, cold, weary seaman addresses to his ship, and in moments of exasperation is disposed to extend to all ships that ever were launched—to the whole everlastingly exacting brood that swims in deep waters. And I have heard curses launched at the unstable element itself, whose fascination, outlasting the accumulated experience of ages, had captured him as it had captured the generations of his forebears.

For all that has been said of the love that certain natures (on shore) have professed to feel for it, for all the celebrations it had been the object of in prose and song, the sea has never been friendly to man. At most it has been the accomplice of human restlessness, and playing the part of dangerous abettor of world-wide ambitions. Faithful to no race after the manner of the kindly earth, receiving no impress from valour and toil and self-sacrifice, recognizing no finality of dominion, the sea has never adopted the cause of its masters like those lands where the victorious nations of mankind have taken root, rocking their cradles and setting up their gravestones. He—man or people—who, putting his trust in the friendship of the sea, neglects the strength and cunning of his right hand, is a fool! As if it were too great, too mighty for common virtues, the ocean has no compassion, no faith, no law, no memory. Its fickleness is to be held true to men's purposes only by an undaunted resolution and by a sleepless, armed, jealous vigilance, in which, perhaps, there has always been more hate than love. *Odi et amo* may well be the confession of those who consciously or blindly have surrendered their existence to the fascination of the sea. All the tempestuous passions of mankind's young days, the love of loot and the love of glory, the love of adventure and the love of danger, with the great love of the unknown and vast dreams of dominion and power, have passed like images reflected from a mirror, leaving no record upon the mysterious face of the sea. Impenetrable and heartless, the sea has given nothing of itself to the suitors for its precarious favours. Unlike the earth, it cannot be subjugated at any cost of patience and toil. For all its fascination that has lured so many to a violent death, its immensity has never been loved as the mountains, the plains, the desert itself, have been loved. Indeed, I suspect that, leaving aside the protestations and tributes of writers who, one is safe in saying, care for little else in the world than the rhythm of their lines and the cadence of their phrase, the love of the sea, to which some men and nations confess so readily, is a complex sentiment wherein pride enters for much, necessity for not a little, and the love of ships—the untiring servants of our hopes and our self-esteem—for the best and most genuine part. For the hundreds who have reviled the sea, beginning with Shakespeare in the line: 'More fell than hunger, anguish, or the sea' down to the last obscure sea-dog of the 'old model', having but few words and still fewer thoughts, there could not be found, I believe, one sailor who has ever coupled a curse with the good or bad name of a ship. If ever his profanity, provoked by the hardships of the sea, went so far as to touch his ship, it would be lightly, as a hand may, without sin, be laid in the way of kindness on a woman.

The love that is given to ships is profoundly different from the love men feel for every other work of their hands—the love they bear to their houses, for instance—because it is untainted by the pride of possession. The pride of skill, the pride of responsibility, the pride of endurance there may be, but otherwise it is a disinterested sentiment. No seaman ever cherished a ship, even if she belonged to him, merely because of the profit she put in his pocket. No one, I think, ever did; for a ship-owner, even of the best, has always been outside the pale of that sentiment embracing in a feeling of intimate, equal fellowship the ship and the man, backing each other against the implacable, if sometimes dissembled, hostility of their world of waters. The sea—this truth must be confessed—has no generosity. No display of manly qualities—courage, hardihood, endurance, faithfulness—has ever been known to touch its irresponsible consciousness of power. The ocean has the conscienceless temper of a savage autocrat spoiled by much adulation. He cannot brook the slightest appearance of defiance, and has remained the irreconcilable enemy of ships and men ever since ships and men had the unheard-of audacity to go afloat together in the face of his frown. From that day he has gone on swallowing up fleets and men without his resentment being glutted by the number of victims—by so many wrecked ships and wrecked lives. Today, as ever, he is ready to beguile and betray, to smash and to drown the incorrigible optimism of men who, backed by the fidelity of ships, are trying to wrest from him the fortune of their house, the dominion of their world, or only a dole of food for their hunger. If not always in the hot mood to smash, he is always stealthily ready for a drowning. The most amazing wonder of the deep is its unfathomable cruelty.

I felt its dread for the first time in mid-Atlantic one day, many years ago, when we took off the crew of a Danish brig homeward bound from the West Indies. A thin, silvery mist softened the calm and majestic splendour of light without shadows—seemed to render the sky less remote and the ocean less immense. It was one of the days, when the might of the sea appears indeed lovable, like the nature of a strong man in moments of quiet intimacy. At sunrise we had made out a black speck to the westward, apparently suspended high up in the void behind a stirring, shimmering veil of silvery blue gauze that seemed at times to stir and float in the breeze which fanned us slowly along. The peace of that enchanting forenoon was so profound, so untroubled, that it seemed that every word pronounced loudly on our deck would penetrate to the very heart of that infinite mystery born from the conjunction of water and sky. We did not raise our voices. 'A water-logged derelict, I think, sir,' said the second officer, quietly, coming down from aloft with the binoculars in their case slung across his shoulders; and our captain, without a word, signed to the helmsman to steer for the black speck. Presently we made out a low, jagged stump sticking up forward—all that remained of her departed masts.

The captain was expatiating in a low conversational tone to the chief mate upon the danger of these derelicts, and upon his dread of coming upon them at night, when suddenly a man forward screamed out, 'There's people on board of her, sir! I see them!' in a most extraordinary voice—a voice never heard before in our ship; the amazing voice of a stranger. It gave the signal for a sudden tumult of shouts. The watch below ran up the forecastle head in a body, the cook dashed out of the galley. Everybody saw the poor fellows now. They were there! And all at once our ship, which had the well-earned name of being without a rival for speed in light winds, seemed to us to have lost the power of motion, as if the sea, becoming viscous, had clung to her sides. And yet she moved. Immensity, the inseparable companion of a ship's life, chose that day to breathe upon her as gently as a sleeping child. The clamour of our excitement had died out, and our living ship, famous for never losing steerage way as long as there was air enough to float a feather, stole, without a ripple, silent and white as a ghost, towards her mutilated and wounded sister, come upon at the point of death in the sunlit haze of a calm day at sea.

With the binoculars glued to his eyes, the captain said in a quavering tone: 'They are waving to us with something aft there.' He put down the glasses on the skylight brusquely, and began to walk about the poop. 'A shirt or a flag', he ejaculated, irritably. 'Can't make it out. . . . Some damn rag or other!' He took a few more turns on the poop, glancing down over the rail now and then to see how fast we were moving. His nervous footsteps rang sharply in the quiet of the ship, where the other men, all looking the same way, had forgotten themselves in a staring immobility. 'This will never do!' he cried out, suddenly. 'Lower the boats at once! Down with them!'

Before I jumped into mine he took me aside, as being an experienced junior, for a word of warning:

'You look out as you come alongside that she doesn't take you down with her. You understand?'

He murmured this confidentially, so that none of the men at the falls should overhear, and I was shocked. 'Heavens! as if in such an emergency one stopped to think of danger!' I exclaimed to myself mentally, in scorn of such cold-blooded caution.

It takes many lessons to make a real seaman, and I got my rebuke at once. My experienced commander seemed in one searching glance to read my thoughts on my ingenuous face.

'What you're going for is to save life, not to drown your boat's crew for nothing,' he growled, severely, in my ear. But as we shoved off he leaned over and cried out: 'It all rests on the power of your arms, men. Give way for life!'

We made a race of it, and I would never have believed that a common boat's crew of a merchantman could keep up so much determined fierceness in the regular swing of their stroke. What our captain had clearly perceived before we left had become plain to all of us since. The issue of our enterprise hung on a hair above that abyss of waters which will not give up its dead till the Day of Judgement. It was a race of two ship's boats matched against Death for a prize of nine men's lives, and Death had a long start. We saw the crew of the brig from afar working at the pumps—still pumping on that wreck, which already had settled so far down that the gentle, low swell, over which our boats rose and fell easily without a check to their speed, welling up almost level with her head-rails, plucked at the ends of broken gear swinging desolately under her naked bowsprit.

We could not, in all conscience, have picked out a better day for our regatta had we had the free choice of all the days that ever dawned upon the lonely struggles and solitary agonies of ships since the Norse rovers first steered to the westward against the run of Atlantic waves. It was a very good race. At the finish there was not an oar's length between the first and second boat, with Death coming in a good third on the top of the very next smooth swell, for all one knew to the contrary. The scuppers of the brig gurgled softly all together when the water rising against her sides subsided sleepily with a low wash, as if playing about an immovable rock. Her bulwarks were gone fore and aft, and one saw her bare deck low-lying like a raft and swept clean of boats, spars, houses—of everything except the ringbolts and the heads of the pumps. I had one dismal glimpse of it as I braced myself up to receive upon my breast the last man to leave her, the captain, who literally let himself fall into my arms.

It had been a weirdly silent rescue—a rescue without a hail, without a single uttered word, without a gesture or a sign, without a conscious exchange of glances. Up to the very last moment those on board stuck to their pumps, which spouted two clear streams of water upon their bare feet. Their brown skin showed through the rents of their shirts; and the two small bunches of half-naked, tattered men went on bowing from the waist to each other in their back-breaking labour, up and down, absorbed, with no time for a glance over the shoulder at the help that was coming to them. As we dashed, unregarded, alongside a voice let out one, only one hoarse howl of command, and then, just as they stood, without caps, with the salt drying grey in the wrinkles and folds of their hairy, haggard faces, blinking stupidly at us their red eyelids, they made a bolt away from the handles, tottering and jostling against each other, and positively flung themselves over upon our very heads. The clatter they made tumbling into the boats had an extraordinarily destructive effect upon the illusion of tragic dignity our self-esteem had thrown over the contests of mankind with the sea. On that exquisite day of gentle breathing peace and veiled sunshine perished my romantic love to what men's imagination had proclaimed the most august aspect of Nature. The cynical indifference of the sea to the merits of human suffering and courage, laid bare in this ridiculous, panic-tainted performance extorted from the dire extremity of nine good and honourable seamen, revolted me. I saw the duplicity of the sea's most tender mood. It was so because it could not help itself, but the awed respect of the early days was gone. I felt ready to smile bitterly at its enchanting charm and glare viciously at its furies. In a moment, before we shoved off, I had looked coolly at the life of my choice. Its illusions were gone, but its fascination remained. I had become a seaman at last.

We pulled hard for a quarter of an hour, then laid on our oars waiting for our ship. She was coming down on us with swelling sails, looking delicately tall and exquisitely noble through the mist. The captain of the brig, who sat in the stern sheets by my side with his face in his hands, raised his head and began to speak with a sort of sombre volubility. They had lost their masts and sprung a leak in a hurricane; drifted for weeks, always at the pumps, met more bad weather; the ships they sighted failed to make them out, the leak gained upon them slowly, and the seas had left them nothing to make a raft of. It was very hard to see ship after ship pass by at a distance, 'as if everybody had agreed that we must be left to drown', he added. But they went on trying to keep the brig afloat as long as possible, and working the pumps constantly on insufficient food, mostly raw, till 'yesterday evening,' he continued, monotonously, 'just as the sun went down, the men's hearts broke'.

He made an almost imperceptible pause here, and went on again with exactly the same intonation:

'They told me the brig could not be saved, and they thought they had done enough for themselves. I said nothing to that. It was true. It was no mutiny. I had nothing to say to them. They lay about aft all night, as still as so many dead men. I did not lie down. I kept a look-out. When the first light came I saw your ship at once. I waited for more light; the breeze began to fail on my face. Then I shouted out as loud as I was able, "Look at that ship!" but only two men got up very slowly and came to me. At first only we three stood alone, for a long time, watching you coming down to us, and feeling the breeze drop to a calm almost; but afterwards others, too, rose, one after another, and by and by I had all my crew behind me. I turned round and said to them that they could see the ship was coming our way, but in this small breeze she might come too late after all, unless we turned to and tried to keep the brig afloat long enough to give you time to save us all. I spoke like that to them, and then I gave the command to man the pumps.'

He gave the command, and gave the example, too, by going himself to the handles, but it seems that these men did actually hang back for a moment, looking at each other dubiously before they followed him. 'He! he! he!' He broke out into a most unexpected, imbecile, pathetic, nervous little giggle. 'Their hearts were broken so! They had been played with too long', he explained, apologetically, lowering his eyes, and became silent.

Twenty-five years is a long time—a quarter of a century is a dim and distant past; but to this day I remember the dark-brown feet, hands, and faces of two of these men whose hearts had been broken by the sea. They were lying very still on their sides on the bottom boards between the thwarts, curled up like dogs. My boat's crew, leaning over the looms of their oars, stared and listened as if at the play. The master of the brig looked up suddenly to ask me what day it was.

They had lost the date. When I told him it was Sunday, the 22nd, he frowned, making some mental calculation, then nodded twice sadly to himself, staring at nothing.

His aspect was miserably unkempt and wildly sorrowful. Had it not been for the unquenchable candour of his blue eyes, whose unhappy, tired glance every moment sought his abandoned, sinking brig, as if it could find rest nowhere else, he would have appeared mad. But he was too simple to go mad, too simple with that manly simplicity which alone can bear men unscathed in mind and body through an encounter with the deadly playfulness of the sea or with its less abominable fury.

Neither angry, nor playful, nor smiling, it enveloped our distant ship growing bigger as she neared us, our boats with the rescued men and the dismantled hull of the brig we were leaving behind, in the large and placid embrace of its quietness, half lost in the fair haze, as if in a dream of infinite and tender clemency. There was no frown, no wrinkle on its face, not a ripple. And the run of the slight swell was so smooth that it resembled the graceful undulation of a piece of shimmering grey silk shot with gleams of green. We pulled an easy stroke; but when the master of the brig, after a glance over his shoulder, stood up with a low exclamation, my men feathered their oars instinctively, without an order, and the boat lost her way.

He was steadying himself on my shoulder with a strong grip, while his other arm, flung up rigidly, pointed a denunciatory finger at the immense tranquillity of the ocean. After his first exclamation, which stopped the swing of our oars, he made no sound, but his whole attitude seemed to cry out an indignant 'Behold!' . . . I could not imagine what vision of evil had come to him. I was startled, and the amazing energy of his immobilized gesture made my heart beat faster with the anticipation of something monstrous and unsuspected. The stillness around us became crushing.

For a moment the succession of silky undulations ran on innocently. I saw each of them swell up the misty line of the horizon, far, far away beyond the derelict brig, and the next moment, with a slight friendly toss of our boat, it had passed under us and was gone. The lulling cadence of the rise and fall, the invariable gentleness of this irresistible force, the great charm of the deep waters, warmed my breast deliciously, like the subtle poison of a love-potion. But all this lasted only a few soothing seconds before I jumped up, too, making the boat roll like the veriest land-lubber.

Something startling, mysterious, hastily confused was taking place. I watched it with incredulous and fascinated awe, as one watches the confused, swift movements of some deed of violence done in the dark. As if at a given signal, the run of the smooth undulations seemed checked suddenly around the brig. By a strange optical delusion the whole sea appeared to rise upon her in one overwhelming heave of its silky surface where in one spot a smother of foam broke out ferociously. And then the effort subsided. It was all over, and the smooth swell ran on as before from the horizon in uninterrupted cadence of motion, passing under us with a slight friendly toss of our boat. Far away where the brig had been, an angry white stain undulating on the surface of steely grey waters, shot with gleams of green, diminished swiftly without a hiss, like a patch of pure snow melting in the sun. And the great stillness after this initiation into the sea's implacable hate seemed full of dread thoughts and shadows of disaster.

'Gone!' ejaculated from the depths of his chest my bowman in a final tone. He spat in his hands, and took a better grip on his oar. The captain of the brig lowered his rigid arm slowly, and looked at our faces in a solemnly conscious silence, which called upon us to share in his simple-minded, marvelling awe. All at once he sat down by my side, and leaned forward earnestly at my boat's crew, who, swinging together in a long, easy stroke, kept their eyes fixed upon him faithfully.

'No ship could have done so well,' he addressed them, firmly, after a moment of strained silence, during which he seemed with trembling lips to seek for words fit to bear such high testimony. 'She was small, but she was good. I had no anxiety. She was strong. Last voyage I had my wife and two children in her. No other ship could have stood so long the weather she had to live through for days and days before we got dismasted a fortnight ago. She was fairly worn out, and that's all. You may believe me. She lasted under us for days and days, but she could not last for ever. It was long enough. I am glad it is over. No better ship was ever left to sink at sea on such a day as this.'

He was competent to pronounce the funereal oration of a ship, this son of ancient sea-folk, whose national existence, so little stained by the excesses of manly virtues, had demanded nothing but the merest foothold from the earth. By the merits of his sea-wise forefathers and by the artlessness of his heart, he was made fit to deliver this excellent discourse. There was nothing wanting in its orderly arrangement— neither piety nor faith, nor the tribute of praise due to the worthy dead, with the edifying recital of their achievement. She had lived, he had loved her; she had suffered, and he was glad she was at rest. It was an excellent discourse. And it was orthodox, too, in its fidelity to the cardinal article of a seaman's faith, of which it was a single-minded confession. 'Ships are all right.' They are. They who live with the sea have got to hold by that creed first and last; and it came to me, as I glanced at him sideways, that some men were not altogether unworthy in honour and conscience to pronounce the funereal eulo-gium of a ship's constancy in life and death.

After this, sitting by my side with his loosely clasped hands hanging between his knees, he uttered no word, made no movement till the shadow of our ship's sails fell on the boat, when, at the loud cheer greeting the return of the victors with their prize, he lifted up his troubled face with a faint smile of pathetic indulgence. This smile of the worthy descendant of the most ancient sea-folk whose audacity and hardihood had left no trace of greatness and glory upon the waters, completed the cycle of my initiation. There was an infinite depth of hereditary wisdom in its pitying sadness. It made the hearty bursts of cheering sound like a childish noise of triumph. Our crew shouted with immense confidence—honest souls! As if anybody could ever make sure of having prevailed against the sea, which has betrayed so many ships of great 'name', so many proud men, so many towering ambitions of fame, power, wealth, greatness!

As I brought the boat under the falls my captain, in high good humour, leaned over, spreading his red and freckled elbows on the rail, and called down to me sarcastically out of the depths of his cynic philosopher's beard:

'So you have brought the boat back after all, have you?'

Sarcasm was 'his way', and the most that can be said for it is that it was natural. This did not make it lovable. But it is decorous and expedient to fall in with one's commander's way. 'Yes. I brought the boat back all right, sir,' I answered. And the good man believed me. It was not for him to discern upon me the marks of my recent initiation. And yet I was not exactly the same youngster who had taken the boat away—all impatience for a race against Death, with the prize of nine men's lives at the end.

Already I looked with other eyes upon the sea. I knew it capable of betraying the generous ardour of youth as implacably as, indifferent to evil and good, it would have betrayed the basest greed or the noblest heroism. My conception of its magnanimous greatness was gone. And I looked upon the true sea—the sea that plays with men till their hearts are broken, and wears stout ships to death. Nothing can touch the brooding bitterness of its soul. Open to all and faithful to none, it exercises its fascination for the undoing of the best. To love it is not well. It knows no bond of plighted troth, no fidelity to misfortune, to long companionship, to long devotion. The promise it holds out perpetually is very great; but the only secret of its possession is strength, strength—the jealous, sleepless strength of a man guarding a coveted treasure within his gates.

The Voyage

WASHINGTON IRVING

> Ships, ships, I will descrie you
>   Amidst the main,
> I will come and try you
> What you are protecting
> And projecting,
>   What's your end and aim.
> One goes abroad for merchandize and trading,
> Another stays to keep his country from invading,
> A third is coming home with rich and wealthy lading.
>   Hallo my fancie, whither wilt thou go?
>
> *Old Poem*

To an American visiting Europe the long voyage he has to make is an excellent preparative. The temporary absence of worldly scenes and employments produces a state of mind peculiarly fitted to receive new and vivid impressions. The vast space of waters, that separates the hemispheres is like a blank page in existence. There is no gradual transition by which as in Europe the features and population of one country blend almost imperceptibly with those of another. From the moment you lose sight of the land you have left, all is vacancy until you step on the opposite shore, and are launched at once into the bustle and novelties of another world.

In travelling by land there is a continuity of scene and a connected succession of persons and incidents, that carry on the story of life, and lessen the effect of absence and separation. We drag, it is true, 'a lengthening chain' at each remove of our pilgrimage; but the chain is unbroken—we can trace it back link by link; and we feel that the last still grapples us to home. But a wide sea voyage severs us at once. It makes us conscious of being cast loose from the secure anchorage of settled life and sent adrift upon a doubtful world. It interposes a gulph, not merely imaginary, but real, between us and our homes—a gulph subject to tempest and fear and uncertainty, rendering distance palpable and return precarious.

Such at least was the case with myself. As I saw the last blue line of my native land fade away like a cloud in the horizon, it seemed as if I had closed one volume of the world and its concerns, and had time for meditation before I opened another. That land too, now vanishing from my view; which contained all that was most dear to me in life; what vicissitudes might occur in it—what changes might take place in me, before I should visit it again. Who can tell when he sets forth to wander, whither he may be driven by the uncertain currents of existence; or when he may return; or whether it may ever be his lot to revisit the scenes of his childhood?

I said that at sea all is vacancy—I should correct the expression. To one given to day dreaming and fond of losing himself in reveries, a sea voyage is full of subjects for meditation: but then they are the wonders of the deep and of the air, and rather tend to abstract the mind from worldly themes. I delighted to loll over the quarter railing or climb to the main top of a calm day, and muse for hours together, on the tranquil bosom of a summer's sea. To gaze upon the piles of golden clouds just peering above the horizon; fancy them some fairy realms and people them with a creation of my own. To watch the gently undulating billows, rolling their silver volumes as if to die away on those happy shores.

There was a delicious sensation of mingled security and awe with which I looked down from my giddy height on the monsters of the deep at their uncouth gambols. Shoals of porpoises tumbling about the bow of the ship; the grampus slowly heaving his huge form above the surface, or the ravenous shark darting like a spectre through the blue waters. My imagination would conjure up all that I had heard or read of the watery world beneath me. Of the finny herds that roam its fathomless valleys; of the shapeless monsters that lurk among the very foundations of the earth and of those wild phantasms that swell the tales of fishermen and sailors.

Sometimes a distant sail, gliding along the edge of the ocean would be another theme of idle speculation. How interesting this fragment of a world, hastening to rejoin the great mass of existence. What a glorious monument of human invention; which has in a manner triumphed over wind and wave; has brought the ends of the earth into communion; has established an interchange of blessings—pouring into the sterile regions of the north all the luxuries of the south; has diffused the light of knowledge and the charities of cultivated life, and has thus bound together those scattered portions of the human race, between which nature seemed to have thrown an insurmountable barrier.

We one day descried some shapeless object drifting at a distance. At sea everything that breaks the monotony of the surrounding expanse attracts attention. It proved to be the mast of a ship that must have been completely wrecked; for there were the remains of handkerchiefs, by which some of the crew had fastened themselves to this spar to prevent their being washed off by the waves. There was no trace by which the name of the ship could be ascertained. The wreck had evidently drifted about for many months: clusters of shell fish had fastened about it; and long sea weeds flaunted at its sides.

But where, thought I, is the crew! Their struggle has long been over—they have gone down amidst the roar of the tempest—their bones lie whitening among the caverns of the deep. Silence—oblivion, like the waves, have closed over them, and no one can tell the story of their end. What sighs have been wafted after that ship; what prayers offered up at the deserted fireside of home. How often has the mistress, the wife, the mother pored over the daily news to catch some casual intelligence of this rover of the deep. How has expectation darkened into anxiety—anxiety into dread and dread into despair. Alas! not one memento may ever return for love to cherish. All that may ever be known is, that she sailed from her port, 'and was never heard of more!'

The sight of this wreck, as usual, gave rise to many dismal anecdotes. This was particularly the case in the evening when the weather, which had hitherto been fair began to look wild and threatening, and gave indications of one of those sudden storms which will sometimes break in upon the serenity of a summer voyage. As we sat round the dull light of a lamp in the cabin, that made the gloom more ghastly, everyone had his tale of shipwreck and disaster. I was peculiarly struck with a short one related by the captain.

'As I was once sailing', said he, 'in a fine stout ship across the banks of Newfoundland, one of those heavy fogs which prevail in those parts rendered it impossible for us to see far ahead even in the daytime; but at night the weather was so thick that we could not distinguish any object at twice the length of the ship. I kept lights at the mast head and a constant watch forward to look out for fishing smacks, which are accustomed to lie at anchor on the banks. The wind was blowing a smacking breeze and we were going at a great rate through the water. Suddenly the watch gave the alarm of "a sail ahead!"—it was scarcely uttered before we were upon her. She was a small schooner at anchor, with the broad side toward us. The crew were all asleep and had neglected to hoist a light. We struck her just amidships. The force, the size and weight of our vessel bore her down below the waves—we passed over her and were hurried on our course. As the crashing wreck was sinking beneath us I had a glimpse of two or three half-naked wretches, rushing from her cabin—they just started from their beds to be swallowed shrieking by the waves. I heard their drowning cry mingling with the wind. The blast that bore it to our ears swept us out of all further hearing—I shall never forget that cry!—It was some time before we could put the ship about; she was under such headway. We returned as nearly as we could guess to the place where the smack had anchored. We cruised about for several hours in the dense fog. We fired signal guns and listened if we might hear the halloo of any survivors; but all was silent—we never saw or heard anything of them more!'

I confess these stories for a time put an end to all my fine fancies. The storm increased with the night. The sea was lashed up into tremendous confusion. There was a fearful sullen sound of rushing waves and broken surges. Deep called unto deep. At times the black volume of clouds overhead seemed rent asunder by flashes of lightning which quivered along the foaming billows, and made the succeeding darkness doubly terrible. The thunders bellowed over the wild waste of waters and were echoed and prolonged by the mountain waves. As I saw the ship staggering and plunging among these roaring caverns, it seemed miraculous that she regained her balance or preserved her buoyancy. Her yards would dip into the water; her bow was almost buried beneath the waves. Sometimes an impending surge appeared ready to overwhelm her, and nothing but a dextrous movement of the helm preserved her from the shock.

When I retired to my cabin the awful scene still followed me. The whistling of the wind through the rigging sounded like funereal wailings. The creaking of the masts; the straining and groaning of bulkheads as the ship laboured in the weltering sea were frightful. As I heard the waves rushing along the side of the ship and roaring in my very ear, it seemed as if death were raging round this floating prison, seeking for his prey—the mere starting of a nail—the yawning of a seam might give him entrance.

A fine day, however, with a tranquil sea and favouring breeze soon put all these dismal reflections to flight. It is impossible to resist the gladdening influence of fine weather and fair wind at sea. When the ship is decked out in all her canvas, every sail swelled, and careering gaily over the curling waves, how lofty, how gallant she appears—how she seems to lord it over the deep!

I might fill a volume with the reveries of a sea voyage, for with me it is almost a continual reverie—but it is time to get to shore.

It was a fine sunny morning when the thrilling cry of Land! was given from the mast head. None but those who have experienced it can form an idea of the delicious throng of sensations which rush into an American's bosom, when he first comes in sight of Europe. There is a volume of associations with the very name. It is the land of promise, teeming with everything of which his childhood has heard, or on which his studious years have pondered.

From that time until the moment of arrival it was all feverish excitement. The ships of war that prowled like guardian giants along the coast—the headlands of Ireland stretching out into the channel—the Welsh mountains towering into the clouds, all were objects of intense interest. As we sailed up the Mersey I reconnoitered the shores with a telescope. My eye dwelt with delight on neat cottages with their trim shrubberies and green grass plots. I saw the mouldering ruin of an abbey overrun with ivy, and the taper spire of a village church rising from the brow of a neighbouring hill—all were characteristic of England.

The tide and wind were so favourable that the ship was enabled to come at once to the pier. It was thronged with people; some idle lookers-on, others eager expectants of friends or relatives. I could distinguish the merchant to whom the ship was consigned. I knew him by his calculating brow and restless air. His hands were thrust into his pockets; he was whistling thoughtfully and walking to and fro, a small space having been accorded him by the crowd in deference to his temporary importance. There were repeated cheerings and salutations interchanged between the shore and the ship, as friends happened to recognize each other. I particularly noticed one young woman of humble dress, but interesting demeanour. She was leaning forward from among the crowd; her eye hurried over the ship as it neared the shore, to catch some wished for countenance. She seemed disappointed and agitated; when I heard a faint voice call her name. It was from a poor sailor who had been ill all the voyage and had excited the sympathy of everyone on board. When the weather was fine his messmates had spread a mattress for him on deck in the shade, but of late his illness had so increased, that he had taken to his hammock, and only breathed a wish that he might see his wife before he died. He had been helped on deck as we came up the river, and was now leaning against the shrouds, with a countenance so wasted, so pale, so ghastly that it was no wonder even the eye of affection did not recognize him. But at the sound of his voice her eye darted on his features—it read at once a whole volume of sorrow—she clasped her hands; uttered a faint shriek and stood wringing them in silent agony.

All now was hurry and bustle. The meetings of acquaintances—the greetings of friends—the consultations of men of business. I alone was solitary and idle. I had no friend to meet, no cheering to receive. I stepped upon the land of my forefathers—but felt that I was a stranger in the land.

A Descent into the Maelström

EDGAR ALLAN POE

The ways of God in Nature, as in Providence, are not as *our* ways;
nor are the models that we frame any way commensurate to the
vastness, profundity, and unsearchableness of His works, *which
have a depth in them greater than the well of Democritus.*

Joseph Glanville

WE had now reached the summit of the loftiest crag. For
some minutes the old man seemed too much exhausted to speak.

'Not long ago,' said he at length, 'and I could have guided you on
this route as well as the youngest of my sons; but, about three years
past, there happened to me an event such as never happened before
to mortal man—or at least such as no man ever survived to tell
of—and the six hours of deadly terror which I then endured have
broken me up body and soul. You suppose me a *very* old man—but
I am not. It took less than a single day to change these hairs from a
jetty black to white, to weaken my limbs, and to unstring my nerves,
so that I tremble at the least exertion, and am frightened at a shadow.
Do you know I can scarcely look over this little cliff without getting
giddy?'

The 'little cliff', upon whose edge he had so carelessly thrown
himself down to rest that the weightier portion of his body hung over
it, while he was only kept from falling by the tenure of his elbow on
its extreme and slippery edge—this 'little cliff' arose, a sheer unobstructed precipice of black shining rock, some fifteen or sixteen hundred feet from the world of crags beneath us. Nothing would have
tempted me to within half a dozen yards of its brink. In truth so
deeply was I excited by the perilous position of my companion, that

I fell at full length upon the ground, clung to the shrubs around me, and dared not even glance upward at the sky—while I struggled in vain to divest myself of the idea that the very foundations of the mountain were in danger from the fury of the winds. It was long before I could reason myself into sufficient courage to sit up and look out into the distance.

'You must get over these fancies,' said the guide, 'for I have brought you here that you might have the best possible view of the scene of that event I mentioned—and to tell you the whole story with the spot just under your eye.'

'We are now,' he continued, in that particularizing manner which distinguished him—'we are now close upon the Norwegian coast—in the sixty-eighth degree of latitude—in the great province of Nordland—and in the dreary district of Lofoden. The mountain upon whose top we sit is Helseggen, the Cloudy. Now raise yourself up a little higher—hold on to the grass if you feel giddy—so—and look out, beyond the belt of vapour beneath us, into the sea.'

I looked dizzily, and beheld a wide expanse of ocean, whose waters wore so inky a hue as to bring at once to my mind the Nubian geographer's account of the *Mare Tenebrarum*. A panorama more deplorably desolate no human imagination can conceive. To the right and left, as far as the eye could reach, there lay outstretched, like ramparts of the world, lines of horridly black and beetling cliff, whose character of gloom was but the more forcibly illustrated by the surf which reared high up against it its white and ghastly crest, howling and shrieking for ever. Just opposite the promontory upon whose apex we were placed, and at a distance of some five or six miles out at sea, there was visible a small, bleak-looking island; or, more properly, its position was discernible through the wilderness of surge in which it was enveloped. About two miles nearer the land, arose another of smaller size, hideously craggy and barren, and encompassed at various intervals by a cluster of dark rocks.

The appearance of the ocean, in the space between the more distant island and the shore, had something very unusual about it. Although, at the time, so strong a gale was blowing landward that a brig in the remote offing lay to under a double-reefed trysail, and constantly plunged her whole hull out of sight, still there was here nothing like a regular swell, but only a short, quick, angry cross dashing of water in every direction—as well in the teeth of the wind as otherwise. Of foam there was little except in the immediate vicinity of the rocks.

'The island in the distance', resumed the old man, 'is called by the Norwegians Vurrgh. The one midway is Moskoe. That a mile to the northward is Ambaaren. Yonder are Iflesen, Hoeyholm, Kieldholm, Suarven, and Buckholm. Further off—between Moskoe and Vurrgh— are Otterholm, Flimen, Sandflesen, and Skarholm. These are the true names of the places—but why it has been thought necessary to name them at all, is more than either you or I can understand. Do you hear anything? Do you see any change in the water?'

We had now been about ten minutes upon the top of Helseggen, to which we had ascended from the interior of Lofoden, so that we had caught no glimpse of the sea until it had burst upon us from the summit. As the old man spoke, I became aware of a loud and grad- ually increasing sound, like the moaning of a vast herd of buffaloes upon an American prairie; and at the same moment I perceived that what seamen term the *chopping* character of the ocean beneath us was rapidly changing into a current which set to the eastward. Even while I gazed, this current acquired a monstrous velocity. Each moment added to its speed—to its headlong impetuosity. In five minutes the whole sea, as far as Vurrgh, was lashed into ungovernable fury; but it was between Moskoe and the coast that the main uproar held its sway. Here the vast bed of the waters, seamed and scarred into a thousand conflicting channels, burst suddenly into phrensied convulsion—heav- ing, boiling, hissing—gyrating in gigantic and innumerable vortices, and all whirling and plunging on to the eastward with a rapidity which water never elsewhere assumes except in precipitous descents.

In a few minutes more, there came over the scene another radical alteration. The general surface grew somewhat more smooth, and the whirlpools, one by one, disappeared, while prodigious streaks of foam became apparent where none had been seen before. These streaks, at length, spreading out to a great distance, and entering into combina- tion, took unto themselves the gyratory motion of the subsided vor- tices, and seemed to form the germ of another more vast. Suddenly— very suddenly—this assumed a distinct and definite existence, in a circle of more than half a mile in diameter. The edge of the whirl was represented by a broad belt of gleaming spray; but no particle of this slipped into the mouth of the terrific funnel, whose interior, as far as the eye could fathom it, was a smooth, shining, and jet-black wall of water, inclined to the horizon at an angle of some forty-five degrees, speeding dizzily round and round with a swaying and sweltering motion, and sending forth to the winds an appalling voice, half shriek, half roar, such as not even the mighty cataract of Niagara ever lifts up in its agony to Heaven.

The mountain trembled to its very base, and the rock rocked. I threw myself upon my face, and clung to the scant herbage in an excess of nervous agitation.

'This,' said I at length, to the old man—'this *can* be nothing else than the great whirlpool of the Maelström.'

'So it is sometimes termed,' said he. 'We Norwegians call it the Moskoe-ström, from the island of Moskoe in the midway.'

The ordinary accounts of this vortex had by no means prepared me for what I saw. That of Jonas Ramus, which is perhaps the most circumstantial of any, cannot impart the faintest conception either of the magnificence, or of the horror of the scene—or of the wild bewildering sense of *the novel* which confounds the beholder. I am not sure from what point of view the writer in question surveyed it, nor at what time; but it could neither have been from the summit of Helseggen, nor during a storm. There are some passages of his description, nevertheless, which may be quoted for their details, although their effect is exceedingly feeble in conveying an impression of the spectacle.

'Between Lofoden and Moskoe', he says, 'the depth of the water is between thirty-six and forty fathoms; but on the other side, toward Ver (Vurrgh) this depth decreases so as not to afford a convenient passage for a vessel, without the risk of splitting on the rocks, which happens even in the calmest weather. When it is flood, the stream runs up the country between Lofoden and Moskoe with a boisterous rapidity; but the roar of its impetuous ebb to the sea is scarce equalled by the loudest and most dreadful cataracts; the noise being heard several leagues off, and the vortices or pits are of such an extent and depth, that if a ship comes within its attraction, it is inevitably absorbed and carried down to the bottom, and there beat to pieces against the rocks; and when the water relaxes, the fragments thereof are thrown up again. But these intervals of tranquillity are only at the turn of the ebb and flood, and in calm weather, and last but a quarter of an hour, its violence gradually returning. When the stream is most boisterous, and its fury heightened by a storm, it is dangerous to come within a Norway mile of it. Boats, yachts, and ships have been carried away by not guarding against it before they were within its reach. It likewise happens frequently, that whales come too near the stream, and are overpowered by its violence; and then it is impossible to describe their howlings and bellowings in their fruitless struggles to disengage themselves. A bear once, attempting to swim from Lofoden to Moskoe, was caught by the stream and borne down, while he roared terribly, so as to be heard on shore. Large stocks of firs and pine trees, after being absorbed by the current, rise again broken and torn to such a degree as if bristles grew upon them. This plainly shows the bottom to consist of craggy rocks, among which they are whirled to and fro. This stream is regulated by the flux and reflux of the sea—it being constantly high and low water every six hours. In the year 1645, early in the morning of Sexagesima Sunday, it raged with such noise and impetuosity that the very stones of the houses on the coast fell to the ground.'

In regard to the depth of the water, I could not see how this could have been ascertained at all in the immediate vicinity of the vortex. The 'forty fathoms' must have reference only to portions of the channel close upon the shore either of Moskoe or Lofoden. The depth in the centre of the Moskoe-ström must be immeasurably greater; and no better proof of this fact is necessary than can be obtained from even the sidelong glance into the abyss of the whirl which may be had from the highest crag of Helseggen. Looking down from this pinnacle upon the howling Phlegethon below, I could not help smiling at the simplicity with which the honest Jonas Ramus records, as a matter difficult of belief, the anecdotes of the whales and the bears; for it appeared to me, in fact, a self-evident thing, that the largest ship of the line in existence, coming within the influence of that deadly attraction, could resist it as little as a feather the hurricane, and must disappear bodily and at once.

The attempts to account for the phenomenon—some of which, I remember, seemed to me sufficiently plausible in perusal—now wore a very different and unsatisfactory aspect. The idea generally received is that this, as well as three smaller vortices among the Ferroe islands, 'have no other cause than the collision of waves rising and falling, at flux and reflux, against a ridge of rocks and shelves, which confines the water so that it precipitates itself like a cataract; and thus the higher the flood rises, the deeper must the fall be, and the natural result of all is a whirlpool or vortex, the prodigious suction of which is sufficiently known by lesser experiments.' These are the words of the *Encyclopædia Britannica*. Kircher and others imagine that in the centre of the channel of the Maelström is an abyss penetrating the globe, and issuing in some very remote part—the Gulf of Bothnia being somewhat decidedly named in one instance. This opinion, idle in itself, was the one to which, as I gazed, my imagination most readily assented; and, mentioning it to the guide, I was rather surprised to hear him say that, although it was the view almost universally entertained of the subject by the Norwegians, it nevertheless was not his own. As to the former notion he confessed his inability to comprehend it; and here I agreed with him—for, however conclusive on paper, it becomes altogether unintelligible, and even absurd, amid the thunder of the abyss.

'You have had a good look at the whirl now,' said the old man, 'and if you will creep round this crag, so as to get in its lee, and deaden the roar of the water, I will tell you a story that will convince you I ought to know something of the Moskoe-ström.'

I placed myself as desired, and he proceeded.

'Myself and my two brothers once owned a schooner-rigged smack of about seventy tons burthen, with which we were in the habit of fishing among the islands beyond Moskoe, nearly to Vurrgh. In all violent eddies at sea there is good fishing, at proper opportunities, if one has only the courage to attempt it; but among the whole of the Lofoden coastmen, we three were the only ones who made a regular business of going out to the islands, as I tell you. The usual grounds are a great way lower down to the southward. There fish can be got at all hours, without much risk, and therefore these places are preferred. The choice spots over here among the rocks, however, not only yield the finest variety, but in far greater abundance; so that we often got in a single day, what the more timid of the craft could not scrape together in a week. In fact, we made it a matter of desperate speculation—the risk of life standing instead of labour, and courage answering for capital.

'We kept the smack in a cove about five miles higher up the coast than this; and it was our practice, in fine weather, to take advantage of the fifteen minutes' slack to push across the main channel of the Moskoe-ström, far above the pool, and then drop down upon anchorage somewhere near Otterholm, or Sandflesen, where the eddies are not so violent as elsewhere. Here we used to remain until nearly time for slackwater again, when we weighed and made for home. We never set out upon this expedition without a steady side wind for going and coming—one that we felt sure would not fail us before our return—and we seldom made a miscalculation upon this point. Twice, during six years, we were forced to stay all night at anchor on account of a dead calm, which is a rare thing indeed just about here; and once we had to remain on the grounds nearly a week, starving to death, owing to a gale which blew up shortly after our arrival, and made the channel too boisterous to be thought of. Upon this occasion we should have been driven out to sea in spite of everything (for the whirlpools threw us round and round so violently, that, at length, we fouled our anchor and dragged it) if it had not been that we drifted into one of the innumerable cross currents—here today and gone tomorrow—which drove us under the lee of Flimen, where, by good luck, we brought up.

'I could not tell you the twentieth part of the difficulties we encountered "on the grounds"—it is a bad spot to be in, even in good weather—but we made shift always to run the gauntlet of the Moskoeström itself without accident; although at times my heart has been in my mouth when we happened to be a minute or so behind or before the slack. The wind sometimes was not as strong as we thought it at starting, and then we made rather less way than we could wish, while the current rendered the smack unmanageable. My eldest brother had a son eighteen years old, and I had two stout boys of my own. These would have been of great assistance at such times, in using the sweeps, as well as afterward in fishing—but, somehow, although we ran the risk ourselves, we had not the heart to let the young ones get into the danger—for, after all is said and done, it *was* a horrible danger, and that is the truth.

'It is now within a few days of three years since what I am going to tell you occurred. It was on the tenth day of July, 18—, a day which the people of this part of the world will never forget—for it was one in which blew the most terrible hurricane that ever came out of the heavens. And yet all the morning, and indeed until late in the afternoon, there was a gentle and steady breeze from the south-west, while the sun shone brightly, so that the oldest seaman among us could not have foreseen what was to follow.

'The three of us—my two brothers and myself—had crossed over to the islands about two o'clock p.m., and had soon nearly loaded the smack with fine fish, which, we all remarked, were more plenty that day than we had ever known them. It was just seven, *by my watch*, when we weighed and started for home, so as to make the worst of the Ström at slack-water, which we knew would be at eight.

'We set out with a fresh wind on our starboard quarter and for some time spanked along at a great rate, never dreaming of danger, for indeed we saw not the slightest reason to apprehend it. All at once we were taken aback by a breeze from over Helseggen. This was most unusual—something that had never happened to us before—and I began to feel a little uneasy, without exactly knowing why. We put the boat on the wind, but could make no headway at all for the eddies and I was upon the point of proposing to return to the anchorage, when, looking astern, we saw the whole horizon covered with a singular copper-coloured cloud that rose with the most amazing velocity.

'In the meantime the breeze that had headed us off fell away, and we were dead becalmed, drifting about in every direction. This state of things, however, did not last long enough to give us time to think about it. In less than a minute the storm was upon us—in less than two the sky was entirely overcast—and what with this and the driving spray, it became suddenly so dark that we could not see each other in the smack.

'Such a hurricane as then blew it is folly to attempt describing. The oldest seaman in Norway never experienced anything like it. We had let our sails go by the run before it cleverly took us; but, at the first puff, both our masts went by the board as if they had been sawed off—the mainmast taking with it my youngest brother, who had lashed himself to it for safety.

'Our boat was the lightest feather of a thing that ever sat upon water. It had a complete flush deck, with only a small hatch near the bow, and this hatch it had always been our custom to batten down when about to cross the Ström, by way of precaution against the chopping seas. But for this circumstance we should have foundered at once—for we lay entirely buried for some moments. How my elder brother escaped destruction I cannot say, for I never had an opportunity of ascertaining. For my part, as soon as I had let the foresail run, I threw myself flat on deck, with my feet against the narrow gunwale of the bow, and with my hands grasping a ring-bolt near the foot of the foremast. It was mere instinct that prompted me to do this—which was undoubtedly the very best thing I could have done—for I was too much flurried to think.

'For some moments we were completely deluged, as I say, and all this time I held my breath, and clung to the bolt. When I could stand it no longer I raised myself upon my knees, still keeping hold with my hands, and thus got my head clear. Presently our little boat gave herself a shake, just as a dog does in coming out of the water, and thus rid herself, in some measure, of the seas. I was now trying to get the better of the stupor that had come over me, and to collect my senses so as to see what was to be done, when I felt somebody grasp my arm. It was my elder brother, and my heart leaped for joy, for I had made sure that he was overboard—but the next moment all this joy was turned into horror—for he put his mouth close to my ear, and screamed out the word "*Moskoe-ström!*"

'No one ever will know what my feelings were at that moment. I shook from head to foot as if I had had the most violent fit of the ague. I knew what he meant by that one word well enough—I knew what he wished to make me understand. With the wind that now drove us on, we were bound for the whirl of the Ström, and nothing could save us!

'You perceive that in crossing the Ström *channel*, we always went a long way up above the whirl, even in the calmest weather, and then had to wait and watch carefully for the slack—but now we were driving right upon the pool itself, and in such a hurricane as this! "To be sure", I thought, "we shall get there just about the slack—there is some little hope in that"—but in the next moment I cursed myself for being so great a fool as to dream of hope at all. I knew very well that we were doomed, had we been ten times a ninety-gun ship.

'By this time the first fury of the tempest had spent itself, or perhaps we did not feel it so much, as we scudded before it, but at all events the seas, which at first had been kept down by the wind, and lay flat and frothing, now got up into absolute mountains. A singular change, too, had come over the heavens. Around in every direction it was still as black as pitch, but nearly overhead there burst out, all at once, a circular rift of clear sky—as clear as I ever saw—and of a deep bright blue—and through it there blazed forth the full moon with a lustre that I never before knew her to wear. She lit up everything about us with the greatest distinctness—but, oh God, what a scene it was to light up!

'I now made one or two attempts to speak to my brother—but, in some manner which I could not understand, the din had so increased that I could not make him hear a single word, although I screamed at the top of my voice in his ear. Presently he shook his head, looking as pale as death, and held up one of his fingers, as if to say "*listen!*"

'At first I could not make out what he meant—but soon a hideous thought flashed upon me. I dragged my watch from its fob. It was not going. I glanced at its face by the moonlight, and then burst into tears as I flung it far away into the ocean. *It had run down at seven o'clock! We were behind the time of the slack, and the whirl of the Ström was in full fury!*

'When a boat is well built, properly trimmed, and not deep laden, the waves in a strong gale, when she is going large, seem always to slip from beneath her—which appears very strange to a landsman—and this is what is called *riding*, in sea phrase. Well, so far we had ridden the swells very cleverly; but presently a gigantic sea happened to take us right under the counter, and bore us with it as it rose—up—up—as if into the sky. I would not have believed that any wave could rise so high. And then down we came with a sweep, a slide, and a plunge, that made me feel sick and dizzy, as if I was falling from some lofty mountain-top in a dream. But while we were up I had thrown a quick glance around—and that one glance was all sufficient. I saw our exact position in an instant. The Moskoe-ström whirlpool was about a quarter of a mile dead ahead—but no more like the everyday Moskoe-ström, than the whirl as you now see it is like a mill-race. If I had not known where we were, and what we had to expect, I should not have recognized the place at all. As it was, I involuntarily closed my eyes in horror. The lids clenched themselves together as if in a spasm.

'It could not have been more than two minutes afterward until we suddenly felt the waves subside, and were enveloped in foam. The boat made a sharp half turn to larboard, and then shot off in its new direction like a thunderbolt. At the same moment the roaring noise of the water was completely drowned in a kind of shrill shriek—such a sound as you might imagine given out by the waste-pipes of many thousand steam-vessels, letting off their steam all together. We were now in the belt of surf that always surrounds the whirl; and I thought, of course, that another moment would plunge us into the abyss—down which we could only see indistinctly on account of the amazing velocity with which we were borne along. The boat did not seem to sink into the water at all, but to skim like an air-bubble upon the surface of the surge. Her starboard side was next the whirl, and on the larboard arose the world of ocean we had left. It stood like a huge writhing wall between us and the horizon.

'It may appear strange, but now, when we were in the very jaws of the gulf, I felt more composed than when we were only approaching it. Having made up my mind to hope no more, I got rid of a great deal of that terror which unmanned me at first. I suppose it was despair that strung my nerves.

'It may look like boasting—but what I tell you is truth—I began to reflect how magnificent a thing it was to die in such a manner, and how foolish it was in me to think of so paltry a consideration as my own individual life, in view of so wonderful a manifestation of God's power. I do believe that I blushed with shame when this idea crossed my mind. After a little while I became possessed with the keenest curiosity about the whirl itself. I positively felt a *wish* to explore its depths, even at the sacrifice I was going to make; and my principal grief was that I should never be able to tell my old companions on shore about the mysteries I should see. These, no doubt, were singular fancies to occupy a man's mind in such extremity—and I have often thought since, that the revolutions of the boat around the pool might have rendered me a little light-headed.

'There was another circumstance which tended to restore my self-possession; and this was the cessation of the wind, which could not reach us in our present situation—for, as you saw yourself, the belt of surf is considerably lower than the general bed of the ocean, and this latter now towered above us, a high, black, mountainous ridge. If you have never been at sea in a heavy gale, you can form no idea of the confusion of mind occasioned by the wind and spray together. They blind, deafen, and strangle you, and take away all power of action or reflection. But we were now, in a great measure, rid of these annoyances—just as death-condemned felons in prison are allowed petty indulgences, forbidden them while their doom is yet uncertain.

'How often we made the circuit of the belt it is impossible to say. We careered round and round for perhaps an hour, flying rather than floating, getting gradually more and more into the middle of the surge, and then nearer and nearer to its horrible inner edge. All this time I had never let go of the ring-bolt. My brother was at the stern, holding on to a large empty water cask which had been securely lashed under the coop of the counter, and was the only thing on deck that had not been swept overboard when the gale first took us. As we approached the brink of the pit he let go his hold upon this, and made for the ring, from which, in the agony of his terror, he endeavoured to force my hands, as it was not large enough to afford us both a secure grasp. I never felt deeper grief than when I saw him attempt this act—although I knew he was a madman when he did it—a raving maniac through sheer fright. I did not care, however, to contest the point with him. I thought it could make no difference whether either of us held on at all; so I let him have the bolt, and went astern to the cask. This there was no great difficulty in doing; for the smack flew round steadily enough, and upon an even keel—only swaying to and fro, with the immense sweeps and swelters of the whirl. Scarcely had I secured myself in my new position, when we gave a wild lurch to starboard, and rushed headlong into the abyss. I muttered a hurried prayer to God, and thought all was over.

'As I felt the sickening sweep of the descent, I had instinctively tightened my hold upon the barrel, and closed my eyes. For some seconds I dared not open them—while I expected instant destruction, and wondered that I was not already in my death-struggles with the water. But moment after moment elapsed. I still lived. The sense of falling had ceased; and the motion of the vessel seemed much as it had been before while in the belt of foam, with the exception that she now lay more along. I took courage, and looked once again upon the scene.

'Never shall I forget the sensations of awe, horror, and admiration with which I gazed about me. The boat appeared to be hanging, as if by magic, midway down, upon the interior surface of a funnel vast in circumference, prodigious in depth, and whose perfectly smooth sides might have been mistaken for ebony, but for the bewildering rapidity with which they spun around, and for the gleaming and ghastly radiance they shot forth, as the rays of the full moon, from that circular rift amid the clouds which I have already described, streamed in a flood of golden glory along the black walls, and far away down into the inmost recesses of the abyss.

'At first I was too much confused to observe anything accurately. The general burst of terrific grandeur was all that I beheld. When I recovered myself a little, however, my gaze fell instinctively downward. In this direction I was able to obtain an unobstructed view, from the manner in which the smack hung on the inclined surface of the pool. She was quite upon an even keel—that is to say, her deck lay in a plane parallel with that of the water—but this latter sloped at an angle of more than forty-five degrees, so that we seemed to be lying upon our beam-ends. I could not help observing, nevertheless, that I had scarcely more difficulty in maintaining my hold and footing in this situation, than if we had been upon a dead level; and this, I suppose, was owing to the speed at which we revolved.

'The rays of the moon seemed to search the very bottom of the profound gulf; but still I could make out nothing distinctly, on account of a thick mist in which everything there was enveloped, and over which there hung a magnificent rainbow, like that narrow and tottering bridge which Mussulmen say is the only pathway between Time and Eternity. This mist, or spray, was no doubt occasioned by the clashing of the great walls of the funnel, as they all met together at the bottom—but the yell that went up to the Heavens from out of that mist, I dare not attempt to describe.

'Our first slide into the abyss itself, from the belt of foam above, had carried us a great distance down the slope; but our further descent was by no means proportionate. Round and round we swept—not with any uniform movement—but in dizzying swings and jerks, that sent us sometimes only a few hundred feet—sometimes nearly the complete circuit of the whirl. Our progress downward, at each revolution, was slow, but very perceptible.

'Looking about me upon the wide waste of liquid ebony on which we were thus borne, I perceived that our boat was not the only object in the embrace of the whirl. Both above and below us were visible fragments of vessels, large masses of building timber and trunks of trees, with many smaller articles, such as pieces of house furniture, broken boxes, barrels and staves. I have already described the unnatural curiosity which had taken the place of my original terrors. It appeared to grow upon me as I drew nearer and nearer to my dreadful doom. I now began to watch, with a strange interest, the numerous things that floated in our company. I *must* have been delirious—for I even sought *amusement* in speculating upon the relative velocities of their several descents toward the foam below. "This fir tree", I found myself at one time saying, "will certainly be the next thing that takes the awful plunge and disappears"—and then I was disappointed to find that the wreck of a Dutch merchant ship overtook it and went down before. At length, after making several guesses of this nature, and being deceived in all—this fact—the fact of my invariable miscalculation—set me upon a train of reflection that made my limbs again tremble, and my heart beat heavily once more.

'It was not a new terror that thus affected me, but the dawn of a more exciting *hope*. This hope arose partly from memory, and partly from present observation. I called to mind the great variety of buoyant matter that strewed the coast of Lofoden, having been absorbed and then thrown forth by the Moskoe-ström. By far the greater number of the articles were shattered in the most extraordinary way—so chafed and roughened as to have the appearance of being stuck full of splinters—but then I distinctly recollected that there were *some* of them which were not disfigured at all. Now I could not account for this difference except by supposing that the roughened fragments were the only ones which had been *completely absorbed*—that the others had entered the whirl at so late a period of the tide, or, for some reason, had descended so slowly after entering, that they did not reach the bottom before the turn of the flood came, or of the ebb, as the case might be. I conceived it possible, in either instance, that they might thus be whirled up again to the level of the ocean, without undergoing the fate of those which had been drawn in more early, or absorbed more rapidly. I made, also, three important observations. The first was, that, as a general rule, the larger the bodies were, the more rapid their descent—the second, that, between two masses of equal extent, the one spherical, and the other *of any other shape*, the superiority in speed of descent was with the sphere—the third, that, between two masses of equal size, the one cylindrical, and the other of any other shape, the cylinder was absorbed the more slowly. Since my escape, I have had several conversations on this subject with an old school-master of the district; and it was from him that I learned the use of the words "cylinder" and "sphere". He explained to me— although I have forgotten the explanation—how what I observed was, in fact, the natural consequence of the forms of the floating fragments—and showed me how it happened that a cylinder, swimming in a vortex, offered more resistance to its suction, and was drawn in with greater difficulty than an equally bulky body, of any form whatever.*

'There was one startling circumstance which went a great way in enforcing these observations, and rendering me anxious to turn them to account, and this was that, at every revolution, we passed something like a barrel, or else the broken yard or the mast of a vessel, while many of these things, which had been on our level when I first opened my eyes upon the wonders of the whirlpool, were now high up above us, and seemed to have moved but little from their original station.

'I no longer hesitated what to do. I resolved to lash myself securely to the water cask upon which I now held, to cut it loose from the counter, and to throw myself with it into the water. I attracted my brother's attention by signs, pointed to the floating barrels that came near us, and did everything in my power to make him understand

* See Archimedes, 'De Incidentibus in Fluido.'—lib. 2.

what I was about to do. I thought at length that he comprehended my design—but, whether this was the case or not, he shook his head despairingly, and refused to move from his station by the ring-bolt. It was impossible to force him; the emergency admitted no delay; and so, with a bitter struggle, I resigned him to his fate, fastened myself to the cask by means of the lashings which secured it to the counter, and precipitated myself with it into the sea, without another moment's hesitation.

'The result was precisely what I had hoped it might be. As it is myself who now tells you this tale—as you see that I *did* escape—and as you are already in possession of the mode in which this escape was effected, and must therefore anticipate all that I have further to say—I will bring my story quickly to conclusion. It might have been an hour, or thereabout, after my quitting the smack, when, having descended to a vast distance beneath me, it made three or four wild gyrations in rapid succession, and, bearing my loved brother with it, plunged headlong, at once and forever, into the chaos of foam below. The cask to which I was attached sank very little further than half the distance between the bottom of the gulf and the spot at which I leaped overboard, before a great change took place in the character of the whirlpool. The slope of the sides of the vast funnel became momently less and less steep. The gyrations of the whirl grew, gradually, less and less violent. By degrees, the froth and the rainbow disappeared, and the bottom of the gulf seemed slowly to uprise. The sky was clear, the winds had gone down, and the full moon was setting radiantly in the west, when I found myself on the surface of the ocean, in full view of the shores of Lofoden, and above the spot where the pool of the Moskoe-ström *had been*. It was the hour of the slack—but the sea still heaved in mountainous waves from the effects of the hurricane. I was borne violently into the channel of the Ström, and in a few minutes was hurried down the coast into the "grounds" of the fishermen. A boat picked me up—exhausted from fatigue—and (now that the danger was removed) speechless from the memory of its horror. Those who drew me on board were my old mates and daily companions—but they knew me no more than they would have known a traveller from the spirit-land. My hair which had been raven-black the day before, was as white as you see it now. They say too that the whole expression of my countenance had changed. I told them my story. They did not believe it. I now tell it to *you*—and I can scarcely expect you to put more faith in it than did the merry fishermen of Lofoden.'

I Have Been Drowned

~~~

TOM HOPKINSON

WHEN I was a boy my mother took me to a gypsy. She was a dirty old woman in a tent. My mother hoped, I suppose, that she would foretell me fame and fortune. I was hoping she would foretell me a pony for my birthday.

The gypsy looked for a long time into a crystal. Then she spoke, and in a clear voice quite unlike the gruff dialect in which she bargained, lied and quarrelled, the gypsy said, 'The boy will meet death by drowning.' Then she added, rather oddly, 'I see him drown; at least I think I do.'

Even at my age I realized that her hesitation set the stamp of truth on what she said. If she had wanted to invent a tale she would have made up a high-sounding one, full of the good fortune and grand events for which my mother longed.

Though I accepted death by drowning as my fate from that day on, I was never in the least haunted by the thought of it. I have always felt in my heart I should prefer to die that way, and now it is clear to me I never shall, I feel almost as though I had been cheated of a promised honour, degraded from my proper destiny, done out of my own death.

Anyone who remembers being a child will recall the overwhelming desires which seized one for an object, not beautiful or useful, which could only be a nuisance to one when one had it. I have been kept awake at night by craving for a white mouse belonging to a friend. I stole the mouse, and was awake next night with excitement of the theft and mad joy of possession.

Something of the sort must have happened to me, I suppose, on the hot June afternoon when I saw *Stella*.

Stella, I could see straightaway, was almost everything a boat ought not to be. She was too fine forward, so that she would thrust her nose into the waves instead of lifting to them. She was too square aft, so that a following sea would smash her from side to side with constant danger of a jibe. She had a long deep keel, making it impossible to run for shelter into the muddy inlets of the coast where I should sail her.

I went into the builder's yard above which I had seen her much too tall mast tower, and looked her over inch by inch. When I had done I hated the very sight of her, hated her with the hatred one can only feel for something to which one is inextricably married, married not by force of circumstances which may change or weaken, but by an unchanging thing implanted in the substance of one's self.

There were some few points in *Stella*'s favour. I had hoped as I looked at her to find her rotten, and so be given excuse for never setting eyes on her again. She was dry as a barrel and tough as a concrete pavement, built of teak.

'She don't make a cupful of water in two years', said the builder, kicking angrily at her side—I could see he hated the boat as much as I did, and would gladly have scored off her by leaving her in his yard to rot unsold—adding in explanation, 'She've got two skins.'

Her enormous keel was of lead—worth money in itself—and she had a prodigious variety of sails, ranging from a tiny trysail and storm-jib to a towering mainsail and pot-bellied spinnaker. At the worst, I thought, I could ride out bad weather with two of the smallest pocket-handkerchiefs. Then I looked at the steepling mast and loaded keel and saw that without a stitch of sail the thresh and strain in any wind and sea would be terrific.

I bought *Stella*, as I had known all along I should. I paid £120 for her. It was nothing for a boat of her size and condition. From the point of view of anyone wanting a pleasant sail, it was money thrown down the drain. Sailing her could only be a nightmare.

As I went out of the yard I looked at the builder. 'She's a maniac's boat,' I said. 'Built by a lunatic, to be sailed by idiots. She ought to be broken up now, before someone injures themselves in her. If I'd any sense, I'd pay you twice the money not to let me have her.'

The dealer looked at me under his wedged-down bowler-hat. The deal was finished. I should not back out. 'She's a drowning boat,' he said.

I took *Stella* away from the yard a fortnight later. She scarcely needed touching. I had her painted black with varnished decks. She was copper-fastened all through and her fittings were brass.

I sailed her down the Medway and across to the Essex coast—where I lived aboard, and got to know her. But 'got to know' are not the words to use of *Stella*. I had known everything about her long before. She played no trick which I had not foreseen and dreaded. Her good qualities I had counted on without her ever proving they were there.

She buried her bows instead of lifting them, as I had known she would. When running before a wind, I had to be 'steering' all the time, spinning the wheel as a following wave crashed against her counter, spinning it back before the boom could fly over and snap the mast off like a hemlock stalk. A good boat steers herself. A good helmsman scarcely uses the wheel; he thinks instead. It was no good thinking at *Stella*. You had to take two hands to her and use all the strength of arms and planted legs and lever body.

As against that, she would sail almost into the wind's eye in defiance of all nature, and with her monstrous spread of canvas would go driving on as if towed by an unseen army of porpoises, when other boats were tossing helplessly up and down, their crews playing nap for occupation.

At first I sailed *Stella* with a paid hand. That was all right for a week or two. But before long I began to find fault with everything he did. He could never set a sail just right. He left specks on the brass-work. At the wheel he let her fall away, shipped more seas than he need. I sacked him. When he went off he looked not at all angrily at me, but sympathetically, as a man might look at another he must leave for the winter in the frozen north.

From that time I ministered to all *Stella*'s wants myself. I would let no one so much as splice a rope for her. By the end of the summer I would not even let other people come aboard. If, in an access of friendship or desire, I asked someone to stay on her for the weekend, I would be certain the day before to send a wire and put them off. My friends all gave me up.

I did not mind: I scarcely noticed. During August and September I won seven firsts and two second prizes with *Stella* in ten races, sailing her with a young fisherman whom I paid for every race and packed off as soon as it was over. He hated the sea, and said so, and he hated me and my boat more than ever he could find words to say.

In return, I ignored his presence. He had no more to do with *Stella* or me than the seagulls overhead. I had no bunk for him. He brought his own food in a red handkerchief. I carried no more than a single lifebelt for myself.

The day on which I was to drown was the day of the Yantlet Regatta. It was a dirty day, and a falling glass showed worse to follow. There was talk of calling the regatta off. All the competitors in my class—we had to sail an eighteen-mile course out of the estuary and back—agreed that, if our race were to be held they would refuse to sail it.

I found a note in the rules of the Club saying that, if one member of a class wished to sail the course, he could do so and claim the race. I declared that I would sail and claim. The others held together and refused. 'It's a game for lunatics,' they said.

I found my fisherman and offered him double money for the race. 'You know what you can do with that,' he said. I offered him £20— double the prize money—for himself. He looked down at his boots, round at the sky, and went to get his oilskins.

There was nobody except the officials of the Club to see us start. The race had been banned by the members. They were watching the dinghies race behind a breakwater.

We tacked out to the buoy, six miles in all, with four of open sea, *Stella* sailing as she had never sailed before. If the other boats had come out they would have looked like carthorses. After that we had to beat to windward—five further miles of open water, as 'open' as any water I've ever known—then we rounded a lightship and set off for home, doing six knots and the boat half lost in spray.

We had covered perhaps a quarter of the way on this last leg when mist came down, not a nice gentle mist, but a foul blinding mist; not a mist that is laid round you like a blanket, but a mist that is flung at you like rough-cast. At the same time the tide turned.

The sea was now running out and the wind blowing in. The waves got bigger every minute. *Stella* began to plunge her nose. Great slews of water flushed along the deck. I had been soaked two hours before. Now the water was battering me solid. It was as much as I could do to stand. 'Better take in sail,' said the fisherman; they called him Jack.

'You'll take in nothing,' I shouted at him through the wind. Jack did not answer. He went forward and begun to reef the foresail.

'Blast him,' I thought. 'I'll show him.'

We were running then over a sand-bar, a spit that would be dried out at low water. Now there were two or three fathoms over it. The waves were pounding and threshing, half-breaking, instead of lifting and then sliding away beneath her keel. There was a big one just ahead and as we came to it, instead of turning *Stella*'s bows away so

that she met it slightly at an angle, I drove them in. She took it solid.
A great belch of water burst along the deck, splitting over every
obstacle, reuniting again the second it was past.

It took Jack waist-high, knocked him off his feet. I thought for a
second he was gone. But it takes more than a wave to drown a
fisherman, and as he swept down the scuppers he grabbed hold of one
of the stays and clung till the burst of water had gone by. That
happened in one second.

In the next there was a crack, a painful crack like the sound of a
living body broken, then an outbreak of smaller cracks, wires and
ropes whipped through the air or sank coiling by my feet, all of the
decks forward and yards of sea on either side were suddenly over-
spread with canvas, through which the water welled and over which
it spread.

The weight and shock of that sea smashing her nose down, while a
gust was bursting and lifting into her sails behind, had snapped the
mainmast short.

I had thought very often of what I should do if *Stella* went down at
sea, and she was going to go down now. Without way on her she
would not steer. The next two or three waves would hammer her
counter round and broach her to. Once broadside on, a couple of
waves would fill her, and the tons of valuable lead on her bottom
would do the rest.

I slithered and bolted forward to where the lifebelt was fixed on top
of a skylight, tore it off, thrust head and arms inside, and went
overboard in a patch of sea clear of sail and cordage. Two minutes
later she was gone.

As soon as I found myself in the sea I knew what I must do. I never
wore sea-boots on *Stella*. I did not trust her well enough. I could keep
afloat with my lifebelt for several hours if I did not exhaust myself in
swimming. A fool would have tried to make coast against the tide.
But to make those four or five miles—an hour's quick walking—I
would have needed to swim twenty. I could not even hold my place
and hope to be washed back when the tide turned in six hours' time.
All I could do was keep afloat and save my strength.

Having a lifebelt, I knew, is not everything. Plenty of drowned
bodies are washed up with lifebelts underneath their arms. You have
got to keep your face out of the water, not easy when the waves are
running high, and you have got to keep yourself from dying of exhaustion, though normally you feel cold after a five minutes' bathe.

I reckoned I had two chances of life. One was to drift down to the lightship we had rounded, and hope to grab one of her chains or rouse her men by shouting. The other was that some of the men at the Yacht Club would set out to look for me.

They would not do that for several hours, because we had been making much faster time than they'd expect, and when we did not turn up they'd only think we had run into a creek for shelter, and would telephone round for a while to see if there was news of us. There was chance of a stray steamer catching sight of me, but it was just as likely in the mist she'd cut me down.

I decided to swim gently to keep the blood flowing, first on my breast, then on my back, shifting the lifebelt slightly as I turned from one position to the other. Once as I was swimming forward it slipped back and caught for a moment on my hips, lifting my body and pressing my face into the water. I rolled over on my back and worked it free. Once as I changed position my arm got pinned inside the belt. With muscles weakening, I had a horror that I could not get it back. I got it back quite easily and went on swimming.

I swam as I had been taught to swim, in long slow sweeping strokes, driving with my legs, which are strong and slow to tire, using my arms for direction and support, drawing great breaths of air upon the upward stroke, blowing it out from my mouth before me through the water as my arms came in and forward. I took as much care with my style as though I had been in for a competition and there were judges pacing beside me in a bath, for in the swim I had embarked on, just one more stroke, or ten or twenty, might serve, I knew, to keep me in the world.

At first, while I was over the bar where the waves were breaking, I thought I should be choked. Water beat up my nose, into my mouth. My throat was full of water and I thought my lungs must be filling too. A man must breathe to live, and every time I took in air I took in sea. Then I drifted into deep water where there was swell instead of breakers, and breath came more easily. I felt for a moment as glad as if I had been rescued. I looked round, and swam on almost gaily through the mist and swell.

I had been in the water for perhaps half an hour when I happened to catch sight of my fingers. Two of the fingers of my left hand and the little finger of my right hand were pale green. I have never been much at home in water, and after a short bathe it takes me half an hour to get my blood flowing properly through my limbs. There was no chance to get it flowing again now.

It was not long before my other fingers had gone dead. I imagined the paralysis moving up like mercury in a thermometer, from wrist to elbow, elbow to shoulder, and then running on down inside my body. It became necessary for me to know if my toes were dead as well, to see if the paralysis were coming from both ends or only one. I did actually raise my feet in the water and tried to take one of my shoes off to find out, but my dead fingers would no longer work.

I cannot say how long I had been in the sea before I saw the lightship. I was conscious one moment of something dark seen out of the corner of my eye. The next instant there she was, not as I'd pictured her, a comfortable and friendly presence with men waiting ready to haul me out and take me in but a great heaving mass rolling her iron belly up with every surge, then crashing back with showers of spray into the swell, a dreadful heap of metal uncontrolled terrifying to a thing of flesh.

Her cables, which I imagined as so steady and accessible, almost like life-chains in a swimming-bath, whipped through the waves as though themselves alive, the links grinding and crashing like goods-trucks in a siding. I kicked away from her with all the strength I had, and as I kicked I shouted.

I shouted once, a high-pitched shriek. There came no sign or answer. I thought the thin sound might seem too like a seagull's cry. I gathered up air into my lungs, and burst it all out again in roars. In the wild clatter and grind of everything on board, in the explosions all round me of the bursting seas, I scarcely heard the sound myself.

As the lightship passed away out of sight I began to expect death. It would not be possible for me much longer to keep my face out of the waves. It was already impossible to keep it from the crests, and I drew breath by gulping in the quiet troughs. Soon whether crests or troughs would be alike to me, my face would lie helplessly forward on the surge, and water would make its way into my lungs, driving the last air out in bubbles from my nose.

My weary arms were becoming with every stroke more difficult to lift. I had to force them serve my purpose, almost shouting my orders to them with my mouth, as they began to disregard the messages sent along sinews from my brain.

I was becoming a prey to bodily fancies that would have stopped me swimming altogether had I heeded them. It was not my arms that weighed so much and were so difficult to move, it was the burden of clothing on my arms which had now grown weightier than I could lift. It was the thickness of the clothing on my body which prevented me dividing the waters like a fish.

So conscious was I of my painful arms that I forgot the existence of my legs, and when for some reason they came into my mind, I could not feel their presence for the cold. I imagined I had swum away from them entirely or else that they had kicked loose from their joints, and was urgent in myself to get them back, feeling I could never make land without their help.

Inside my body I endured a feeling of extreme and painful cold, as though my entrails had been taken out and ice sewn in. The sole remaining patch of warmth, it seemed to me, was around my heart and my upper chest, between ribs and shoulder-blades.

As I swam on, these various pains lifted and moved away from my outlying parts, becoming centred on my controlling, guiding head. The driving rain, the flung spume lashing from the waves against my face, the constant muscular effort to peer my eyes out of their sockets and lose no chance of sighting ship or boat, combined with the opposite effort to draw them right in beneath their brows for shelter, had caused inside my head a deadly pain, as though the metal of my forehead were on fire.

At the time these things were happening to me in the body, quite other things were happening in my mind.

I had no experience at all of that delightful coloured-cinema phenomenon in which the whole of one's past life is said to unroll progressively before one's eyes, and at the same time, to be presented to one in a flash. Yet certain moments in my life resurged before me with a more than natural vividness, growing sharper in image or receding, in time to the rising and falling of the waves.

I had, first, a picture of myself as infant, bundled together within my mother's womb, drawing life through a tube as a diver draws air down from the surface.

I watched myself grow from a small swimming fish into a dwarf, enormous-headed, sightless-eyed, his useless limbs tucked round him for convenience, the whole creature shaken and quivering from the pulsing of his own determined heart.

This, the first vision that I saw, was curiously without colour. It had the grey appearance of a photograph, the nebulous vagueness of an X-ray picture.

Then, with no intervening stages, I was become a small boy of seven or eight, constructing for myself a house of packing-cases in the garden. No house, it seemed to me, is complete without store of food, and I was scratching with trowel a hole in the turf floor, to let in a biscuit-tin larder for odd crusts and cakes, a green apple and a bottle which had once held lemonade.

And now, a few years older, I was making of my bicycle a sort of moving home, covering handlebars and frame with fittings, lashing a tiny tent behind the saddle, delighted to show myself independent of the world on those two travelling wheels.

At seventeen, a senior boy at school, I owned a study. It had no door, consisted of no more than a slab of wood for desk, a seat close up against the slab, a bookshelf running round the sides. No bride ever lavished more care on her first home of love than I on that small wooden stall. Its sides I drapped with flowered cloth, hung or pinned pictures over that, bargained with other boys for ornaments that caught my fancy.

My study and my youth washed by me on a wave.

Now, a young man, I walked through a town where I once lived. It was a seaport. Tall, gabled warehouses darkened the street through which I passed.

Suddenly from a doorway ran a cat, a small and common tabby. At the sight I was filled with such passion of tenderness I stood, rooted, to follow with clenched hands its progress up the street. The cat turned off the road, crept low with flattened paunch beneath a warehouse door. My breath ran slowly out in a long sigh. The cat was safe. No wheel would break its back, no boot its belly.

I stopped there in the street to marvel at myself. What was the cat to me—that I should be its loving lord and father? And even as I stopped and asked, I saw the answer. The cat was pregnant. Two days before my love had told me there was a child of mine inside her body.

Over all these pictures as they dawned and faded, I experienced no emotion or regret. There was nothing I wished to bring back of what was gone, no untried course I wished that I had rather followed. I did not even wish or unwish the experience of seeing what I saw. The visions simply came before my eyes, and glowed, and died.

Between the vanishing of one and the appearance of the next were some few moments of torment when I thought, not that my soul was separating from my body, but that my body was being wrenched by violence from my living soul.

Illusions of size and shape obsessed me. Now I traced all my sufferings to the battering of the waves which had compressed my frame
to the substance of a tiny pellet, so that all I was bearing had been
concentrated and rendered more intense by the small space in me
available for suffering.

Now, now, it was the dreadful opposite. My racked and elongated
body sprawled over so vast an acreage of water, that not one only, but
a thousand waves attacked it from all sides. The tides both ebbed and
flowed along its length. A million screaming gulls let fall their fishy
droppings on my freezing back.

Again I was suffering, not from my size but from my shape. I had
become a sponge. A hundred broad and narrow inlets carried the
coldness of the sea into the very centre of my being.

Through all this time, as a man may feel in a limb which is lost, so
I swam on with strength that was long since spent. I had not died,
and till I died my legs and arms could not refuse their work. But I
had now no plan, no hope. I did not know if I was moving towards,
or from, the shore, or whether perhaps there was no shore at all.

When I first entered the sea the world had withdrawn into two
elements. There was the water upon which I rose and fell, and there
was the mist that drove above me, and from which I sucked in the
dry air needed by any human lungs. But now the distinction between
these elements was vanishing away. It seemed to me that the world
as I swam had somehow been turned over. Water was now on top,
and air beneath. The breath I was drawing from above had become
too thick and watery to feed my lungs. I should do better, I saw, to
draw breath from below.

With this discovery, that I could draw breath from the sea, there
succeeded a strangely happy mood. I had done the most that could
be expected of me. I had swum quite truly all I could. Exhausted but
not terrified, I should swim now straight on from this life to the next.
The dead would see me swimming as I entered upon the tideways of
their world. I should come upon my new life, not drifting like a
coward or amoeba, but swimming like a man. The thought gave
comfort to me.

In life I had always loved and admired sailors above other men.
Many sailors, I thought, must have passed from one life to the other
in this way, upon cold bellies and with working arms and legs. I was
proud and glad of the company I kept. I took a deep breath of water
to sustain me, and swam on.

There were two bodies lying on the beach; I was aware of this long before I understood that one of them was mine. Two bodies, and a crowd of men round each. The men were shouting, sometimes it seemed to me that they were arguing. I was too far from the ears which I now saw to be my own for any sound to reach me through them. At last the men all turned from one body and worked upon the other. I saw now what they were up to: they were trying to bring me back to life.

A scorch of fire passed through my frame, and I sat up, shaking, crying, coughing, chattering. These were no angels by my side, but ordinary men. The channels of my ears were suddenly unstopped, and sound came through: the voices of the fishermen came to me from far away, like the cries of travellers calling across an estuary at evening for a ferry.

'Drowned,' they said, 'drowned'—but it was of Jack they said it, not of me.

Mocha Dick

Or The White Whale of the Pacific:
A Leaf from a Manuscript Journal.

～～～

J. N. REYNOLDS

WE expected to find the island of Santa Maria still more remarkable for the luxuriance of its vegetation, than even the fertile soil of Mocha; and the disappointment arising from the unexpected shortness of our stay at the latter place, was in some degree relieved, by the prospect of our remaining for several days in safe anchorage at the former. Mocha lies upon the coast of Chili, in lat. 38° 28′ south, twenty leagues north of Mono del Bonifacio, and opposite the Imperial river, from which it bears w. s. w. During the last century, this island was inhabited by the Spaniards, but it is at present, and has been for some years, entirely deserted. Its climate is mild, with little perceptible difference of temperature between the summer and winter seasons. Frost is unknown on the lowlands, and snow is rarely seen, even on the summits of the loftiest mountains.

It was late in the afternoon, when we left the schooner; and while we bore up for the north, she stood away for the southern extremity of the island. As evening was gathering around us, we fell in with a vessel, which proved to be the same whose boats, a day or two before, we had seen in the act of taking a whale. Aside from the romantic and stirring associations it awakened, there are few objects in themselves more picturesque or beautiful, than a whale-ship, seen from a distance of three or four miles, on a pleasant evening, in the midst of the great Pacific. As she moves gracefully over the water, rising and falling on the gentle undulations peculiar to this sea; her sails glowing in the quivering light of the fires that flash from below, and a thick volume of smoke ascending from the midst, and curling away in dark masses upon the wind; it requires little effort of the fancy, to imagine one's self gazing upon a floating volcano.

As we were both standing to the north, under easy sail, at nine o'clock at night we had joined company with the stranger. Soon after, we were boarded by his whale-boat, the officer in command of which bore us the compliments of the captain, together with a friendly invitation to partake the hospitalities of his cabin. Accepting, without hesitation, a courtesy so frankly tendered, we proceeded, in company with Captain Palmer, on board, attended by the mate of the Penguin, who was on his way to St Mary's to repair his boat, which had some weeks before been materially injured in a storm.

We found the whaler a large, well-appointed ship, owned in New York, and commanded by such a man as one might expect to find in charge of a vessel of this character; plain, unassuming, intelligent, and well informed upon all the subjects relating to his peculiar calling. But what shall we say of his first mate, or how describe him? To attempt his portrait by a comparison, would be vain, for we have never looked upon his like; and a detailed description, however accurate, would but faintly shadow forth the *tout ensemble* of his extraordinary figure. He had probably numbered about thirty-five years. We arrived at this conclusion, however, rather from the untamed brightness of his flashing eye, than the general appearance of his features, on which torrid sun and polar storm had left at once the furrows of more advanced age, and a tint swarthy as that of the Indian. His height, which was a little beneath the common standard, appeared almost dwarfish, from the immense breadth of his overhanging shoulders; while the unnatural length of the loose, dangling arms which hung from them, and which, when at rest, had least the appearance of ease, imparted to his uncouth and muscular frame an air of grotesque awkwardness, which defies description. He made few pretensions as a sailor, and had never aspired to the command of a ship. But he would not have exchanged the sensations which stirred his blood, when steering down upon a school of whales, for the privilege of treading, as master, the deck of the noblest liner that ever traversed the Atlantic. According to the admeasurement of his philosophy, whaling was the most dignified and manly of all sublunary pursuits. Of this he felt perfectly satisfied, having been engaged in the noble vocation for upward of twenty years, during which period, if his own assertions were to be received as evidence, no man in the American spermaceti fleet had made so many captures, or met with such wild adventures, in the exercise of his perilous profession. Indeed, so completely were all his propensities, thoughts, and feelings, identified with his occupation; so intimately did he seem acquainted with the habits and instincts of the objects of his pursuit, and so little conversant with the ordinary affairs of life; that one felt less inclined to class him in the genus *homo*, than as a sort of intermediate something between man and the cetaceous tribe.

Soon after the commencement of his nautical career, in order to prove that he was not afraid of a whale, a point which it is essential for the young whaleman to establish beyond question, he offered, upon a wager, to run his boat 'bows on' against the side of an 'old bull', leap from the 'cuddy' to the back of the fish, sheet his lance home, and return on board in safety. This feat, daring as it may be considered, he undertook and accomplished; at least so it was chronicled in his log, and he was ready to bear witness, on oath, to the veracity of the record. But his conquest of the redoubtable MOCHA DICK, unquestionably formed the climax of his exploits.

Before we enter into the particulars of this triumph, which, through their valorous representative, conferred so much honour on the lancers of Nantucket, it may be proper to inform the reader who and what Mocha Dick was; and thus give him a posthumous introduction to one who was, in his day and generation, so emphatically among fish the 'Stout Gentleman' of his latitudes. The introductory portion of his history we shall give, in a condensed form, from the relation of the mate. Substantially, however, it will be even as he rendered it; and as his subsequent narrative, though not deficient in rude eloquence, was coarse in style and language, as well as unnecessarily diffuse, we shall assume the liberty of altering the expression; of adapting the phraseology to the occasion; and of presenting the whole matter in a shape more succinct and connected. In this arrangement, however, we shall leave our adventurer to tell his *own story*, although not always in his own words, and shall preserve the person of the original.

But to return to Mocha Dick—which, it may be observed, few were solicitous to do, who had once escaped from him. This renowned monster, who had come off victorious in a hundred fights with his pursuers, was an old bull whale, of prodigious size and strength. From the effect of age, or more probably from a freak of nature, as exhibited in the case of the Ethiopian Albino, a singular consequence had resulted—*he was white as wool!* Instead of projecting his spout obliquely forward, and puffing with a short, convulsive effort, accompanied by a snorting noise, as usual with his species, he flung the water from his nose in a lofty, perpendicular, expanded volume, at regular and somewhat distant intervals; its expulsion producing a continuous roar, like that of vapour struggling from the safety-valve of a powerful steam engine. Viewed from a distance, the practised eye of the sailor only could decide, that the moving mass, which constituted this enormous animal, was not a white cloud sailing along the horizon. On the spermaceti whale, barnacles are rarely discovered; but upon the head of this *lusus naturæ*, they had clustered, until it became absolutely rugged with the shells. In short, regard him as you would, he was a most extraordinary fish; or, in the vernacular of Nantucket, 'a genuine old sog', of the first water.

Opinions differ as to the time of his discovery. It is settled, however, that previous to the year 1810, he had been seen and attacked near the island of Mocha. Numerous boats are known to have been shattered by his immense flukes, or ground to pieces in the crush of his powerful jaws; and, on one occasion, it is said that he came off victorious from a conflict with the crews of three English whalers, striking fiercely at the last of the retreating boats, at the moment it was rising from the water, in its hoist up to the ship's davits. It must not be supposed, howbeit, that through all this desperate warfare, our leviathan passed scathless. A back serried with irons, and from fifty to a hundred yards of line trailing in his wake, sufficiently attested, that though unconquered, he had not proved invulnerable. From the period of Dick's first appearance, his celebrity continued to increase, until his name seemed naturally to mingle with the salutations which whalemen were in the habit of exchanging, in their encounters upon the broad Pacific; the customary interrogatories almost always closing with, 'Any news from Mocha Dick?' Indeed, nearly every whaling captain who rounded Cape Horn, if he possessed any professional ambition, or valued himself on his skill in subduing the monarch of the seas, would lay his vessel along the coast, in the hope of having an opportunity to try the muscle of this doughty champion, who was never known to shun his assailants. It was remarked, nevertheless, that the old fellow seemed particularly careful as to the portion of his body which he exposed to the approach of the boat-steerer; generally presenting, by some well-timed manœuvre, his back to the harpooner; and dexterously evading every attempt to plant an iron under his fin, or a spade on his 'small'. Though naturally fierce, it was not customary with

Dick, while unmolested, to betray a malicious disposition. On the contrary, he would sometimes pass quietly round a vessel, and occasionally swim lazily and harmlessly among the boats, when armed with full craft, for the destruction of his race. But this forbearance gained him little credit, for if no other cause of accusation remained to them, his foes would swear they saw a lurking deviltry in the long, careless sweep of his flukes. Be this as it may, nothing is more certain, than that all indifference vanished with the first prick of the harpoon; while cutting the line, and a hasty retreat to their vessel, were frequently the only means of escape from destruction, left to his discomfited assaulters.

Thus far the whaleman had proceeded in his story, and was about commencing the relation of his own individual encounters with its subject, when he was cut short by the mate of the Penguin, to whom allusion has already been made, and who had remained, up to this point, an excited and attentive listener. Thus he would have continued, doubtless, to the end of the chapter, notwithstanding his avowed contempt for every other occupation than sealing, had not an observation escaped the narrator, which tended to arouse his professional jealousy. The obnoxious expression we have forgotten. Probably it involved something of boasting or egotism; for no sooner was it uttered, than our sealer sprang from his seat, and planting himself in front of the unconscious author of the insult, exclaimed:

'*You!*—you whale-killing, blubber-hunting, light-gathering varmint!— *you* pretend to manage a boat better than a Stonington sealer! A Nantucket whaleman', he continued, curling his lip with a smile of supreme disdain, 'presume to teach a Stonington sealer how to manage a boat! Let all the small craft of your South Sea fleet range among the rocks and breakers where I have been, and if the whales would not have a peaceful time of it, for the next few years, may I never strip another jacket, or book another skin! What's taking a whale? Why, I could reeve a line through one's blow-hole, make it fast to a thwart, and then beat his brains out with my seal-club!'

Having thus given play to the first ebullition of his choler, he proceeded with more calmness to institute a comparison between whaling and sealing. 'A whaler', said he, 'never approaches land, save when he enters a port to seek fresh grub. Not so the sealer. *He* thinks that his best fortune, which leads him where the form of man has never before startled the game he's after; where a quick eye, steady nerve, and stout heart, are his only guide and defence, in difficulty and danger. Where the sea is roughest, the whirlpool wildest, and the surf roars and dashes madly among the jagged cliffs, there—I was going to say there *only*—are the peak-nosed, black-eyed rogues we hunt for, to be found, gambolling in the white foam, and there must the sealer follow them. Were I to give you an account of my adventures about the Falkland Isles; off the East Keys of Staten Land; through the South Shetlands; off the Cape, where we lived on salt pork and seal's flippers; and finally, the story of a season spent with a single boat's crew on Diego Ramirez,* you would not make such a fuss about your Mocha Dick. As to the straits of Magellen, Sir, they are as familiar to me, as Broadway to a New-York dandy; though *it* should strut along that fashionable promenade twelve dozen times a day.'

Our son of the sea would have gone on to particularize his 'hair-breadth 'scapes and moving accidents', had we not interposed, and insisted that the remainder of the night should be devoted to the conclusion of Dick's history; at the same time assuring the 'knight of the club' that so soon as we met at Santa Maria, he should have an entire evening expressly set apart, on which he might glorify himself and his calling. To this he assented, with the qualification, that his compliance with the general wish, in thus yielding precedence to his rival, should not be construed into an admission, that Nantucket whalemen were the best boatmen in the world, or that sealing was not as honorable and as pretty a business for coining a penny, as the profession of 'blubber-hunting' ever was.

The whaler now resumed. 'I will not weary you', said he, 'with the uninteresting particulars of a voyage to Cape Horn. Our vessel, as capital a ship as ever left the little island of Nantucket, was finely manned and commanded, as well as thoroughly provided with every requisite for the peculiar service in which she was engaged. I may here observe, for the information of such among you as are not familiar with these things, that soon after a whale-ship from the United States is fairly at sea, the men are summoned aft; then boats' crews are selected by the captain and first mate, and a ship-keeper, at the same time, is usually chosen. The place to be filled by this individual is an important one; and the person designated should be a careful and sagacious man. His duty is, more particularly, to superintend the vessel while the boats are away, in chase of fish; and at these times, the cook and steward are perhaps his only crew. His station, on these

* Diego Ramirez is a small island, lying s.w. from Cape Horn.

occasions, is at the mast-head, except when he is wanted below, to assist in working the ship. While aloft, he is to look out for whales, and also to keep a strict and tireless eye upon the absentees, in order to render them immediate assistance, should emergency require it. Should the game rise to windward of their pursuers, and they be too distant to observe personal signs, he must run down the jib. If they rise to leeward, he should haul up the spanker; continuing the little black signal-flag at the mast, so long as they remain on the surface. When the 'school' turn flukes, and go down, the flag is to be struck, and again displayed when they are seen to ascend. When circumstances occur which require the return of the captain on board, the colours are to be hoisted at the mizzen peak. A ship-keeper must further be sure that provisions are ready for the men, on their return from the chase, and that drink be amply furnished, in the form of a bucket of 'switchel'. 'No whale, no switchel,' is frequently the rule; but *I* am inclined to think that, whale or no whale, a little rum is not amiss, after a lusty pull.

'I have already said, that little of interest occurred, until after we had doubled Cape Horn. We were now standing in upon the coast of Chili, before a gentle breeze from the south, that bore us along almost imperceptibly. It was a quiet and beautiful evening, and the sea glanced and glistened in the level rays of the descending sun, with a surface of waving gold. The western sky was flooded with amber light, in the midst of which, like so many islands, floated immense clouds, of every conceivable brilliant dye; while far to the north-east, looming darkly against a paler heaven, rose the conical peak of Mocha. The men were busily employed in sharpening their harpoons, spades, and lances, for the expected fight. The look-out at the mast-head, with cheek on his shoulder, was dreaming of the 'dangers he had passed', instead of keeping watch for those which were to come; while the captain paced the quarter-deck with long and hasty stride, scanning the ocean in every direction, with a keen, expectant eye. All at once, he stopped, fixed his gaze intently for an instant on some object to leeward, that seemed to attract it, and then, in no very conciliating tone, hailed the mast-head:

' "Both ports shut?" he exclaimed, looking aloft, and pointing back-ward, where a long white bushy spout was rising, about a mile off the larboard bow, against the glowing horizon. "Both ports shut? I say, you leaden-eyed lubber! Nice lazy son of a sea-cook *you* are, for a look-out! Come down, Sir!"

' "There she blows!—sperm whale—old sog, sir," said the man, in a deprecatory tone, as he descended from his nest in the air. It was at once seen that the creature was companionless; but as a lone whale is generally an old bull, and of unusual size and ferocity, more than ordinary sport was anticipated, while unquestionably more than ordinary honour was to be won from its successful issue.

'The second mate and I were ordered to make ready for pursuit; and now commenced a scene of emulation and excitement, of which the most vivid description would convey but an imperfect outline, unless you have been a spectator or an actor on a similar occasion. Line-tubs, water-kegs, and wafe-poles, were thrown hurriedly into the boats; the irons were placed in the racks, and the necessary evolutions of the ship gone through, with a quickness almost magical; and this too, amidst what to a landsman would have seemed inextricable confusion, with perfect regularity and precision; the commands of the officers being all but forestalled by the enthusiastic eagerness of the men. In a short time, we were as near the object of our chase, as it was considered prudent to approach.

' "Back the main-top-s'l!" shouted the captain. "There she blows! there she blows!—there she blows!"—cried the look-out, who had taken the place of his sleepy shipmate, raising the pitch of his voice with each announcement, until it amounted to a downright yell. 'Right ahead. Sir!—spout as long an 's thick as the mainyard!'

' "Stand by to lower!" exclaimed the captain; "all hands; cook, steward, cooper—every d—d one of ye, stand by to lower!"

'An instantaneous rush from all quarters of the vessel answered this appeal, and every man was at his station, almost before the last word had passed the lips of the skipper.

' "Lower away!"—and in a moment the keels splashed in the water. "Follow down the crews; jump in my boys; ship the crotch; line your oars; now pull, as if the d—l was in your wake!" were the successive orders, as the men slipped down the ship's side, took their places in the boats, and began to give way.

'The second mate had a little the advantage of me in starting. The stern of his boat grated against the bows of mine, at the instant I grasped my steering-oar, and gave the word to shove off. One sweep of my arm, and we sprang foaming in his track. Now came the tug-of-war. To become a first-rate oarsman, you must understand, requires a natural gift. My crew were not wanting in the proper qualification; every mother's son of them pulled as if he had been born with an oar in his hand; and as they stretched every sinew for the glory of darting the first iron it did my heart good to see the boys spring. At every stroke, the tough blades bent like willow wands, and quivered like tempered steel in the warm sunlight, as they sprang forward from the tension of the retreating wave. At the distance of half a mile, and directly before us, lay the object of our emulation and ambition, heaving his huge bulk in unwieldly gambols, as though totally unconscious of our approach.

' "There he blows! An old bull, by Jupiter! Eighty barrels, boys, waiting to be towed alongside! Long and quick—shoot ahead! Now she feels it; waist-boat never could beat us; now she feels the touch!— now she walks through it! Again—*now!*" Such were the broken exclamations and adjurations with which I cheered my rowers to their toil, as, with renewed vigor, I plied my long steering-oar. In another moment, we were alongside our competitor. The shivering blades flashed forward and backward, like sparks of light. The waters boiled under our prow, and the trenched waves closed, hissing and whirling, in our wake, as we swept, I might almost say were *lifted*, onward in our arrowy course.

'We were coming down upon our fish, and could hear the roar of his spouting above the rush of the sea, when my boat began to take the lead.

' "Now, my fine fellows," ' I exclaimed, in triumph, "now we'll show them our stern—only spring! Stand ready, harpooner, but don't dart, till I give the word."

' "Carry me on, and his name's *Dennis!*"* cried the boat-steerer, in a confident tone. We were perhaps a hundred feet in advance of the waist-boat, and within fifty of the whale, about an inch of whose hump only was to be seen above the water, when, heaving slowly into view a pair of flukes some eighteen feet in width, he went down. The men lay on their oars. "There he blows, again!" cried the tub-oarsman, as a lofty, perpendicular spout sprang into the air, a few furlongs away on the starboard side. Presuming from his previous movement, that the old fellow had been "gallied" by other boats, and might probably be jealous of our purpose, I was about ordering the men to pull away as softly and silently as possible, when we received fearful intimation that he had no intention of balking our inclination, or even yielding us the honour of the first attack. Lashing the sea with his enormous

---

* A whale's name is 'Dennis', when he spouts blood.

tail, until he threw about him a cloud of surf and spray, he came down, at full speed, "jaws on", with the determination, apparently, of doing battle in earnest. As he drew near, with his long curved back looming occasionally above the surface of the billows, we perceived that it was *white as the surf around him*; and the men stared aghast at each other, as they uttered, in a suppressed tone, the terrible name of MOCHA DICK!

' "Mocha Dick or the d—l," said I, "this boat never sheers off from anything that wears the shape of a whale. Pull easy; just give her way enough to steer." As the creature approached, he somewhat abated his frenzied speed, and, at the distance of a cable's length, changed his course to a sharp angle with our own.

' "Here he comes!" I exclaimed. "Stand up, harpooner! Don't be hasty—don't be flurried. Hold your iron higher—firmer. Now!" I shouted, as I brought our bows within a boat's length of the immense mass which was wallowing heavily by. *"Now!—give it to him solid!"*

'But the leviathan plunged on, unharmed. The young harpooner, though ordinarily as fearless as a lion, had imbibed a sort of superstitious dread of Mocha Dick, from the exaggerated stories of that prodigy, which he had heard from his comrades. He regarded him, as he had heard him described in many a tough yarn during the middle watch, rather as some ferocious fiend of the deep, than a regular-built, legitimate whale! Judge then of his trepidation, on beholding a creature, answering the wildest dreams of his fancy, and sufficiently formidable, without any superadded terrors, bearing down upon him with thrashing flukes and distended jaws! He stood erect, it cannot be denied. He planted his foot—he grasped the coil—he poised his weapon. But his knee shook, and his sinewy arm wavered. The shaft was hurled, but with unsteady aim. It just grazed the back of the monster, glanced off, and darted into the sea beyond. A second, still more abortive, fell short of the mark. The giant animal swept on for a few rods, and then, as if in contempt of our fruitless and childish attempt to injure him, flapped a storm of spray in our faces with his broad tail, and dashed far down into the depths of the ocean, leaving our little skiff among the waters where he sank, to spin and duck in the whirlpool.

'Never shall I forget the choking sensation of disappointment which came over me at that moment. My glance fell on the harpooner. "Clumsy lubber!" I vociferated, in a voice hoarse with passion; *"you a* whaleman! You are only fit to spear eels! Cowardly spawn! Curse me, if you are not *afraid* of a whale!"

'The poor fellow, mortified at his failure, was slowly and thoughtfully hauling in his irons. No sooner had he heard me stigmatize him as "afraid of a whale", than he bounded upon his thwart, as if bitten by a serpent. He stood before me for a moment, with a glowing cheek and flashing eye; then, dropping the iron he had just drawn in, without uttering a word, he turned half round, and sprang head-foremost into the sea. The tub-oarsman, who was re-coiling the line in the after part of the boat, saw his design just in season to grasp him by the heel, as he made his spring. But he was not to be dragged on board again without a struggle. Having now become more calm, I endeavoured to soothe his wounded pride with kind and flattering words; for I knew him to be a noble-hearted fellow, and was truly sorry that my hasty reproaches should have touched so fine a spirit so deeply.

'Night being now at hand, the captain's signal was set for our return to the vessel; and we were soon assembled on her deck, discussing the mischances of the day, and speculating on the prospect of better luck on the morrow.

'We were at breakfast next morning, when the watch at the foretop-gallant head sung out merrily, "There she breaches!" In an instant everyone was on his feet. "Where away?" cried the skipper, rushing from the cabin, and upsetting in his course the steward, who was returning from the caboose with a replenished biggin of hot coffee. "Not loud but deep" were the grumblings and groans of that functionary, as he rubbed his scalded shins, and danced about in agony; but had they been far louder, they would have been drowned in the tumult of vociferation which answered the announcement from the mast-head.

' "Where away?" repeated the captain, as he gained the deck.

' "Three points off the leeward bow."

' "How far?"

' "About a league, Sir; heads same as we do. There she blows!" added the man, as he came slowly down the shouds, with his eyes fixed intently upon the spouting herd.

' "Keep her off two points! Steady!—steady, as she goes!"

' "Steady it is, Sir," answered the helmsman.

' "Weather braces, a small pull. Loose to'-gallant-s'ls! Bear a hand, my boys! Who knows but we may tickle their ribs at this rising?"

'The captain had gone aloft, and was giving these orders from the main-to'-gallant-cross-trees. "There she top-tails! there she blows!" added he, as, after taking a long look at the sporting shoal, he glided down the back stay. "Sperm whale, and a thundering big school of 'em!" was his reply to the rapid and eager enquiries of the men. "See the lines in the boats," he continued; "get in the craft; swing the cranes!"

'By this time the fish had gone down and every eye was strained to catch the first intimation of their reappearance.

' "There she *spouts!*" screamed a young greenhorn in the main chains, "close by; a mighty big whale, Sir!"

' "We'll know that better at the trying out, my son", said the third mate, drily.

' "Back the main-top-s'l!" was now the command. The ship had little headway at the time, and in a few minutes we were as motionless as if lying at anchor.

' "Lower away, all hands!" And in a twinkling, and together, the starboard, larboard, and waist-boats struck the water. Each officer leaped into his own; the crews arranged themselves at their respective stations; the boat-steerers began to adjust their "craft"; and we left the ship's side in company; the captain, in laconic phrase, bidding us to "get up and get fast", as quickly as possible.

'Away we dashed, in the direction of our prey, who were frolicking, if such a term can be applied to their unwieldly motions, on the surface of the waves. Occasionally, a huge, shapeless body would flounce out of its proper element, and fall back with a heavy splash; the effort forming about as ludicrous a caricature of agility, as would the attempt of some over-fed alderman to execute the Highland fling.

'We were within a hundred rods of the herd, when, as if from a common impulse, or upon some preconcerted signal, they all suddenly disappeared. "Follow me!" I shouted, waving my hand to the men in the other boats; "I see their track under water; they swim fast, but we'll be among them when they rise. Lay back," I continued, addressing myself to my own crew, "back to the thwarts! Spring *hard!* We'll be in the thick of 'em when they come up; only *pull!*"

'And they did pull, manfully. After rowing for about a mile, I ordered them to "lie". The oars were peaked, and we rose to look out for the first "noddle-head" that should break water. It was at this time a dead calm. Not a single cloud was passing over the deep blue of the heavens, to vary their boundless transparency, or shadow for a moment the gleaming ocean which they spanned. Within a short distance lay our noble ship, with her idle canvas hanging in drooping festoons from her yards; while she seemed resting on her inverted image, which, distinct and beautiful as its original, was glassed in the smooth expanse beneath. No sound disturbed the general silence, save our own heavy breathings, the low gurgle of the water against the side of the boat, or the noise of flapping wings, as the albatross wheeled sleepily along through the stagnant atmosphere. We had remained quiet for about five minutes, when some dark object was descried ahead, moving on the surface of the sea. It proved to be a small "calf", playing in the sunshine.

' "Pull up and strike it," said I to the third mate; "it may bring up the old one—perhaps the whole school."

'And so it did, with a vengeance! The sucker was transpierced, after a short pursuit; but hardly had it made its first agonized plunge, when an enormous cow-whale rose close beside her wounded offspring. Her first endeavour was to take it under her fin, in order to bear it away; and nothing could be more striking than the maternal tenderness she manifested in her exertions to accomplish this object. But the poor thing was dying, and while she vainly tried to induce it to accompany her, it rolled over, and floated dead at her side. Perceiving it to be beyond the reach of her caresses, she turned to wreak her vengeance on its slayers, and made directly for the boat, crashing her vast jaws the while, in a paroxysm of rage. Ordering his boat-steerer aft, the mate sprang forward, cut the line loose from the calf, and then snatched from the crotch the remaining iron, which he plunged with his gathered strength into the body of the mother, as the boat sheered off to avoid her onset. I saw that the work was well done, but had no time to mark the issue; for at that instant, a whale "breached" at the distance of about a mile from us, on the starboard quarter. The glimpse I caught of the animal in his descent, convinced me that I once more beheld my old acquaintance, Mocha Dick. That falling mass was white as a snow-drift!

'One might have supposed the recognition mutual, for no sooner was his vast square head lifted from the sea, than he charged down upon us, scattering the billows into spray as he advanced, and leaving a wake of foam a rod in width, from the violent lashing of his flukes.

' "He's making for the bloody water!" cried the men, as he cleft his way toward the very spot where the calf had been killed. "Here, harpooner, steer the boat, and let me dart!" I exclaimed, as I leaped into the bows. "May the '*Goneys*' eat me, if he dodge us *this* time, though he were Beelzebub himself! Pull for the red water!"

'As I spoke, the fury of the animal seemed suddenly to die away. He paused in his career, and lay passive on the waves, with his arching back thrown up like the ridge of a mountain. "The old sog's lying to!" I cried, exultingly. "Spring, boys! spring *now*, and we have him! All my clothes, tobacco, everything I've got, shall be yours, only lay me 'longside that whale before another boat comes up! My *grimky!* what a hump! Only look at the irons in his back! No, don't *look*—PULL! Now, boys, if you care about seeing your sweethearts and wives in old Nantuck!—if you love Yankee-land—if you love *me*—pull ahead, *won't* ye? Now then, to the thwarts! Lay back, my boys! I feel ye, my hearties! Give her the touch! Only five seas off! *Not* five seas off! One minute—*half* a minute more! Softly—no noise! Softly with your oars! That will do——"

'And as the words were uttered, I raised the harpoon above my head, took a rapid but no less certain aim, and sent it, hissing, deep into his thick white side!

' "Stern all! for your lives!" I shouted; for at the instant the steel quivered in his body, the wounded leviathan plunged his head beneath the surface, and whirling around with great velocity, smote the sea violently, with fin and fluke, in a convulsion of rage and pain.

'Our little boat flew dancing back from the seething vortex around him, just in season to escape being overwhelmed or crushed. He now started to run. For a short time, the line rasped, smoking, through the chocks. A few turns round the loggerhead then secured it; and with oars a-peak, and bows tilted to the sea, we went leaping onward in the wake of the tethered monster. Vain were all his struggles to break from our hold. The strands were too strong, the barbed iron too deeply fleshed, to give way. So that whether he essayed to dive or breach, or dash madly forward, the frantic creature still felt that he was held in check. At one moment, in impotent rage, he reared his immense blunt head, covered with barnacles, high above the surge; while his jaws fell together with a crash that almost made me shiver; then the upper outline of his vast form was dimly seen, gliding amidst showers of sparkling spray; while streaks of crimson on the white surf that boiled in his track, told that the shaft had been driven home.

'By this time, the whole "school" was about us; and spouts from a hundred spiracles, with a roar that almost deafened us, were raining on every side; while in the midst of a vast surface of chafing sea, might be seen the black shapes of the rampant herd, tossing and plunging, like a legion of maddened demons. The second and third mates were in the very centre of this appalling commotion.

'At length, Dick began to lessen his impetuous speed. "Now, my boys," cried I, "haul me on; wet the line, you second oarsman, as it comes in. Haul away, ship-mates!—why the devil don't you haul? Leeward side—*leeward!* I tell you! Don't you know how to approach a whale?"

'The boat brought fairly up upon his broadside as I spoke, and I gave him the lance just under the shoulder blade. At this moment, just as the boat's head was laid off; and I was straitening for a second lunge, my lance, which I had "boned" in the first, a piercing cry from the boat-steerer drew my attention quickly aft, and I saw the waist-boat, or more properly a fragment of it, falling through the air, and underneath, the dusky forms of the struggling crew, grasping at the oars, or clinging to portions of the wreck; while a pair of flukes, descending in the midst of the confusion, fully accounted for the catastrophe. The boat had been struck and shattered by a whale!

' "Good heaven!" I exclaimed, with impatience, and in a tone which I fear showed me rather mortified at the interruption, than touched with proper feeling for the sufferers; "good heavens!—hadn't they sense enough to keep out of the red water! And I must lose this glorious prize, through their infernal stupidity!" This was the first outbreak of my selfishness.

' "But we must not see them drown, boys," I added, upon the instant; "cut the line!" The order had barely passed my lips, when I caught sight of the captain, who had seen the accident from the quarter-deck, bearing down with oar and sail to the rescue.

' "Hold on!" I thundered, just as the knife's edge touched the line; "for the glory of old Nantuck, hold on! The captain will pick them up, and Mocha Dick will be ours, after all!"

'This affair occurred in half the interval I have occupied in the relation. In the mean time, with the exception of a slight shudder, which once or twice shook his ponderous frame, Dick lay perfectly quiet upon the water. But suddenly, as though goaded into exertion by some fiercer pang, he started from his lethargy with apparently augmented power. Making a leap toward the boat, he darted perpen-dicularly downward, hurling the after oarsman, who was helmsman at the time, ten feet over the quarter, as he struck the long steering-oar in his descent. The unfortunate seaman fell, with his head forward, just upon the flukes of the whale, as he vanished, and was drawn down by suction of the closing waters, as if he had been a feather. After being carried to a great depth, as we inferred from the time he remained below the surface, he came up, panting and exhausted, and was dragged on board, amidst the hearty congratulations of his comrades.

'By this time two hundred fathoms of line had been carried spinning through the chocks, with an impetus that gave back in steam the water cast upon it. Still the gigantic creature bored his way downward, with undiminished speed. Coil after coil went over, and was swallowed up. There remained but three flakes in the tub!

' "Cut!" I shouted; "cut quick, or he'll take us down!" But as I spoke, the hissing line flew with trebled velocity through the smoking wood, jerking the knife he was in the act of applying to the heated strands out of the hand of the boat-steerer. The boat rose on end, and her bows were buried in an instant; a hurried ejaculation, at once shriek and prayer, rose to the lips of the bravest, when, unexpected mercy! the whizzing cord lost its tension, and our light bark, half filled with water, fell heavily back on her keel. A tear was in every eye, and I believe every heart bounded with gratitude, at this unlooked-for deliverance.

'Overpowered by his wounds, and exhausted by his exertions and the enormous pressure of the water above him, the immense creature was compelled to turn once more upward, for a fresh supply of air. And upward he came, indeed; shooting twenty feet of his gigantic length above the waves, by the impulse of his ascent. He was not disposed to be idle. Hardly had we succeeded in bailing out our swamping boat, when he again darted away, as it seemed to me with renewed energy. For a quarter of a mile, we parted the opposing waters as though they had offered no more resistance than air. Our game then abruptly brought to, and lay as if paralysed, his massy frame quivering and twitching, as if under the influence of galvanism. I gave the word to haul on; and seizing a boat-spade, as we came near him, drove it twice into his small; no doubt partially disabling him by the vigour and certainty of the blows. Wheeling furiously around, he answered this salutation, by making a desperate dash at the boat's quarter. We were so near him, that to escape the shock of his onset, by any practicable manœuvre, was out of the question. But at the critical moment, when we expected to be crushed by the collision, his powers seemed to give way. The fatal lance had reached the seat of life. His strength failed him in mid-career, and sinking quietly beneath our keel, grazing it as he wallowed along, he rose again a few rods from us, on the side opposite that where he went down.

' "Lay around, my boys, and let us set on him!" I cried, for I saw his spirit was broken at last. But the lance and spade were needless now. The work was done. The dying animal was struggling in a whirlpool of bloody foam, and the ocean far around was tinted with crimson. "Stern all!" I shouted, as he commenced running impetuously in a circle, beating the water alternately with his head and flukes, and smiting his teeth ferociously into their sockets with a crashing sound, in the strong spasms of dissolution. "Stern all! or we shall be stove!"

'As I gave the command, a stream of black, clotted gore rose in a thick spout above the expiring brute, and fell in a shower around, bedewing, or rather drenching us, with a spray of blood.

' "*There's the flag!*" I exclaimed; "there! thick as tar! Stern! every soul of ye! He's going in his flurry!" And the monster, under the convulsive influence of his final paroxysm, flung his huge tail into the air, and then, for the space of a minute, thrashed the waters on either side of him with quick and powerful blows; the sound of the concussions resembling that of the rapid discharge of artillery. He then turned slowly and heavily on his side, and lay a dead mass upon the sea through which he had so long ranged a conqueror.

' "He's fin up at last!" I screamed, at the very top of my voice. "Hurrah! hurrah! hurrah!" And snatching off my cap, I sent it spinning aloft, jumping at the same time from thwart to thwart, like a madman.

'We now drew alongside our floating spoil; and I seriously question if the brave commodore who first, and so nobly, broke the charm of British invincibility, by the capture of the Guerriere, felt a warmer rush of delight, as he beheld our national flag waving over the British ensign, in assurance of his victory, than I did, as I leaped upon the quarter-deck of Dick's back, planted my wafe-pole in the midst, and saw the little canvas flag, that tells so important and satisfactory a tale to the whaleman, fluttering above my hard-earned prize.

'The captain and second mate, each of whom had been fortunate enough to kill his fish, soon after pulled up, and congratulated me on my capture. From them I learned the particulars of the third mate's disaster. He had fastened, and his fish was sounding, when another whale suddenly rose, almost directly beneath the boat, and with a single blow of his small, absolutely cut it in twain, flinging the bows, and those who occupied that portion of the frail fabric, far into the air. Rendered insensible, or immediately killed by the shock, two of the crew sank without a struggle, while a third, unable in his confusion to disengage himself from the flakes of the tow-line, with which he had become entangled, was, together with the fragment to which the warp was attached, borne down by the harpooned whale, and was seen no more! The rest, some of them severely bruised, were saved from drowning by the timely assistance of the captain.

'To get the harness on Dick, was the work of an instant; and as the ship, taking every advantage of a light breeze which had sprung up within the last hour, had stood after us, and was now but a few rods distant, we were soon under her stern. The other fish, both of which were heavy fellows, lay floating near; and the tackle being affixed to one of them without delay, all hands were soon busily engaged in cutting in. Mocha Dick was the longest whale I ever looked upon. He measured more than seventy feet from his noddle to the tips of his flukes; and yielded one hundred barrels of clear oil, with a proportionate quantity of "head-matter". It may emphatically be said, that "the scars of his old wounds were near his new", for not less than twenty harpoons did we draw from his back; the rusted mementos of many a desperate rencounter.'

The mate was silent. His yarn was reeled off. His story was told; and with far better tact than is exhibited by many a modern orator, he had the modesty and discretion to stop with its termination. In response, a glass of 'o-be-joyful' went merrily round; and this tribute having been paid to courtesy, the vanquisher of Mocha Dick was unanimously called upon for a song. Too sensible and too good-natured to wait for a second solicitation, when he had the power to oblige, he took a 'long pull' and a strong, at the grog as an appropriate overture to the occasion, and then, in a deep, sonorous tone, gave us the following professional ballad, accompanied by a superannuated hand-organ, which constituted the musical portion of the cabin furniture:

I.

'Don't bother my head about catching of seals!
To me there's more glory in catching of eels;
Give me a tight ship, and under snug sail,
And I ask for no more, 'long side the sperm whale,
  In the Indian Ocean,
  Or Pacific Ocean,

No matter *what* ocean;
    Pull ahead, yo heave O!

II.

'When our anchor's a-peak, sweethearts and wives
Yield a warm drop at parting, breathe a prayer for our lives;
With hearts full of promise, they kiss off the tear
From the eye that grows rarely dim—never with fear!
    Then for the ocean, boys,
    The billow's commotion, boys,
    That's our devotion, boys,
      Pull ahead, yo heave O!

III.

'Soon we hear the glad cry of "Town O!—there she blows!"
Slow as night, my brave fellows, to leeward she goes:
Hard up! square the yards! then steady, lads, so!
Cries the captain, "My maiden lance soon shall she know!"
    "Now we get near, boys,
    In with the gear, boys,
    Swing the cranes clear, boys;
      Pull ahead, yo heave O!"

IV.

'Our boat's in the water, each man at his oar
Bends strong to the sea, while his bark bounds before,
As the fish of all sizes, still flouncing and blowing,
With fluke and broad fin, scorn the best of hard rowing:
    "Hang to the oar, boys,
    Another stroke more, boys;
    Now line the oar, boys;
      Pull ahead, yo heave O!"

V.

'Then rises long Tom, who never knew fear;
Cries the captain, "Now nail her, my bold harpooner!"
He speeds home his lance, then exclaims, "I am fast!"
While blood, in a torrent, leaps high as the mast:
    "Starn! starn! hurry, hurry, boys!
    She's gone in her flurry, boys,
    She'll soon be in 'gurry', boys!
      Pull ahead, yo heave O!"

VI.

'Then give me a whaleman, wherever he be,
Who fears not a fish that can swim the salt sea;
Then give me a tight ship, and under snug sail,
And last lay me 'side of the noble sperm whale;
  "In the Indian ocean,
  Or Pacific ocean,
  No matter *what* ocean;
  Pull ahead, yo heave O!" '

The song 'died away into an echo', and we all confessed ourselves
delighted with it—save and except the gallant knight of the seal-club.
He indeed allowed the lay and the music to be well enough, considering the subject; but added: 'If you want to hear genuine, heart-stirring
harmony, you must listen to a rookery of fur seal. For many an hour,
on the rocks round Cape Horn, have I sat thus, listening to these
gentry, as they clustered on the shelving cliffs above me; the surf
beating at my feet, while——'

'Come, come, my old fellow!' exclaimed the captain, interrupting
the loquacious sealer; 'you forget the evening you are to have at Santa
Maria. It is three o'clock in the morning, and more.' Bidding farewell
to our social and generous entertainers, we were soon safely on board
our ship, when we immediately made all sail to the north.

To me, the evening had been one of singular enjoyment. Doubtless
the particulars of the tale were in some degree highly coloured, from
the desire of the narrator to present his calling in a prominent light,
and especially one that should eclipse the occupation of sealing. But
making every allowance for what, after all, may be considered a
natural embellishment, the facts presented may be regarded as a fair
specimen of the adventures which constitute so great a portion of the
romance of a whaler's life; a life which, viewing all the incidents that
seem inevitably to grow out of the enterprise peculiar to it, can be
said to have no parallel. Yet vast as the field is, occupied by this class
of our resolute seamen, how little can we claim to know of the
particulars of a whaleman's existence! That our whale ships leave port,
and usually return, in the course of three years, with full cargoes, to
swell the fund of national wealth, is nearly the sum of our knowledge
concerning them. Could we comprehend, at a glance, the mighty
surface of the Indian or Pacific seas, what a picture would open upon
us of unparalleled industry and daring enterprise! What scenes of toil
along the coast of Japan, up the straits of Mozambique, where the

dangers of the storm, impending as they may be, are less regarded than the privations and sufferings attendant upon exclusion from all intercourse with the shore! Sail onward, and extend your view around New-Holland, to the coast of Guinea; to the eastern and western shores of Africa; to the Cape of Good Hope; and south, to the waters that lash the cliffs of Kergulan's Land, and you are ever upon the whaling-ground of the American seaman. Yet onward, to the vast expanse of the two Pacifics, with their countless summer isles, and your course is still over the common arena and highway of our whalers. The varied records of the commercial world can furnish no precedent, can present no comparison, to the intrepidity, skill, and fortitude, which seem the peculiar prerogatives of this branch of our marine. These characteristics are not the growth of forced exertion; they are incompatible with it. They are the natural result of the ardour of a free people; of a spirit of fearless independence, generated by free institutions. Under such institutions alone, can the human mind attain its fullest expansion, in the various departments of science, and the multiform pursuits of busy life.

The Chase

~~~

HERMAN MELVILLE

T HAT night, in the mid-watch, when the old man—as his wont at intervals—stepped forth from the scuttle in which he leaned, and went to his pivot-hole, he suddenly thrust out his face fiercely, snuffing up the sea air as a sagacious ship's dog will, in drawing nigh to some barbarous isle. He declared that a whale must be near. Soon that peculiar odour, sometimes to a great distance given forth by the living sperm whale, was palpable to all the watch; nor was any mariner surprised when, after inspecting the compass, and then the dog-vane, and then ascertaining the precise bearing of the odour as nearly as possible, Ahab rapidly ordered the ship's course to be slightly altered, and the sail to be shortened.

The acute policy dictating these movements was sufficiently vindicated at daybreak, by the sight of a long sleek on the sea directly and lengthwise ahead, smooth as oil, and resembling in the pleated watery wrinkles bordering it, the polished metallic-like marks of some swift tide-rip, at the mouth of a deep, rapid stream.

'Man the mast-heads! Call all hands!'

Thundering with the butts of three clubbed handspikes on the forecastle deck, Daggoo roused the sleepers with such judgement claps that they seemed to exhale from the scuttle, so instantaneously did they appear with their clothes in their hands.

'What d'ye see?' cried Ahab, flattening his face to the sky.

'Nothing, nothing, sir!' was the sound hailing down in reply.

'T'gallant sails!—stunsails! alow and aloft, and on both sides!'

All sail being set, he now cast loose the life-line, reserved for swaying him to the main royal-mast head; and in a few moments they were hoisting him thither, when, while but two-thirds of the way aloft, and while peering ahead through the horizontal vacancy between the main-top-sail and top-gallant-sail, he raised a gull-like cry in the air, 'There she blows!—there she blows! A hump like a snow-hill! It is Moby Dick!'

Fired by the cry which seemed simultaneously taken up by the three look-outs, the men on deck rushed to the rigging to behold the famous whale they had so long been pursuing. Ahab had now gained his final perch, some feet above the other look-outs, Tashtego standing just beneath him on the cap of the top-gallant-mast, so that the Indian's head was almost on a level with Ahab's heel. From this height the whale was now seen some mile or so ahead, at every roll of the sea revealing his high sparkling hump, and regularly jetting his silent spout into the air. To the credulous mariners it seemed the same silent spout they had so long ago beheld in the moonlit Atlantic and Indian Oceans.

'And did none of ye see it before?' cried Ahab, hailing the perched men all around him.

'I saw him almost that same instant, sir, that Captain Ahab did, and I cried out,' said Tashtego.

'Not the same instant; not the same—no, the doubloon is mine, Fate reserved the doubloon for me. *I* only; none of ye could have raised the White Whale first. There she blows! there she blows!—there she blows! There again!—there again!' he cried, in long-drawn, lingering, methodic tones, attuned to the gradual prolongings of the whale's visible jets. 'He's going to sound! In stunsails! Down top-gallant-sails! Stand by three boats. Mr Starbuck, remember, stay on board, and keep the ship. Helm there! Luff, luff a point! So; steady, man, steady! There go flukes! No, no; only black water! All ready the boats there? Stand by, stand by! Lower me, Mr Starbuck; lower, lower—quick, quicker!' and he slid through the air to the deck.

'He is heading straight to leeward, sir,' cried Stubb, 'right away from us; cannot have seen the ship yet.'

'Be dumb, man! Stand by the braces! Hard down the helm!—brace up! Shiver her!—shiver her! So; well that! Boats, boats!'

Soon all the boats but Starbuck's were dropped; all the boat-sails set—all the paddles plying; with rippling swiftness, shooting to leeward; and Ahab heading the onset. A pale, death-glimmer lit up Fedallah's sunken eyes; a hideous motion gnawed his mouth.

Like noiseless nautilus shells, their light prows sped through the sea; but only slowly they neared the foe. As they neared him, the ocean grew still more smooth; seemed drawing a carpet over its waves; seemed a noon-meadow, so serenely it spread. At length the breathless hunter came so nigh his seemingly unsuspecting prey, that his entire dazzling hump was distinctly visible, sliding along the sea as if an isolated thing, and continually set in a revolving ring of finest, fleecy, greenish foam. He saw the vast, involved wrinkles of the slightly projecting head beyond. Before it, far out on the soft Turkish-rugged waters, went the glistening white shadow from his broad, milky forehead, a musical rippling playfully accompanying the shade; and behind, the blue waters interchangeably flowed over into the moving valley of his steady wake; and on either hand bright bubbles arose and danced by his side. But these were broken again by the light toes of hundreds of gay fowl softly feathering the sea, alternate with their fitful flight; and like to some flag-staff rising from the painted hull of an argosy, the tall but shattered pole of a recent lance projected from the white whale's back; and at intervals one of the cloud of soft-toed fowls hovering, and to and-fro skimming like a canopy over the fish, silently perched and rocked on this pole, the long tail feathers streaming like pennons.

A gentle joyousness—a mighty mildness of repose in swiftness, invested the gliding whale. Not the white bull Jupiter swimming away with ravished Europa clinging to his graceful horns; his lovely, leering eyes sideways intent upon the maid; with smooth bewitching fleetness, rippling straight for the nuptial bower in Crete; not Jove, not that great majesty Supreme! did surpass the glorified White Whale as he so divinely swam.

On each soft side—coincident with the parted swell, that but once laving him, then flowed so wide away—on each bright side, the whale shed off enticings. No wonder there had been some among the hunters who namelessly transported and allured by all this serenity, had ventured to assail it; but had fatally found that quietude but the vesture of tornadoes. Yet calm, enticing calm, oh, whale! thou glidest on, to all who for the first time eye thee, no matter how many in that same way thou may'st have bejuggled and destroyed before.

And thus, through the serene tranquillities of the tropical sea, among waves whose hand-clappings were suspended by exceeding rapture, Moby Dick moved on, still withholding from sight the full terrors of his submerged trunk, entirely hiding the wrenched hideousness of his jaw. But soon the fore part of him slowly rose from the water; for an instant his whole marbleized body formed a high arch, like Virginia's Natural Bridge, and warningly waving his bannered flukes in the air, the grand god revealed himself, sounded, and went out of sight. Hoveringly halting, and dipping on the wing, the white sea-fowls longingly lingered over the agitated pool that he left.

With oars apeak, and paddles down, the sheets of their sails adrift, the three boats now stilly floated, awaiting Moby Dick's reappearance.

'An hour,' said Ahab, standing rooted in his boat's stern; and he gazed beyond the whale's place, towards the dim blue spaces and wide wooing vacancies to leeward. It was only an instant; for again his eyes seemed whirling round in his head as he swept the watery circle. The breeze now freshened; the sea began to swell.

'The birds!—the birds!' cried Tashtego.

In long Indian file, as when herons take wing, the white birds were now all flying towards Ahab's boat; and when within a few yards began fluttering over the water there, wheeling round and round, with joyous, expectant cries. Their vision was keener than man's; Ahab could discover no sign in the sea. But suddenly as he peered down and down into its depths, he profoundly saw a white living spot no bigger than a white weasel, with wonderful celerity uprising, and magnifying as it rose, till it turned, and then there were plainly revealed two long crooked rows of white, glistening teeth, floating up from the undiscoverable bottom. It was Moby Dick's open mouth and scrolled jaw; his vast, shadowed bulk still half blending with the blue of the sea. The glittering mouth yawned beneath the boat like an open-doored marble tomb; and giving one sidelong sweep with his steering oar, Ahab whirled the craft aside from this tremendous apparition. Then, calling upon Fedallah to change places with him, went forward to the bows, and seizing Perth's harpoon, commanded his crew to grasp their oars and stand by to stern.

Now, by reason of this timely spinning round the boat upon its axis, its bow, by anticipation, was made to face the whale's head while yet under water. But as if perceiving this stratagem, Moby Dick, with that malicious intelligence ascribed to him, sidelingly transplanted himself, as it were, in an instant, shooting his pleated head lengthwise beneath the boat.

Through and through; through every plank and each rib, it thrilled for an instant, the whale obliquely lying on his back, in the manner of a biting shark, slowly and feelingly taking its bows full within his mouth, so that the long, narrow, scrolled lower jaw curled high up into the open air, and one of the teeth caught in a row-lock. The bluish pearl-white of the inside of the jaw was within six inches of Ahab's head, and reached higher than that. In this attitude the White Whale now shook the slight cedar as a mildly cruel cat her mouse. With unastonished eyes Fedallah gazed, and crossed his arms; but the tiger-yellow crew were tumbling over each other's heads to gain the uttermost stern.

And now, while both elastic gunwales were springing in and out, as the whale dallied with the doomed craft in this devilish way; and from his body being submerged beneath the boat, he could not be darted at from the bows, for the bows were almost inside of him, as it were; and while the other boats involuntarily paused, as before a quick crisis impossible to withstand, then it was that monomaniac Ahab, furious with this tantalizing vicinity of his foe, which placed him all alive and helpless in the very jaws he hated; frenzied with all this, he seized the long bone with his naked hands, and wildly strove to wrench it from its gripe. As now he thus vainly strove, the jaw slipped from him; the frail gunwales bent in, collapsed, and snapped, as both jaws, like an enormous shears, sliding further aft, bit the craft completely in twain, and locked themselves fast again in the sea, midway between the two floating wrecks. These floated aside, the broken ends drooping, the crew at the stern-wreck clinging to the gunwales, and striving to hold fast to the oars to lash them across.

At that preluding moment, ere the boat was yet snapped, Ahab, the first to perceive the whale's intent, by the crafty upraising of his head, a movement that loosed his hold for the time; at that moment his hand had made one final effort to push the boat out of the bite. But only slipping further into the whale's mouth, and tilting over sideways as it slipped, the boat had shaken off his hold on the jaw; spilled him out of it, as he leaned to the push; and so he fell flat-faced upon the sea.

Ripplingly withdrawing from his prey, Moby Dick now lay at a little distance, vertically thrusting his oblong white head up and down in the billows; and at the same time slowly revolving his whole spindled body; so that when his vast wrinkled forehead rose—some twenty or more feet out of the water—the now rising swells, with all their confluent waves, dazzlingly broke against it; vindictively tossing their

shivered spray still higher into the air.* So, in a gale, the but half
baffled Channel billows only recoil from the base of the Eddystone,
triumphantly to overleap its summit with their scud.

But soon resuming his horizontal attitude, Moby Dick swam swiftly
round and round the wrecked crew; sideways churning the water in
his vengeful wake, as if lashing himself up to still another and more
deadly assault. The sight of the splintered boat seemed to madden
him, as the blood of grapes and mulberries cast before Antiochus'
elephants in the book of Maccabees. Meanwhile Ahab half smothered
in the foam of the whale's insolent tail, and too much of a cripple to
swim—though he could still keep afloat, even in the heart of such a
whirlpool as that; helpless Ahab's head was seen, like a tossed bubble
which the least chance shock might burst. From the boat's fragmentary stern, Fedallah incuriously and mildly eyed him; the clinging crew,
at the other drifting end, could not succour him; more than enough
was it for them to look to themselves. For so revolvingly appalling
was the White Whale's aspect, and so planetarily swift the ever-
contracting circles he made, that he seemed horizontally swooping
upon them. And though the other boats, unharmed, still hovered hard
by; still they dared not pull into the eddy to strike, lest that should
be the signal for the instant destruction of the jeopardized castaways,
Ahab and all; nor in that case could they themselves hope to escape.
With straining eyes, then, they remained on the outer edge of the
direful zone, whose centre had now become the old man's head.

Meantime, from the beginning all this had been descried from the
ship's mast-heads; and squaring her yards, she had borne down upon
the scene; and was now so nigh, that Ahab in the water hailed
her—'Sail on the'—but that moment a breaking sea dashed on him
from Moby Dick, and whelmed him for the time. But struggling out
of it again, and chancing to rise on a towering crest, he shouted—'Sail
on the whale!—Drive him off!'

The Pequod's prow was pointed; and breaking up the charmed
circle, she effectually parted the white whale from his victim. As he
sullenly swam off, the boats flew to the rescue.

Dragged into Stubb's boat with blood-shot, blinded eyes, the white
brine caking in his wrinkles; the long tension of Ahab's bodily

---

* This motion is peculiar to the sperm whale. It receives its designation (pitchpoling)
from its being likened to that preliminary up-and-down poise of the whale-lance, in the
exercise called pitchpoling, previously described. By this motion the whale must best
and most comprehensively view whatever objects may be encircling him.

strength did crack, and helplessly he yielded to his body's doom: for a time, lying all crushed in the bottom of Stubb's boat, like one trodden under foot of herds of elephants. Far inland, nameless wails came from him, as desolate sounds from out ravines.

But this intensity of his physical prostration did but so much the more abbreviate it. In an instant's compass, great hearts sometimes condense to one deep pang, the sum total of those shallow pains kindly diffused through feebler men's whole lives. And so, such hearts, though summary in each one suffering; still, if the gods decree it, in their life-time aggregate a whole age of woe, wholly made up of instantaneous intensities; for even in their pointless centres, those noble natures contain the entire circumferences of inferior souls.

'The harpoon,' said Ahab, half way rising, and draggingly leaning on one bended arm—'is it safe?'

'Aye, sir, for it was not darted; this is it', said Stubb, showing it.

'Lay it before me—any missing men?'

'One, two, three, four, five—there were five oars, sir, and here are five men.'

'That's good. Help me, man; I wish to stand. So, so, I see him! there! there! going to leeward still; what a leaping spout! Hands off from me! The eternal sap runs up in Ahab's bones again! Set the sail; out oars; the helm!'

It is often the case that when a boat is stove, its crew, being picked up by another boat, help to work that second boat; and the chase is thus continued with what is called double-banked oars. It was thus now. But the added power of the boat did not equal the added power of the whale, for he seemed to have treble-banked his every fin; swimming with a velocity which plainly showed, that if now, under these circumstances, pushed on, the chase would prove an indefinitely prolonged, if not a hopeless one; nor could any crew endure for so long a period, such an unintermitted, intense straining at the oar; a thing barely tolerable only in some one brief vicissitude. The ship itself, then, as it sometimes happens, offered the most promising intermediate means of overtaking the chase. Accordingly, the boats now made for her, and were soon swayed up to their cranes—the two parts of the wrecked boat having been previously secured by her—and then hoisting everything to her side, and stacking her canvas high up, and sideways outstretching it with stun-sails, like the double-jointed wings of an albatross; the Pequod bore down in the leeward wake of Moby Dick. At the well-known, methodic intervals, the whale's glittering spout was regularly announced from the manned mast-heads; and when he would be reported as just gone down, Ahab would take the time, and then pacing the deck, binnacle-watch in hand, so soon as the last second of the allotted hour expired, his voice was heard. 'Whose is the doubloon now? D'ye see him?' and if the reply was, No, sir! straightway he commanded them to lift him to his perch. In this way the day wore on; Ahab, now aloft and motionless; anon, unrestingly pacing the planks.

As he was thus walking, uttering no sound, except to hail the men aloft, or to bid them hoist a sail still higher, or to spread one to a still greater breadth—thus to and fro pacing, beneath his slouched hat, at every turn he passed his own wrecked boat, which had been dropped upon the quarter-deck, and lay there reversed; broken bow to shattered stern. At last he paused before it; and as in an already over-clouded sky fresh troops of clouds will sometimes sail across, so over the old man's face there now stole some such added gloom as this.

Stubb saw him pause; and perhaps intending, not vainly, though, to evince his own unabated fortitude, and thus keep up a valiant place in his Captain's mind, he advanced, and eyeing the wreck exclaimed 'The thistle the ass refused; it pricked his mouth too keenly, sir; ha! ha!'

'What soulless thing is this that laughs before a wreck? Man, man! did I not know thee brave as fearless fire (and as mechanical) I could swear thou wert a poltroon. Groan nor laugh should be heard before a wreck.'

'Aye sir,' said Starbuck drawing near, ' 'tis a solemn sight; an omen, and an ill one.'

'Omen? omen? the dictionary! If the gods think to speak outright to man, they will honourably speak outright; not shake their heads, and give an old wives' darkling hint. Begone! Ye two are the opposite poles of one thing; Starbuck is Stubb reversed, and Stubb is Starbuck; and ye two are all mankind; and Ahab stands alone among the millions of the peopled earth, nor gods nor men his neighbours! Cold, cold—I shiver!—How now? Aloft there! D'ye see him? Sing out for every spout, though he spout ten times a second!'

The day was nearly done; only the hem of his golden robe was rustling. Soon, it was almost dark, but the look-out men still remained unset.

'Can't see the spout now, sir; too dark', cried a voice from the air.

'How heading when last seen?'

'As before, sir, straight to leeward.'

'Good! he will travel slower now 'tis night. Down royals and top-gallant stun-sails, Mr Starbuck. We must not run over him before morning; he's making a passage now, and may heave-to a while. Helm there! keep her full before the wind!—Aloft! come down!—Mr Stubb, send a fresh hand to the fore-mast head, and see it manned till morning.' Then advancing towards the doubloon in the main-mast—'Men, this gold is mine, for I earned it; but I shall let it abide here till the White Whale is dead; and then, whosoever of ye first raises him, upon the day he shall be killed, this gold is that man's; and if on that day I shall again raise him, then, ten times its sum shall be divided among all of ye! Away now!—the deck is thine, sir.'

And so saying, he placed himself half way within the scuttle, and slouching his hat, stood there till dawn, except when at intervals rousing himself to see how the night wore on.

SECOND DAY

At day-break, the three mast-heads were punctually manned afresh.

'D'ye see him?' cried Ahab, after allowing a little space for the light to spread.

'See nothing, sir.'

'Turn up all hands and make sail! he travels faster than I thought for; the top-gallant sails! Aye, they should have been kept on her all night. But no matter—'tis but resting for the rush.'

Here be it said, that this pertinacious pursuit of one particular whale, continued through day into night, and through night into day, is a thing by no means unprecedented in the South sea fishery. For such is the wonderful skill, prescience of experience, and invincible confidence acquired by some great natural geniuses among the Nantucket commanders; that from the simple observation of a whale when last descried, they will, under certain given circumstances, pretty accurately foretell both the direction in which he will continue to swim for a time, while out of sight, as well as his probable rate of progression during that period. And, in these cases, somewhat as a pilot, when about losing sight of a coast, whose general trending he well knows, and which he desires shortly to return to again, but at some further point; like as this pilot stands by his compass, and takes the precise bearing of the cape at present visible, in order the more certainly to hit aright the remote, unseen headland, eventually to be visited: so does the fisherman, at his compass, with the whale; for after being chased, and diligently marked, through several hours of daylight, then, when night obscures the fish, the creature's future wake through the darkness is almost as established to the sagacious mind of the hunter, as the pilot's coast is to him. So that to this hunter's wondrous skill, the proverbial evanescence of a thing writ in water, a wake, is to all desired purposes well nigh as reliable as the steadfast land.

And as the mighty iron Leviathan of the modern railway is so familiarly known in its every pace, that, with watches in their hands, men time his rate as doctors that of a baby's pulse; and lightly say of it, the up train or the down train will reach such or such a spot, at such or such an hour; even so, almost, there are occasions when these Nantucketers time that other Leviathan of the deep, according to the observed humour of his speed; and say to themselves, so many hours hence this whale will have gone two hundred miles, will have about reached this or that degree of latitude or longitude. But to render this acuteness at all successful in the end, the wind and the sea must be the whaleman's allies; for of what present avail to the becalmed or windbound mariner is the skill that assures him he is exactly ninety-three leagues and a quarter from his port? Inferable from these statements, are many collateral subtile matters touching the chase of whales.

The ship tore on; leaving such a furrow in the sea as when a cannon-ball, missent, becomes a plough-share and turns up the level field.

'By salt and hemp!' cried Stubb, 'but this swift motion of the deck creeps up one's legs and tingles at the heart. This ship and I are two brave fellows!—Ha! ha! Some one take me up, and launch me, spine-wise, on the sea—for by live-oaks! my spine's a keel. Ha, ha! we go the gait that leaves no dust behind!'

'There she blows—she blows!—she blows!—right ahead!' was now the mast-head cry.

'Aye, aye!' cried Stubb, 'I knew it—ye can't escape—blow on and split your spout, O whale! the mad fiend himself is after ye! blow your trump—blister your lungs!—Ahab will dam off your blood, as a miller shuts his water-gate upon the stream!'

And Stubb did but speak out for well nigh all that crew. The frenzies of the chase had by this time worked them bubblingly up, like old wine worked anew. Whatever pale fears and forebodings some of them might have felt before; these were not only now kept out of sight through the growing awe of Ahab, but they were broken up, and on all sides routed, as timid prairie hares that scatter before the bounding bison. The hand of Fate had snatched all their souls; and by the stirring perils of the previous day; the rack of the past night's suspense; the fixed, unfearing, blind, reckless way in which their wild craft went plunging towards its flying mark; by all these things, their hearts were bowled along. The wind that made great bellies of their sails, and rushed the vessel on by arms invisible as irresistible; this seemed the symbol of that unseen agency which so enslaved them to the race.

They were one man, not thirty. For as the one ship that held them all; though it was put together of all contrasting things—oak, and maple, and pine wood; iron, and pitch, and hemp—yet all these ran into each other in the one concrete hull, which shot on its way, both balanced and directed by the long central keel; even so, all the individualities of the crew, this man's valour, that man's fear; guilt and guiltlessness, all varieties were welded into oneness, and were all directed to that fatal goal which Ahab their one lord and keel did point to.

The rigging lived. The mast-heads, like the tops of tall palms, were outspreadingly tufted with arms and legs. Clinging to a spar with one hand, some reached forth the other with impatient wavings; others, shading their eyes from the vivid sunlight, sat far out on the rocking yards; all the spars in full bearing of mortals, ready and ripe for their fate. Ah! how they still strove through that infinite blueness to seek out the thing that might destroy them!

'Why sing ye not out for him, if ye see him?' cried Ahab, when, after the lapse of some minutes since the first cry, no more had been heard. 'Sway me up, men; ye have been deceived; not Moby Dick casts one odd jet that way, and then disappears.'

It was even so; in their headlong eagerness, the men had mistaken some other thing for the whale-spout, as the event itself soon proved; for hardly had Ahab reached his perch; hardly was the rope belayed to its pin on deck, when he struck the keynote to an orchestra, that made the air vibrate as with the combined discharges of rifles. The triumphant halloo of thirty buckskin lungs was heard, as—much nearer to the ship than the place of the imaginary jet, less than a mile ahead—Moby Dick bodily burst into view! For not by any calm and indolent spoutings; not by the peaceable gush of that mystic fountain in his head, did the White Whale now reveal his vicinity; but by the far more wondrous phenomenon of breaching. Rising with his utmost velocity from the furthest depths, the Sperm Whale thus booms his entire bulk into the pure element of air, and piling up a mountain of dazzling foam, shows his place to the distance of seven miles and more. In those moments, the torn, enraged waves he shakes off, seem his mane; in some cases, this breaching is his act of defiance.

'There she breaches! there she breaches!' was the cry, as in his immeasurable bravadoes the White Whale tossed himself salmon-like to Heaven. So suddenly seen in the blue plain of the sea, and relieved against the still bluer margin of the sky, the spray that he raised, for the moment, intolerably glittered and glared like a glacier; and stood there gradually fading and fading away from its first sparkling intensity, to the dim mistiness of an advancing shower in a vale.

'Aye, breach your last to the sun, Moby Dick!' cried Ahab, 'thy hour and thy harpoon are at hand!—Down! down all of ye, but one man at the fore. The boats!—stand by!'

Unmindful of the tedious rope-ladders of the shrouds, the men, like shooting stars, slid to the deck, by the isolated back-stays and halyards; while Ahab, less dartingly, but still rapidly was dropped from his perch.

'Lower away', he cried, so soon as he had reached his boat—a spare one, rigged the afternoon previous. 'Mr Starbuck, the ship is thine—keep away from the boats, but keep near them. Lower, all!'

As if to strike a quick terror into them, by this time being the first assailant himself, Moby Dick had turned, and was now coming for the three crews. Ahab's boat was central; and cheering his men, he told them he would take the whale head-and-head—that is, pull straight up to his forehead—a not uncommon thing; for when within a certain limit, such a course excludes the coming onset from the whale's sidelong vision. But ere that close limit was gained, and while yet all three boats were plain as the ship's three masts to his eye; the White Whale churning himself into furious speed, almost in an instant as it were, rushing among the boats with open jaws, and a lashing tail, offered appalling battle on every side; and heedless of the irons darted at him from every boat, seemed only intent on annihilating each separate plank of which those boats were made. But skilfully manœuvred, incessantly wheeling like trained chargers in the field; the boats for a while eluded him; though, at times, but by a plank's breadth; while all the time, Ahab's unearthly slogan tore every other cry but his to shreds.

But at last in his untraceable evolutions, the White Whale so crossed and recrossed, and in a thousand ways entangled the slack of the three lines now fast to him, that they foreshortened, and, of themselves, warped the devoted boats towards the planted irons in him; though now for a moment the whale drew aside a little, as if to rally for a more tremendous charge. Seizing that opportunity, Ahab first paid out more line: and then was rapidly hauling and jerking in upon it again—hoping that way to disencumber it of some snarls— when lo!—a sight more savage than the embattled teeth of sharks!

Caught and twisted—corkscrewed in the mazes of the line, loose harpoons and lances, with all their bristling barbs and points, came flashing and dripping up to the chocks in the bows of Ahab's boat. Only one thing could be done. Seizing the boat-knife, he critically reached within—through—and then, without—the rays of steel; dragged in the line beyond, passed it, inboard, to the bowsman, and then, twice sundering the rope near the chocks—dropped the inter- cepted fagot of steel into the sea; and was all fast again. That instant, the White Whale made a sudden rush among the remaining tangles of the other lines; by so doing, irresistibly dragged the more involved boats of Stubb and Flask towards his flukes; dashed them together like two rolling husks on a surf-beaten beach, and then, diving down into the sea, disappeared in a boiling maelstrom, in which, for a space, the odorous cedar chips of the wrecks danced round and round, like the grated nutmeg in a swiftly stirred bowl of punch.

While the two crews were yet circling in the waters, reaching out after the revolving line-tubs, oars, and other floating furniture, while aslope little Flask bobbed up and down like an empty vial, twitching his legs upwards to escape the dreaded jaws of sharks; and Stubb was lustily singing out for someone to ladle him up; and while the old man's line—now parting—admitted of his pulling into the creamy pool to rescue whom he could; in that wild simultaneousness of a thousand concreted perils, Ahab's yet unstricken boat seemed drawn up towards Heaven by invisible wires—as, arrow-like, shooting per- pendicularly from the sea, the White Whale dashed his broad fore- head against its bottom, and sent it, turning over and over, into the air; till it fell again—gun-wale downwards—and Ahab and his men struggled out from under it, like seals from a seaside cave.

The first uprising momentum of the whale—modifying its direction as he struck the surface—involuntarily launched him along it, to a little distance from the centre of the destruction he had made; and with his back to it, he now lay for a moment slowly feeling with his flukes from side to side; and whenever a stray oar, bit of plank, the least chip or crumb of the boats touched his skin, his tail swiftly drew back, and came sideways smiting the sea. But soon, as if satisfied that his work for that time was done, he pushed his pleated forehead through the ocean, and trailing after him the intertangled lines, continued his leeward way at a traveller's methodic pace.

As before, the attentive ship having descried the whole fight, again came bearing down to the rescue, and dropping a boat, picked up the floating mariners, tubs, oars, and whatever else could be caught at, and safely landed them on her decks. Some sprained shoulders, wrists, and ankles; livid contusions; wrenched harpoons and lances; inextricable intricacies of rope; shattered oars and planks; all these were there; but no fatal or even serious ill seemed to have befallen any one. As with Fedallah the day before, so Ahab was now found grimly clinging to his boat's broken half, which afforded a comparatively easy float; nor did it so exhaust him as the previous day's mishap.

But when he was helped to the deck, all eyes were fastened upon him; as instead of standing by himself he still half-hung upon the shoulder of Starbuck, who had thus far been the foremost to assist him. His ivory leg had been snapped off, leaving but one short sharp splinter.

'Aye aye, Starbuck, 'tis sweet to lean sometimes, be the leaner who he will; and would old Ahab had leaned oftener than he has.'

'The ferrule has not stood, sir,' said the carpenter, now coming up; 'I put good work into that leg.'

'But no bones broken, sir, I hope,' said Stubb with true concern.

'Aye! and all splintered to pieces, Stubb!—d'ye see it. But even with a broken bone, old Ahab is untouched; and I account no living bone of mine one jot more me, than this dead one that's lost. Nor white whale, nor man, nor fiend, can so much as graze old Ahab in his own proper and inaccessible being. Can any lead touch yonder floor, any mast scrape yonder roof?—Aloft there! which way?'

'Dead to leeward, sir.'

'Up helm, then; pile on the sail again, ship keepers! down the rest of the spare boats and rig them—Mr Starbuck away, and muster the boats' crews.'

'Let me first help thee towards the bulwarks, sir.'

'Oh, oh, oh! how this splinter gores me now! Accursed fate! that the unconquerable captain in the soul should have such a craven mate!'

'Sir?'

'My body, man, not thee. Give me something for a cane—there, that shivered lance will do. Muster the men. Surely I have not seen him yet. By heaven it cannot be!—missing?—quick! call them all.'

The old man's hinted thought was true. Upon mustering the company, the Parsee was not there.

'The Parsee!' cried Stubb,—'he must have been caught in——'

'The black vomit wrench thee!—run all of ye above, alow, cabin, forecastle—find him—not gone—not gone!'

But quickly they returned to him with the tidings that the Parsee was nowhere to be found.

'Aye, sir,' said Stubb—'caught among the tangles of your line—I thought I saw him dragging under.'

'*My* line! *my* line? Gone?—gone? What means that little word?—What death-knell rings in it, that old Ahab shakes as if he were the belfry. The harpoon, too!—toss over the litter there,—d'ye see it?—the forged iron, men, the white whale's—no, no, no,—blistered fool! this hand did dart it!—'tis in the fish!—Aloft there! Keep him nailed—Quick!—all hands to the rigging of the boats—collect the oars—harpooneers! the irons, the irons!—hoist the royals higher—a pull on all the sheets!—helm there! steady, steady for your life! I'll ten times girdle the unmeasured globe; yea and dive straight through it, but I'll slay him yet!'

'Great God! but for one single instant show thyself,' cried Starbuck; 'never, never wilt thou capture him, old man—In Jesus' name no more of this, that's worse than devil's madness. Two days chased; twice stove to splinters; thy very leg once more snatched from under thee; thy evil shadow gone—all good angels mobbing thee with warnings: what more wouldst thou have? Shall we keep chasing this murderous fish till he swamps the last man? Shall we be dragged by him to the bottom of the sea? Shall we be towed by him to the infernal world? Oh, oh, Impiety and blasphemy to hunt him more!'

'Starbuck, of late I've felt strangely moved to thee; ever since that hour we both saw—thou know'st what, in one another's eyes. But in this matter of the whale, be the front of thy face to me as the palm of this hand—a lipless, unfeatured blank. Ahab is for ever Ahab, man. This whole act's immutably decreed. 'Twas rehearsed by thee and me a billion years before this ocean rolled. Fool! I am the Fates' lieutenant; I act under orders. Look thou, underling! that thou obeyest mine. Stand round me, men. Ye see an old man cut down to the stump; leaning on a shivered lance; propped up on a lonely foot. 'Tis Ahab—his body's part; but Ahab's soul's a centipede, that moves upon a hundred legs. I feel strained, half stranded, as ropes that tow dismasted frigates in a gale; and I may look so. But ere I break, ye'll hear me crack; and till ye hear *that*, know that Ahab's hawser tows his purpose yet. Believe ye, men, in the things called omens? Then laugh aloud, and cry encore! For ere they drown, drowning things will twice rise to the surface; then rise again, to sink for evermore. So with Moby Dick— two days he's floated—tomorrow will be the third. Aye, men, he'll rise once more, but only to spout his last! D'ye feel brave, men, brave?'

'As fearless fire,' cried Stubb.

'And as mechanical', muttered Ahab. Then as the men went forward, he muttered on: 'The things called omens! And yesterday I talked the same to Starbuck there, concerning my broken boat. Oh! how valiantly I seek to drive out of others' hearts what's clinched so fast in mine!—The Parsee—the Parsee!—gone, gone? and he was to go before:—but still was to be seen again ere I could perish—How's that?—There's a riddle now might baffle all the lawyers backed by the ghosts of the whole line of judges: like a hawk's beak it pecks my brain. *I'll, I'll* solve it, though!'

When dusk descended, the whale was still in sight to leeward.

So once more the sail was shortened, and everything passed nearly as on the previous night; only, the sound of hammers, and the hum of the grindstone was heard till nearly daylight, as the men toiled by lanterns in the complete and careful rigging of the spare boats and sharpening their fresh weapons for the morrow. Meantime, of the broken keel of Ahab's wrecked craft the carpenter made him another leg; while still as on the night before, slouched Ahab stood fixed within his scuttle; his hid, heliotrope glance anticipatingly gone backward on its dial; set due eastward for the earliest sun.

THIRD DAY

The morning of the third day dawned fair and fresh, and once more the solitary night-man at the fore-mast-head was relieved by crowds of the daylight look-outs, who dotted every mast and almost every spar.

'D'ye see him?' cried Ahab; but the whale was not yet in sight.

'In his infallible wake, though; but follow that wake, that's all. Helm there; steady, as thou goest, and hast been going. What a lovely day again! were it a new-made world, and made for a summer-house to the angels, and this morning the first of its throwing open to them, a fairer day could not dawn upon that world. Here's food for thought, had Ahab time to think; but Ahab never thinks; he only feels, feels, feels; *that*'s tingling enough for mortal man! to think's audacity. God only has that right and privilege. Thinking is, or ought to be, a coolness and a calmness; and our poor hearts throb, and our poor brains beat too much for that. And yet, I've sometimes thought my brain was very calm—frozen calm, this old skull cracks so, like a glass in which the contents turn to ice, and shiver it. And still this hair is growing now; this moment growing, and heat must breed it; but no, it's like that sort of common grass that will grow anywhere, between the earthy clefts of Greenland ice or in Vesuvius lava. How the wild winds blow it; they whip it about me as the torn shreds of split sails lash the tossed ship they cling to. A vile wind that has no doubt blown ere this through prison corridors and cells, and wards of hospitals, and ventilated them, and now comes blowing hither as innocent as fleeces. Out upon it!—it's tainted. Were I the wind, I'd blow no more on such a wicked, miserable world. I'd crawl somewhere to a cave, and slink there. And yet, 'tis a noble and heroic thing, the wind! who ever conquered it? In every fight it has the last and bitterest blow. Run tilting at it, and you but run through it. Ha! a coward wind that strikes stark naked men, but will not stand to receive a single blow. Even Ahab is a braver thing—a nobler thing than *that*. Would now the wind but had a body; but all the things that most exasperate and outrage mortal man, all these things are bodiless, but only bodiless as objects, not as agents. There's a most special, a most cunning, oh, a most malicious difference! And yet, I say again, and swear it now, that there's something all glorious and gracious in the wind. These warm Trade Winds, at least, that in the clear heavens blow straight on, in strong and steadfast, vigorous mildness; and veer not from their mark, however the baser currents of the sea may turn and tack, and mightiest Mississippies of the land swift and swerve about, uncertain where to go at last. And by the eternal Poles! these same Trades that so directly blow my good ship on; these Trades, or something like them—something so unchangeable, and full as strong, blow my keeled soul along! To it! Aloft there! What d'ye see?'

'Nothing, sir.'

'Nothing! and noon at hand! The doubloon goes a-begging! See the sun! Aye, aye, it must be so. I've oversailed him. How, got the start? Aye, he's chasing *me* now; not I, *him*—that's bad; I might have known it, too. Fool! the lines—the harpoons he's towing. Aye, aye, I have run him by last night. About! about! Come down, all of ye, but the regular look outs! Man the braces!'

Steering as she had done, the wind had been somewhat on the Pequod's quarter, so that now being pointed in the reverse direction, the braced ship sailed hard upon the breeze as she rechurned the cream in her own white wake.

'Against the wind he now steers for the open jaw,' murmured Starbuck to himself, as he coiled the new-hauled mainbrace upon the rail. 'God keep us, but already my bones feel damp within me, and from the inside wet my flesh. I misdoubt me that I disobey my God in obeying him!'

'Stand by to sway me up!' cried Ahab, advancing to the hempen basket. 'We should meet him soon.'

'Aye, aye, sir,' and straightway Starbuck did Ahab's bidding, and once more Ahab swung on high.

A whole hour now passed; gold-beaten out to ages. Time itself now held long breaths with keen suspense. But at last, some three points off the weather bow, Ahab descried the spout again, and instantly from the three mast-heads three shrieks went up as if the tongues of fire had voiced it.

'Forehead to forehead I meet thee, this third time, Moby Dick! On deck there!—brace sharper up; crowd her into the wind's eye. He's too far off to lower yet, Mr Starbuck. The sails shake! Stand over that helmsman with a top-maul! So, so; he travels fast, and I must down. But let me have one more good round look aloft here at the sea; there's time for that. An old, old sight, and yet somehow so young; aye, and not changed a wink since I first saw it, a boy, from the sand-hills of Nantucket! The same!—the same!—the same to Noah as to me. There's a soft shower to leeward. Such lovely leewardings! They must lead somewhere—to something else than common land, more palmy than the palms. Leeward! the white whale goes that way; look to windward, then; the better if the bitterer quarter. But good-bye, goodbye, old mast-head! What's this?—green? aye, tiny mosses in these warped cracks. No such green weather stains on Ahab's head! There's the difference now between man's old age and matter's. But aye, old mast, we both grow old together; sound in our hulls, though, are we not, my ship? Aye, minus a leg, that's all. By heaven this dead wood has the better of my live flesh every way. I can't compare with it; and I've known some ships made of dead trees outlast the lives of men made of the most vital stuff of vital fathers. What's that he said? he should still go before me, my pilot; and yet to be seen again? But where? Will I have eyes at the bottom of the sea, supposing I descend those endless stairs? and all night I've been sailing from him, wherever he did sink to. Aye, aye, like many more thou told'st direful truth as touching thyself, O Parsee; but, Ahab, there thy shot fell short. Good-bye, mast-head—keep a good eye upon the whale, the while I'm gone. We'll talk tomorrow, nay, tonight, when the white whale lies down there, tied by head and tail.'

He gave the word; and still gazing round him, was steadily lowered through the cloven blue air to the deck.

In due time the boats were lowered; but as standing in his shallop's stern, Ahab just hovered upon the point of the descent, he waved to the mate—who held one of the tackle-ropes on deck—and bade him pause.

'Starbuck!'

'Sir?'

'For the third time my soul's ship starts upon this voyage, Starbuck.'

'Aye, sir, thou wilt have it so.'

'Some ships sail from their ports, and ever afterwards are missing, Starbuck!'

'Truth, sir: saddest truth.'

'Some men die at ebb tide; some at low water; some at the full of the flood; and I feel now like a billow that's all one crested comb, Starbuck. I am old; shake hands with me, man.'

Their hands met; their eyes fastened; Starbuck's tears the glue.

'Oh, my captain, my captain!—noble heart—go not—go not!—see, it's a brave man that weeps; how great the agony of the persuasion then!'

'Lower away!', cried Ahab, tossing the mate's arm from him. 'Stand by the crew!'

In an instant the boat was pulling round close under the stern.

'The sharks! the sharks!' cried a voice from the low cabin-window there; 'O master, my master, come back!'

But Ahab heard nothing; for his own voice was high-lifted then; and the boat leaped on.

Yet the voice spake true; for scarce had he pushed from the ship, when numbers of sharks, seemingly rising from out the dark waters beneath the hull, maliciously snapped at the blades of the oars, every time they dipped in the water; and in this way accompanied the boat with their bites. It is a thing not uncommonly happening to the whale-boats in those swarming seas; the sharks at times apparently following them in the same prescient way that vultures hover over the banners of marching regiments in the east. But these were the first sharks that had been observed by the Pequod since the White Whale had been first descried; and whether it was that Ahab's crew were all such tiger-yellow barbarians, and therefore their flesh more musky to the senses of the sharks—a matter sometimes well known to affect them—however it was, they seemed to follow that one boat without molesting the others.

'Heart of wrought steel!' murmured Starbuck gazing over the side, and following with his eyes the receding boat, 'canst thou yet ring boldly to that sight?—lowering thy keel among ravening sharks, and followed by them, open-mouthed to the chase; and this the critical third day? For when three days flow together in one continuous intense pursuit; be sure the first is the morning, the second the noon, and the third the evening and the end of that thing—be that end what it may. Oh! my God! what is this that shoots through me, and leaves me so deadly calm, yet expectant, fixed at the top of a shudder! Future things swim before me, as in empty outlines and skeletons; all the past is somehow grown dim. Mary, girl! thou fadest in pale glories behind me; boy! I seem to see but thy eyes grown wondrous blue. Strangest problems of life seem clearing; but clouds sweep between— Is my journey's end coming? My legs feel faint; like his who has footed it all day. Feel thy heart—beats it yet? Stir thyself, Starbuck!— stave it off—move, move! speak aloud!—Mast-head there! See ye my boy's hand on the hill?—Crazed;—aloft there!—keep thy keenest eye upon the boats:—mark well the whale!—Ho! again!—drive off that hawk! see! he pecks—he tears the vane'—pointing to the red flag flying at the main-truck—'Ha! he soars away with it!—Where's the old man now? sees't thou that sight, oh Ahab!—shudder, shudder!'

The boats had not gone very far, when by a signal from the mast-heads—a downward pointed arm—Ahab knew that the whale had sounded; but intending to be near him at the next rising, he held on his way a little sideways from the vessel; the becharmed crew maintaining the profoundest silence, as the head-beat waves hammered and hammered against the opposing bow.

'Drive, drive in your nails, oh ye waves! to their uttermost heads drive them in! ye but strike a thing without a lid; and no coffin and no hearse can be mine: and hemp only can kill me! Ha! ha!'

Suddenly the waters around them slowly swelled in broad circles; then quickly upheaved, as if sideways sliding from a submerged berg of ice, swiftly rising to the surface. A low rumbling sound was heard; a subterraneous hum; and then all held their breaths; as bedraggled with trailing ropes, and harpoons, and lances, a vast form shot lengthwise, but obliquely from the sea. Shrouded in a thin drooping veil of mist, it hovered for a moment in the rainbowed air; and then fell swamping back into the deep. Crushed thirty feet upwards, the waters flashed for an instant like heaps of fountains, then brokenly sank in a shower of flakes, leaving the circling surface creamed like new milk round the marble trunk of the whale.

'Give way!' cried Ahab to the oarsmen, and the boats darted forward to the attack; but maddened by yesterday's fresh irons that corroded in him, Moby Dick seemed combinedly possessed by all the angels that fell from heaven. The wide tiers of welded tendons overspreading his broad white forehead, beneath the transparent skin, looked knitted together; as head on, he came churning his tail among the boats; and once more flailed them apart; spilling out the irons and lances from the two mates' boats, and dashing in one side of the upper part of their bows, but leaving Ahab's almost without a scar.

While Daggoo and Tashtego were stopping the strained planks; and as the whale swimming out from them, turned, and showed one entire flank as he shot by them again; at that moment a quick cry went up. Lashed round and round to the fish's back; pinioned in the turns upon turns in which, during the past night, the whale had reeled the involutions of the lines around him, the half torn body of the Parsee was seen; his sable raiment frayed to shreds; his distended eyes turned full upon old Ahab.

The harpoon dropped from his hand.

'Befooled, befooled!'—drawing in a long lean breath—'Aye, Parsee! I see thee again. Aye, and thou goest before; and this, *this* then is the hearse that thou didst promise. But I hold thee to the last letter of thy word. Where is the second hearse? Away, mates, to the ship! those boats are useless now; repair them if ye can in time, and return to me;

if not, Ahab is enough to die—Down, men! the first thing that but offers to jump from this boat I stand in, that thing I harpoon. Ye are not other men, but my arms and my legs; and so obey me. Where's the whale? gone down again?'

But he looked too nigh the boat; for as if bent upon escaping with the corpse he bore, and as if the particular place of the last encounter had been but a stage in his leeward voyage, Moby Dick was now again steadily swimming forward; and had almost passed the ship, which thus far had been sailing in the contrary direction to him, though for the present her headway had been stopped. He seemed swimming with his utmost velocity, and now only intent upon pursuing his own straight path in the sea.

'Oh! Ahab,' cried Starbuck, 'not too late is it, even now, the third day, to desist. See! Moby Dick seeks thee not. It is thou, thou, that madly seekest him!'

Setting sail to the rising wind, the lonely boat was swiftly impelled to leeward, by both oars and canvas. And at last when Ahab was sliding by the vessel, so near as plainly to distinguish Starbuck's face as he leaned over the rail, he hailed him to turn the vessel about, and follow him, not too swiftly, at a judicious interval. Glancing upwards, he saw Tashtego, Queequeg, and Daggoo, eagerly mounting to the three mast-heads; while the oarsmen were rocking in the two staved boats which had but just been hoisted to the side, and were busily at work in repairing them. One after the other, through the port-holes, as he sped, he also caught flying glimpses of Stubb and Flask, busying themselves on deck among bundles of new irons and lances. As he saw all this; as he heard the hammers in the broken boats; far other hammers seemed driving a nail into his heart. But he rallied. And now marking that the vane or flag was gone from the main-mast-head, he shouted to Tashtego, who had just gained that perch, to descend again for another flag, and a hammer and nails, and so nail it to the mast.

Whether fagged by the three days' running chase, and the resistance to his swimming in the knotted hamper he bore; or whether it was some latent deceitfulness and malice in him: whichever was true, the White Whale's way now began to abate, as it seemed, from the boat so rapidly nearing him once more; though indeed the whale's last start had not been so long a one as before. And still as Ahab glided over the waves the unpitying sharks accompanied him; and so pertinaciously stuck to the boat; and so continually bit at the plying oars, that the blades became jagged and crunched, and left small splinters in the sea, at almost every dip.

'Heed them not! those teeth but give new rowlocks to your oars. Pull on! 'tis the better rest, the shark's jaw than the yielding water.'

'But at every bite, sir, the thin blades grow smaller and smaller!'

'They will last long enough! pull on! But who can tell', he muttered, 'whether these sharks swim to feast on the whale or on Ahab? But pull on! Aye, all alive, now—we near him. The helm! take the helm; let me pass', and so saying, two of the oarsmen helped him forward to the bows of the still flying boat.

At length as the craft was cast to one side, and ran ranging along with the White Whale's flank, he seemed strangely oblivious of its advance—as the whale sometimes will—and Ahab was fairly within the smoky mountain mist, which, thrown off from the whale's spout, curled round his great, Monadnock hump; he was even thus close to him; when, with body arched back, and both arms lengthwise high-lifted to the poise, he darted his fierce iron, and his far fiercer curse into the hated whale. As both steel and curse sank to the socket, as if sucked into a morass, Moby Dick sideways writhed; spasmodically rolled his nigh flank against the bow, and, without staving a hole in it, so suddenly canted the boat over, that had it not been for the elevated part of the gunwale to which he then clung, Ahab would once more have been tossed into the sea. As it was, three of the oarsmen—who foreknew not the precise instant of the dart, and were therefore unprepared for its effects—these were flung out; but so fell, that, in an instant two of them clutched the gunwale again, and rising to its level on a combing wave, hurled themselves bodily inboard again; the third man helplessly dropping astern, but still afloat and swimming.

Almost simultaneously, with a mighty volition of ungraduated, instantaneous swiftness, the White Whale darted through the weltering sea. But when Ahab cried out to the steersman to take new turns with the line, and hold it so; and commanded the crew to turn round on their seats, and tow the boat up to the mark; the moment the treacherous line felt that double strain and tug, it snapped in the empty air!

'What breaks in me? Some sinew cracks!—'tis whole again; oars! oars! Burst in upon him!'

Hearing the tremendous rush of the sea-crashing boat, the whale wheeled round to present his blank forehead at bay; but in that

evolution, catching sight of the nearing black hull of the ship; seemingly seeing in it the source of all his persecutions; bethinking it—it
may be—a larger and nobler foe; of a sudden, he bore down upon its
advancing prow, smiting his jaws amid fiery showers of foam.

Ahab staggered; his hand smote his forehead. 'I grow blind; hands!
stretch out before me that I may yet grope my way. Is't night?'

'The whale! The ship!' cried the cringing oarsmen.

'Oars! oars! Slope downwards to thy depths, O sea, that ere it be
for ever too late, Ahab may slide this last, last time upon his mark! I
see: the ship! the ship! Dash on, my men! Will ye not save my ship?'

But as the oarsmen violently forced their boat through the sledge-
hammering seas, the before whale-smitten bow-ends of two planks
burst through, and in an instant almost, the temporarily disabled boat
lay nearly level with the waves; its half-wading, splashing crew, trying
hard to stop the gap and bale out the pouring water.

Meantime, for that one beholding instant, Tashtego's mast-head
hammer remained suspended in his hand; and the red flag, half-wrap-
ping him as with a plaid, then streamed itself straight out from him,
as his own forward-flowing heart; while Starbuck and Stubb, standing
upon the bowsprit beneath, caught sight of the down-coming monster just as soon as he.

'The whale, the whale! Up helm, up helm! Oh, all ye sweet powers
of air, now hug me close! Let not Starbuck die, if die he must, in a
woman's fainting fit. Up helm, I say—ye fools, the jaw! the jaw! Is
this the end of all my bursting prayers? all my life-long fidelities? Oh,
Ahab, Ahab, lo, thy work. Steady! helmsman, steady. Nay, nay! Up
helm again! He turns to meet us! Oh, his unappeasable brow drives
on towards one, whose duty tells him he cannot depart. My God,
stand by me now!'

'Stand not by me, but stand under me, whoever you are that will
now help Stubb; for Stubb, too, sticks here. I grin at thee, thou
grinning whale! Who ever helped Stubb, or kept Stubb awake, but
Stubb's own unwinking eye? And now poor Stubb goes to bed upon
a mattress that is all too soft; would it were stuffed with brushwood!
I grin at thee, thou grinning whale! Look ye, sun, moon, and stars! I
call ye assassins of as good a fellow as ever spouted up his ghost. For
all that, I would yet ring glasses with ye, would ye but hand the cup!
Oh, oh! oh, oh! thou grinning whale, but there'll be plenty of gulping
soon! Why fly ye not, O Ahab! For me, off shoes and jacket to it; let
Stubb die in his drawers! A most mouldy and over-salted death,

though; cherries! cherries! cherries! oh, Flask, for one red cherry ere we die!'

'Cherries? I only wish that we were where they grow. Oh, Stubb, I hope my poor mother's drawn my part-pay ere this; if not, few coppers will now come to her, for the voyage is up.'

From the ship's bows, nearly all the seamen now hung inactive; hammers, bits of plank, lances, and harpoons, mechanically retained in their hands, just as they had darted from their various employments; all their enchanted eyes intent upon the whale, which from side to side strangely vibrating his predestinating head, sent a broad band of overspreading semicircular foam before him as he rushed. Retribution, swift vengeance, eternal malice were in his whole aspect, and spite of all that mortal man could do, the solid white buttress of his forehead smote the ship's starboard bow, till men and timbers reeled. Some fell flat upon their faces. Like dislodged trucks, the heads of the harpooneers aloft shook on their bull-like necks. Through the breach, they heard the waters pour, as mountain torrents down a flume.

'The ship! The hearse!—the second hearse!' cried Ahab from the boat; 'its wood could only be American!'

Diving beneath the settling ship, the whale ran quivering along its keel; but turning under water, swiftly shot to the surface again, far off the other bow, but within a few yards of Ahab's boat, where, for a time, he lay quiescent.

'I turn my body from the sun. What ho, Tashtego! let me hear thy hammer. Oh! ye three unsurrendered spires of mine; thou uncracked keel; and only god-bullied hull; thou firm deck, and haughty helm, and Pole-pointed prow, death-glorious ship! must ye then perish, and without me? Am I cut off from the last fond pride of meanest shipwrecked captains? Oh, lonely death on lonely life! Oh, now I feel my topmost greatness lies in my topmost grief. Ho, ho! from all your furthest bounds, pour ye now in, ye bold billows of my whole foregone life, and top this one piled comber of my death! Towards thee I roll, thou all-destroying but unconquering whale; to the last I grapple with thee; from hell's heart I stab at thee; for hate's sake I spit my last breath at thee. Sink all coffins and all hearses to one common pool! and since neither can be mine, let me then tow to pieces, while still chasing thee, though tied to thee, thou damned whale! *Thus*, I give up the spear!'

The harpoon was darted; the stricken whale flew forward; with igniting velocity the line ran through the groove; ran foul. Ahab stooped to clear it; he did clear it; but the flying turn caught him round the neck, and voicelessly as Turkish mutes bowstring their victim, he was shot out of the boat, ere the crew knew he was gone. Next instant, the heavy eye-splice in the rope's final end flew out of the stark-empty tub, knocked down an oarsman, and smiting the sea, disappeared in its depths.

For an instant, the tranced boat's crew stood still; then turned. 'The ship? Great God, where is the ship?' Soon they through dim, bewildering mediums saw her sidelong fading phantom, as in the gaseous Fata Morgana; only the uppermost masts out of water; while fixed by infatuation, or fidelity, or fate, to their once lofty perches, the pagan harpooneers still maintained their sinking lookouts on the sea. And now, concentric circles seized the lone boat itself, and all its crew, and each floating oar, and every lance-pole, and spinning, animate and inanimate, all round and round in one vortex, carried the smallest chip of the Pequod out of sight.

But as the last whelmings intermixingly poured themselves over the sunken head of the Indian at the mainmast, leaving a few inches of the erect spar yet visible, together with long streaming yards of the flag, which calmly undulated, with ironical coincidings, over the destroying billows they almost touched; at that instant, a red arm and a hammer hovered backwardly uplifted in the open air, in the act of nailing the flag faster and yet faster to the subsiding spar. A sky-hawk that tauntingly had followed the main-truck downwards from its natural home among the stars, pecking at the flag, and incommoding Tashtego there; this bird now chanced to intercept its broad fluttering wing between the hammer and the wood; and simultaneously feeling that etherial thrill, the submerged savage beneath, in his death-grasp, kept his hammer frozen there; and so the bird of heaven, with archangelic shrieks, and his imperial beak thrust upwards, and his whole captive form folded in the flag of Ahab, went down with his ship, which, like Satan, would not sink to hell till she had dragged a living part of heaven along with her, and helmeted herself with it.

Now small fowls flew screaming over the yet yawning gulf; a sullen white surf beat against its steep sides; then all collapsed, and the great shroud of the sea rolled on as it rolled five thousand years ago.

A Tragedy of Error

~~~

HENRY JAMES

I

A LOW English phaeton was drawn up before the door of the post office of a French seaport town. In it was seated a lady, with her veil down and her parasol held closely over her face. My story begins with a gentleman coming out of the office and handing her a letter.

He stood beside the carriage a moment before getting in. She gave him her parasol to hold, and then lifted her veil, showing a very pretty face. This couple seemed to be full of interest for the passers-by, most of whom stared hard and exchanged significant glances. Such persons as were looking on at the moment saw the lady turn very pale as her eyes fell on the direction of the letter. Her companion saw it too, and instantly stepping into the place beside her, took up the reins, and drove rapidly along the main street of the town, past the harbour, to an open road skirting the sea. Here he slackened pace. The lady was leaning back, with her veil down again, and the letter lying open in her lap. Her attitude was almost that of unconsciousness, and he could see that her eyes were closed. Having satisfied himself of this, he hastily possessed himself of the letter, and read as follows:

SOUTHAMPTON, *July 16th*, 18—.
MY DEAR HORTENSE: You will see by my postmark that I am a thousand leagues nearer home than when I last wrote, but I have hardly time to explain the change. M. P—— has given me a most unlooked-for *congé*. After so many months of separation, we shall be able to spend a few weeks together. God be praised! We got in here from New York this morning, and I have had the good luck to find a vessel, the *Armorique*, which sails straight for H——. The mail leaves directly, but we shall probably be detained a few hours by the tide;

so this will reach you a day before I arrive: the master calculates we shall get in early Thursday morning. Ah, Hortense! how the time drags! Three whole days! If I did not write from New York, it is because I was unwilling to torment you with an expectancy which, as it is, I venture to hope, you will find long enough. Farewell. To a warmer greeting!

Your devoted
C.B.

When the gentleman replaced the paper on his companion's lap, his face was almost as pale as hers. For a moment he gazed fixedly and vacantly before him, and a half-suppressed curse escaped his lips. Then his eyes reverted to his neighbour. After some hesitation, during which he allowed the reins to hang so loose that the horse lapsed into a walk, he touched her gently on the shoulder.

'Well, Hortense,' said he, in a very pleasant tone, 'what's the matter; have you fallen asleep?'

Hortense slowly opened her eyes, and, seeing that they had left the town behind them, raised her veil. Her features were stiffened with horror.

'Read that,' said she, holding out the open letter.

The gentleman took it, and pretended to read it again.

'Ah! M. Bernier returns. Delightful!' he exclaimed.

'How, delightful?' asked Hortense; 'we mustn't jest at so serious a crisis, my friend.'

'True,' said the other, 'it will be a solemn meeting. Two years of absence is a great deal.'

'O Heaven! I shall never dare to face him,' cried Hortense, bursting into tears.

Covering her face with one hand, she put out the other toward that of her friend. But he was plunged in so deep a reverie, that he did not perceive the movement. Suddenly he came to, aroused by her sobs.

'Come, come,' said he, in the tone of one who wishes to coax another into mistrust of a danger before which he does not himself feel so secure but that the sight of a companion's indifference will give him relief. 'What if he does come? He need learn nothing. He will stay but a short time, and sail away again as unsuspecting as he came.'

'Learn nothing! You surprise me. Every tongue that greets him, if only to say *bon jour*, will wag to the tune of a certain person's misconduct.'

'Bah! People don't think about us quite as much as you fancy. You and I, *n'est-ce-pas?* we have little time to concern ourselves about our neighbours' failings. Very well, other people are in the same box, better or worse. When a ship goes to pieces on those rocks out at sea, the poor devils who are pushing their way to land on a floating spar, don't bestow many glances on those who are battling with the waves beside them. Their eyes are fastened to the shore, and all their care is for their own safety. In life we are all afloat on a tumultuous sea; we are all struggling toward some *terra firma* of wealth or love or leisure. The roaring of the waves we kick up about us and the spray we dash into our eyes deafen and blind us to the sayings and doings of our fellows. Provided we climb high and dry, what do we care for them?'

'Ay, but if we don't? When we've lost hope ourselves, we want to make others sink. We hang weights about their necks, and dive down into the dirtiest pools for stones to cast at them. My friend, you don't feel the shots which are not aimed at you. It isn't of you the town talks, but of me: a poor woman throws herself off the pier yonder, and drowns before a kind hand has time to restrain her, and her corpse floats over the water for all the world to look at. When her husband comes up to see what the crowd means, is there any lack of kind friends to give him the good news of his wife's death?'

'As long as a woman is light enough to float, Hortense, she is not counted drowned. It's only when she sinks out of sight that they give her up.'

Hortense was silent a moment, looking at the sea with swollen eyes.

'Louis,' she said at last, 'we were speaking metaphorically: I have half a mind to drown myself literally.'

'Nonsense!' replied Louis, 'an accused pleads *not guilty*, and hangs himself in prison. What do the papers say? People talk, do they? Can't you talk as well as they? A woman is in the wrong from the moment she holds her tongue and refuses battle. And that you do too often. That pocket handkerchief is always more or less of a flag of truce.'

'I'm sure I don't know,' said Hortense indifferently; 'perhaps it is.'

There are moments of grief in which certain aspects of the subject of our distress seem as irrelevant as matters entirely foreign to it. Her eyes were still fastened on the sea. There was another silence. 'O my poor Charles!' she murmured, at length, 'to what a hearth do you return!'

'Hortense,' said the gentleman, as if he had not heard her, although, to a third person, it would have appeared that it was because he had done so that he spoke: 'I do not need to tell you that it will never happen to me to betray our secret. But I will answer for it that so long as M. Bernier is at home no mortal shall breathe a syllable of it.'

'What of that?' sighed Hortense. 'He will not be with me ten minutes without guessing it.'

'Oh, as for that,' said her companion, drily, 'that's your own affair.'

'Monsieur de Meyrau!' cried the lady.

'It seems to me,' continued the other, 'that in making such a guarantee, I have done my part of the business.'

'Your part of the business!' sobbed Hortense.

M. de Meyrau made no reply, but with a great cut of the whip sent the horse bounding along the road. Nothing more was said. Hortense lay back in the carriage with her face buried in her handkerchief, moaning. Her companion sat upright, with contracted brows and firmly set teeth, looking straight before him, and by an occasional heavy lash keeping the horse at a furious pace. A wayfarer might have taken him for a ravisher escaping with a victim worn out with resistance. Travellers to whom they were known would perhaps have seen a deep meaning in this accidental analogy. So, by a *détour*, they returned to the town.

When Hortense reached home, she went straight up to a little boudoir on the second floor, and shut herself in. This room was at the back of the house, and her maid, who was at that moment walking in the long garden which stretched down to the water, where there was a landing place for small boats, saw her draw in the window blind and darken the room, still in her bonnet and cloak. She remained alone for a couple of hours. At five o'clock, some time after the hour at which she was usually summoned to dress her mistress for the evening, the maid knocked at Hortense's door, and offered her services. Madame called out, from within, that she had a *migraine*, and would not be dressed.

'Can I get anything for madame?' asked Josephine; 'a *tisane*, a warm drink, something?'

'Nothing, nothing.'

'Will madame dine?'

'No.'

'Madame had better not go wholly without eating.'

'Bring me a bottle of wine—of brandy.'

Josephine obeyed. When she returned, Hortense was standing in the doorway, and as one of the shutters had meanwhile been thrown open, the woman could see that, although her mistress's hat had been tossed upon the sofa, her cloak had not been removed, and that her face was very pale. Josephine felt that she might not offer sympathy nor ask questions.

'Will madame have nothing more?' she ventured to say, as she handed her the tray.

Madame shook her head, and closed and locked the door.

Josephine stood a moment vexed, irresolute, listening. She heard no sound. At last she deliberately stooped down and applied her eye to the keyhole.

This is what she saw:

Her mistress had gone to the open window, and stood with her back to the door, looking out at the sea. She held the bottle by the neck in one hand, which hung listlessly by her side; the other was resting on a glass half filled with water, standing, together with an open letter, on a table beside her. She kept this position until Josephine began to grow tired of waiting. But just as she was about to arise in despair of gratifying her curiosity, madame raised the bottle and glass, and filled the latter full. Josephine looked more eagerly. Hortense held it a moment against the light, and then drained it down.

Josephine could not restrain an involuntary whistle. But her surprise became amazement when she saw her mistress prepare to take a second glass. Hortense put it down, however, before its contents were half gone, as if struck by a sudden thought, and hurried across the room. She stooped down before a cabinet, and took out a small opera glass. With this she returned to the window, put it to her eyes, and again spent some moments in looking seaward. The purpose of this proceeding Josephine could not make out. The only result visible to her was that her mistress suddenly dropped the lorgnette on the table, and sank down on an armchair, covering her face with her hands.

Josephine could contain her wonderment no longer. She hurried down to the kitchen.

'Valentine,' said she to the cook, 'what on earth can be the matter with Madame? She will have no dinner, she is drinking brandy by the glassful, a moment ago she was looking out to sea with a lorgnette, and now she is crying dreadfully with an open letter in her lap.'

The cook looked up from her potato-peeling with a significant wink.

'What can it be,' said she, 'but that monsieur returns?'

II

At six o'clock, Josephine and Valentine were still sitting together, discussing the probable causes and consequences of the event hinted at by the latter. Suddenly Madame Bernier's bell rang. Josephine was only too glad to answer it. She met her mistress descending the stairs, combed, cloaked, and veiled, with no traces of agitation, but a very pale face.

'I am going out,' said Madame Bernier; 'if M. le Vicomte comes, tell him I am at my mother-in-law's, and wish him to wait till I return.'

Josephine opened the door, and let her mistress pass; then stood watching her as she crossed the court.

'Her mother-in-law's,' muttered the maid; 'she has the face!'

When Hortense reached the street, she took her way, not through the town, to the ancient quarter where that ancient lady, her husband's mother, lived, but in a very different direction. She followed the course of the quay, beside the harbour, till she entered a crowded region, chiefly the residence of fishermen and boatmen. Here she raised her veil. Dusk was beginning to fall. She walked as if desirous to attract as little observation as possible, and yet to examine narrowly the population in the midst of which she found herself. Her dress was so plain that there was nothing in her appearance to solicit attention; yet, if for any reason a passer-by had happened to notice her, he could not have helped being struck by the contained intensity with which she scrutinized every figure she met. Her manner was that of a person seeking to recognize a long-lost friend, or perhaps, rather, a long-lost enemy, in a crowd. At last she stopped before a flight of steps, at the foot of which was a landing place for half a dozen little boats, employed to carry passengers between the two sides of the port, at times when the drawbridge above was closed for the passage of vessels. While she stood she was witness of the following scene:

A man, in a red woollen fisherman's cap, was sitting on the top of the steps, smoking the short stump of a pipe, with his face to the water. Happening to turn about, his eye fell on a little child, hurrying along the quay toward a dingy tenement close at hand, with a jug in its arms.

'Hullo, youngster!' cried the man; 'what have you got there? Come here.'

The little child looked back, but, instead of obeying, only quickened its walk.

'The devil take you, come here!' repeated the man angrily, 'or I'll wring your beggarly neck. You won't obey your own uncle, eh?'

The child stopped, and ruefully made its way to its relative, looking around several times toward the house, as if to appeal to some counter authority.

'Come, make haste!' pursued the man, 'or I shall go and fetch you. Move!'

The child advanced to within half a dozen paces of the steps, and then stood still, eyeing the man cautiously, and hugging the jug tight.

'Come on, you little beggar, come up close.'

The youngster kept a stolid silence, however, and did not budge. Suddenly its self-styled uncle leaned forward, swept out his arm, clutched hold of its little sunburned wrist, and dragged it toward him.

'Why didn't you come when you were called?' he asked, running his disengaged hand into the infant's frowsy mop of hair, and shaking its head until it staggered. 'Why didn't you come, you unmannerly little brute, eh?—eh?—eh?' accompanying every interrogation with a renewed shake.

The child made no answer. It simply and vainly endeavoured to twist its neck around under the man's gripe, and transmit some call for succour to the house.

'Come, keep your head straight. Look at me, and answer me. What's in that jug? Don't lie.'

'Milk.'

'Who for?'

'Granny.'

'Granny be hanged.'

The man disengaged his hands, lifted the jug from the child's feeble grasp, tilted it toward the light, surveyed its contents, put it to his lips, and exhausted them. The child, although liberated, did not retreat. It stood watching its uncle drink until he lowered the jug. Then, as he met its eyes, it said:

'It was for the baby.'

For a moment the man was irresolute. But the child seemed to have a foresight of the parental resentment, for it had hardly spoken when it darted backward and scampered off, just in time to elude a blow from the jug, which the man sent clattering at its heels. When it was out of sight, he faced about to the water again, and replaced the pipe between his teeth with a heavy scowl and a murmur that sounded to Madame Bernier very like—'I wish the baby'd choke.'

Hortense was a mute spectator of this little drama. When it was over, she turned around, and retraced her steps twenty yards with her hand to her head. Then she walked straight back, and addressed the man.

'My good man,' she said, in a very pleasant voice, 'are you the master of one of these boats?'

He looked up at her. In a moment the pipe was out of his mouth, and a broad grin in its place. He rose, with his hand to his cap.

'I am, madame, at your service.'

'Will you take me to the other side?'

'You don't need a boat; the bridge is closed,' said one of his comrades at the foot of the steps, looking that way.

'I know it,' said Madame Bernier; 'but I wish to go to the cemetery, and a boat will save me half a mile walking.'

'The cemetery is shut at this hour.'

'*Allons*, leave madame alone,' said the man first spoken to. 'This way, my lady.'

Hortense seated herself in the stern of the boat. The man took the sculls.

'Straight across?' he asked.

Hortense looked around her. 'It's a fine evening,' said she; 'suppose you row me out to the lighthouse, and leave me at the point nearest the cemetery on our way back.'

'Very well', rejoined the boatman; 'fifteen sous', and began to pull lustily.

'*Allez*, I'll pay you well,' said Madame.

'Fifteen sous is the fare,' insisted the man.

'Give me a pleasant row, and I'll give you a hundred,' said Hortense.

Her companion said nothing. He evidently wished to appear not to have heard her remark. Silence was probably the most dignified manner of receiving a promise too munificent to be anything but a jest.

For some time this silence was maintained, broken only by the trickling of the oars and the sounds from the neighbouring shores and vessels. Madame Bernier was plunged in a sidelong scrutiny of her ferryman's countenance. He was a man of about thirty-five. His face was dogged, brutal, and sullen. These indications were perhaps exaggerated by the dull monotony of his exercise. The eyes lacked a certain rascally gleam which had appeared in them when he was so *empressé* with the offer of his services. The face was better then—that is, if vice is better than ignorance. We say a countenance is 'lit up' by a smile; and indeed that momentary flicker does the office of a candle in a dark room. It sheds a ray upon the dim upholstery of our souls. The visages of poor men, generally, know few alternations. There is a large class of human beings whom fortune restricts to a single change of expression, or, perhaps, rather to a single expression. Ah me! the faces which wear either nakedness or rags; whose repose is stagnation, whose activity vice; ignorant at their worst, infamous at their best!

'Don't pull too hard,' said Hortense at last. 'Hadn't you better take breath a moment?'

'Madame is very good,' said the man, leaning upon his oars. 'But if you had taken me by the hour,' he added, with a return of the vicious grin, 'you wouldn't catch me loitering.'

'I suppose you work very hard,' said Madame Bernier.

The man gave a little toss of his head, as if to intimate the inadequacy of any supposition to grasp the extent of his labours.

'I've been up since four o'clock this morning, wheeling bales and boxes on the quay, and plying my little boat. Sweating without five minutes' intermission. *C'est comme ça.* Sometimes I tell my mate I think I'll take a plunge in the basin to dry myself. Ha! ha! ha!'

'And of course you gain little,' said Madame Bernier.

'Worse than nothing. Just what will keep me fat enough for starvation to feed on.'

'How? you go without your necessary food?'

'Necessary is a very elastic word, madame. You can narrow it down, so that in the degree above nothing it means luxury. My necessary food is sometimes thin air. If I don't deprive myself of that, it's because I can't.'

'Is it possible to be so unfortunate?'

'Shall I tell you what I have eaten today?'

'Do', said Madame Bernier.

'A piece of black bread and a salt herring are all that have passed my lips for twelve hours.'

'Why don't you get some better work?'

'If I should die tonight', pursued the boatman, heedless of the question, in the manner of a man whose impetus on the track of self-pity drives him past the signal flags of relief, 'what would there be left to bury me? These clothes I have on might buy me a long box. For the cost of this shabby old suit, that hasn't lasted me a twelve-month, I could get one that I wouldn't wear out in a thousand years. *La bonne idée!'*

'Why don't you get some work that pays better?' repeated Hortense. The man dipped his oars again.

'Work that pays better? I must work for work. I must earn that too. Work is wages. I count the promise of the next week's employment the best part of my Saturday night's pocketings. Fifty casks rolled from the ship to the storehouse mean two things: thirty sous and fifty more to roll the next day. Just so a crushed hand, or a dislocated shoulder, mean twenty francs to the apothecary and *bon jour* to my business.'

'Are you married?' asked Hortense.

'No, I thank you. I'm not cursed with that blessing. But I've an old mother, a sister, and three nephews, who look to me for support. The old woman's too old to work; the lass is too lazy, and the little ones are too young. But they're none of them too old or young to be hungry, *allez.* I'll be hanged if I'm not a father to them all.'

There was a pause. The man had resumed rowing. Madame Bernier sat motionless, still examining her neighbour's physiognomy. The sinking sun, striking full upon his face, covered it with an almost lurid glare. Her own features being darkened against the western sky, the direction of them was quite indistinguishable to her companion.

'Why don't you leave the place?' she said at last.

'Leave it! how?' he replied, looking up with the rough avidity with which people of his class receive proposals touching their interests, extending to the most philanthropic suggestions that mistrustful eagerness with which experience has taught them to defend their own side of a bargain—the only form of proposal that she has made them acquainted with.

'Go somewhere else,' said Hortense.

'Where, for instance!'

'To some new country—America.'

The man burst into a loud laugh. Madame Bernier's face bore more evidence of interest in the play of his features than of that discomfiture which generally accompanies the consciousness of ridicule.

'There's a lady's scheme for you! If you'll write for furnished apartments, *là-bas,* I don't desire anything better. But no leaps in the dark for me. America and Algeria are very fine words to cram into an empty stomach when you're lounging in the sun, out of work, just as you stuff tobacco into your pipe and let the smoke curl around your head. But they fade away before a cutlet and a bottle of wine. When the earth grows so smooth and the air so pure that you can see the American coast from the pier yonder, then I'll make up my bundle. Not before.'

'You're afraid, then, to risk anything?'

'I'm afraid of nothing, *moi*. But I am not a fool either. I don't want to kick away my *sabots* till I am certain of a pair of shoes. I can go barefoot here. I don't want to find water where I counted on land. As for America, I've been there already.'

'Ah! you've been there?'

'I've been to Brazil and Mexico and California and the West Indies.'

'Ah!'

'I've been to Asia, too.'

'Ah!'

'*Pardio*, to China and India. Oh, I've seen the world! I've been three times around the Cape.'

'You've been a seaman then?'

'Yes, ma'am; fourteen years.'

'On what ship?'

'Bless your heart, on fifty ships.'

'French?'

'French and English and Spanish; mostly Spanish.'

'Ah?'

'Yes, and the more fool I was.'

'How so?'

'Oh, it was a dog's life. I'd drown any dog that would play half the mean tricks I used to see.'

'And you never had a hand in any yourself?'

'*Pardon*, I gave what I got. I was as good a Spaniard and as great a devil as any. I carried my knife with the best of them, and drew it as quickly, and plunged it as deep. I've got scars, if you weren't a lady. But I'd warrant to find you their mates on a dozen Spanish hides!'

He seemed to pull with renewed vigour at the recollection. There was a short silence.

'Do you suppose,' said Madame Bernier, in a few moments—'do you remember—that is, can you form any idea whether you ever killed a man?'

There was a momentary slackening of the boatman's oars. He gave a sharp glance at his passenger's countenance, which was still so shaded by her position, however, as to be indistinguishable. The tone of her interrogation had betrayed a simple, idle curiosity. He hesitated a moment, and then gave one of those conscious, cautious, dubious smiles, which may cover either a criminal assumption of more than the truth or a guilty repudiation of it.

'*Mon Dieu!*' said he, with a great shrug, 'there's a question! . . . I never killed one without a reason.'

'Of course not', said Hortense.

'Though a reason in South America, *ma foi!*' added the boatman, 'wouldn't be a reason here.'

'I suppose not. What would be a reason there?'

'Well, if I killed a man in Valparaiso—I don't say I did, mind—it's because my knife went in further than I intended.'

'But why did you use it at all?'

'I didn't. If I had, it would have been because he drew his against me.'

'And why should he have done so?'

'*Ventrebleu!* for as many reasons as there are craft in the harbour.'

'For example?'

'Well, that I should have got a place in a ship's company that he was trying for.'

'Such things as that? is it possible?'

'Oh, for smaller things. That a lass should have given me a dozen oranges she had promised him.'

'How odd!' said Madame Bernier, with a shrill kind of laugh. 'A man who owed you a grudge of this kind would just come up and stab you, I suppose, and think nothing of it?'

'Precisely. Drive a knife up to the hilt into your back, with an oath, and slice open a melon with it, with a song, five minutes afterwards.'

'And when a person is afraid, or ashamed, or in some way unable to take revenge himself, does he—or it may be a woman—does she, get someone else to do it for her?'

'*Parbleu!* Poor devils on the lookout for such work are as plentiful all along the South American coast as *commissionaires* on the street corners here.' The ferryman was evidently surprised at the fascination possessed by this infamous topic for so lady-like a person; but having, as you see, a very ready tongue, it is probable that his delight in being able to give her information and hear himself talk were still greater. 'And then down there', he went on, 'they'll never forget a grudge. If a fellow doesn't serve you one day, he'll do it another. A Spaniard's hatred is like lost sleep—you can put it off for a time, but it will gripe you in the end. The rascals always keep their promises to themselves.

... An enemy on shipboard is jolly fun. It's like bulls tethered in the same field. You can't stand still half a minute except against a wall. Even when he makes friends with you, his favours never taste right. Messing with him is like drinking out of a pewter mug. And so it is everywhere. Let your shadow once flit across a Spaniard's path, and he'll always see it there. If you've never lived in any but these damned clockworky European towns, you can't imagine the state of things in a South American seaport—one half the population waiting round the corner for the other half. But I don't see that it's so much better here, where every man's a spy on every other. There you meet an assassin at every turn, here a *sergent de ville* ... At all events, the life *là bas* used to remind me, more than anything else, of sailing in a shallow channel, where you don't know what infernal rock you may ground on. Every man has a standing account with his neighbour, just as madame has at her *fournisseur*'s; and, *ma foi*, those are the only accounts they settle. The master of the *Santiago* may pay me one of these days for the pretty names I heaved after him when we parted company, but he'll never pay me my wages.'

A short pause followed this exposition of the virtues of the Spaniard.

'You yourself never put a man out of the world, then?' resumed Hortense.

'Oh, *que si!* ... Are you horrified?'

'Not at all. I know that the thing is often justifiable.'

The man was silent a moment, perhaps with surprise, for the next thing he said was:

'Madame is Spanish?'

'In that, perhaps, I am,' replied Hortense.

Again her companion was silent. The pause was prolonged. Madame Bernier broke it by a question which showed that she had been following the same train of thought.

'What is sufficient ground in this country for killing a man?'

The boatman sent a loud laugh over the water. Hortense drew her cloak closer about her.

'I'm afraid there is none.'

'Isn't there a right of self-defence?'

'To be sure there is—it's one I ought to know something about. But it's one that *ces messieurs* at the Palais make short work with.'

'In South America and those countries, when a man makes life insupportable to you, what do you do?'

'*Mon Dieu!* I suppose you kill him.'

'And in France?'

'I suppose you kill yourself. Ha! ha! ha!'

By this time they had reached the end of the great breakwater, terminating in a lighthouse, the limit, on one side, of the inner harbour. The sun had set.

'Here we are at the lighthouse,' said the man; 'it's growing dark. Shall we turn?'

Hortense rose in her place a few moments, and stood looking out to sea. 'Yes,' she said at last, 'you may go back—slowly.' When the boat had headed round she resumed her old position, and put one of her hands over the side, drawing it through the water as they moved, and gazing into the long ripples.

At last she looked up at her companion. Now that her face caught some of the lingering light of the west, he could see that it was deathly pale.

'You find it hard to get along in the world,' said she: 'I shall be very glad to help you.'

The man started, and stared a moment. Was it because this remark jarred upon the expression which he was able faintly to discern in her eyes? The next, he put his hand to his cap.

'Madame is very kind. What will you do?'

Madame Bernier returned his gaze.

'I will trust you.'

'Ah!'

'And reward you.'

'Ah? Madame has a piece of work for me?'

'A piece of work,' Hortense nodded.

The man said nothing, waiting apparently for an explanation. His face wore the look of lowering irritation which low natures feel at being puzzled.

'Are you a bold man?'

Light seemed to come in this question. The quick expansion of his features answered it. You cannot touch upon certain subjects with an inferior but by the sacrifice of the barrier which separates you from him. There are thoughts and feelings and glimpses and foreshadowings of thoughts which level all inequalities of station.

'I'm bold enough', said the boatman, 'for anything *you* want me to do.'

'Are you bold enough to commit a crime?'

'Not for nothing.'

'If I ask you to endanger your peace of mind, to risk your personal safety for me, it is certainly not as a favour. I will give you ten times the weight in gold of every grain by which your conscience grows heavier in my service.'

The man gave her a long, hard look through the dim light.

'I know what you want me to do,' he said at last.

'Very well', said Hortense; 'will you do it?'

He continued to gaze. She met his eyes like a woman who has nothing more to conceal.

'State your case.'

'Do you know a vesel named the *Armorique*, a steamer?'

'Yes, it runs from Southampton.'

'It will arrive tomorrow morning early. Will it be able to cross the bar?'

'No; not till noon.'

'I thought so. I expect a person by it—a man.'

Madame Bernier appeared unable to continue, as if her voice had given way.

'Well, well?' said her companion.

'He's the person'—she stopped again.

'The person who—?'

'The person whom I wish to get rid of.'

For some moments nothing was said. The boatman was the first to speak again.

'Have you formed a plan?'

Hortense nodded.

'Let's hear it.'

'The person in question', said Madame Bernier, 'will be impatient to land before noon. The house to which he returns will be in view of the vessel if, as you say, she lies at anchor. If he can get a boat, he will be sure to come ashore. *Eh bien!*—but you understand me.'

'Aha! you mean my boat—*this* boat?'

'O God!'

Madame Bernier sprang up in her seat, threw out her arms, and sank down again, burying her face in her knees. Her companion hastily shipped his oars, and laid his hands on her shoulders.

'*Allons donc*, in the devil's name, don't break down,' said he; 'we'll come to an understanding.'

Kneeling in the bottom of the boat, and supporting her by his grasp, he succeeded in making her raise herself, though her head still drooped.

'You want me to finish him in the boat?'

No answer.

'Is he an old man?'

Hortense shook her head faintly.

'My age?'

She nodded.

'*Sapristi!* it isn't so easy.'

'He can't swim,' said Hortense, without looking up; 'he—he is lame.'

'*Nom de Dieu!*' The boatman dropped his hands. Hortense looked up quickly. Do you read the pantomime?

'Never mind,' added the man at last, 'it will serve as a sign.'

'*Mais oui.* And besides that, he will ask to be taken to the Maison Bernier, the house with its back to the water, on the extension of the great quay. *Tenez*, you can almost see it from here.'

'I know the place,' said the boatman, and was silent, as if asking and answering himself a question.

Hortense was about to interrupt the train of thought which she apprehended he was following, when he forestalled her.

'How am I to be sure of my affair?' asked he.

'Of your reward? I've thought of that. This watch is a pledge of what I shall be able and glad to give you afterwards. There are two thousand francs' worth of pearls in the case.'

'*Il faut fixer la somme*,' said the man, leaving the watch untouched.

'That lies with you.'

'Good. You know that I have the right to ask a high price.'

'Certainly. Name it.'

'It's only on the supposition of a large sum that I will so much as consider your proposal. *Songez donc*, that it's a MURDER you ask of me.'

'The price—the price?'

'*Tenez*,' continued the man, 'poached game is always high. The pearls in that watch are costly because it's worth a man's life to get at them. You want me to be your pearl diver. Be it so. You must guarantee me a safe descent—it's a descent, you know—ha!—you must furnish me the armour of safety; a little gap to breathe through while I'm at my work—the thought of a capful of Napoleons!'

'My good man, I don't wish to talk to you or to listen to your sallies. I wish simply to know your price. I'm not bargaining for a pair of chickens. Propose a sum.'

The boatman had by this time resumed his seat and his oars. He stretched out for a long, slow pull, which brought him closely face to face with his temptress. This position, his body bent forward, his eyes fixed on Madame Bernier's face, he kept for some seconds. It was perhaps fortunate for Hortense's purpose at that moment—it had often aided her purposes before—that she was a pretty woman.* A plain face might have emphasized the utterly repulsive nature of the negotiation. Suddenly, with a quick, convulsive movement, the man completed the stroke.

'*Pas si bête!* propose one yourself.'

'Very well,' said Hortense, 'if you wish it. *Voyons:* I'll give you what I can. I have fifteen thousand francs' worth of jewels. I'll give you them, or, if they will get you into trouble, their value. At home, in a box I have a thousand francs in gold. You shall have those. I'll pay your passage and outfit to America. I have friends in New York. I'll write to them to get you work.'

'And you'll give your washing to my mother and sister *hein?* Ha! ha! Jewels, fifteen thousand francs; one thousand more makes sixteen; passage to America—first class—five hundred francs; outfit—what does Madame understand by that?'

'Everything needful for your success *là-bas*.'

'A written denial that I am an assassin? *Ma foi*, it were better not to remove the impression. It's served me a good turn, on this side of the water at least. Call it twenty-five thousand francs.'

'Very well; but not a sous more.'

'Shall I trust you?'

'Am I not trusting you? It is well for you that I do not allow myself to think of the venture I am making.'

'Perhaps we're even there. We neither of us can afford to make account of certain possibilities. Still, I'll trust you, too. . . .*Tiens!*' added the boatman, 'here we are near the quay.' Then with a mock-solemn touch of his cap, 'Will Madame still visit the cemetery?'

'Come, quick, let me land,' said Madame Bernier, impatiently.

---

* I am told that there was no resisting her smile; and that she had at her command, in moments of grief, a certain look of despair which filled even the roughest hearts with sympathy, and won over the kindest to the cruel cause.

'We *have* been among the dead, after a fashion,' persisted the boatman, as he gave her his hand.

III

It was more than eight o'clock when Madame Bernier reached her own house.

'Has M. de Meyrau been here?' she asked of Josephine.

'Yes, ma'am; and on learning that Madame was out, he left a note, *chez monsieur.*'

Hortense found a sealed letter on the table in her husband's old study. It ran as follows:

I was desolated at finding you out. I had a word to tell you. I have accepted an invitation to sup and pass the night at C——, thinking it would look well. For the same reason I have resolved to take the bull by the horns, and go aboard the steamer on my return, to welcome M. Bernier home—the privilege of an old friend. I am told the *Armorique* will anchor off the bar by daybreak. What do you think? But it's too late to let me know. Applaud my *savoir faire*—you will, at all events, in the end. You will see how it will smoothe matters.

'Baffled! baffled!' hissed Madame, when she had read the note; 'God deliver me from my friends!' She paced up and down the room several times, and at last began to mutter to herself, as people often do in moments of strong emotion: 'Bah! but he'll never get up by daybreak. He'll oversleep himself, especially after tonight's supper. The other will be before him. . . . Oh, my poor head, you've suffered too much to fail in the end!'

Josephine reappeared to offer to remove her mistress's things. The latter, in her desire to reassure herself, asked the first question that occurred to her.

'Was M. le Vicomte alone?'

'No madame; another gentleman was with him—M. de Saulges, I think. They came in a hack, with two portmanteaus.'

Though I have judged best, hitherto, often from an exaggerated fear of trenching on the ground of fiction, to tell you what this poor lady did and said, rather than what she thought, I may disclose what passed in her mind now:

'Is he a coward? is he going to leave me? or is he simply going to pass these last hours in play and drink? He might have stayed with me. Ah! my friend, you do little for me, who do so much for you; who commit murder, and—Heaven help me!—suicide for you! . . . But I suppose he knows best. At all events, he will make a night of it.'

When the cook came in late that evening, Josephine, who had sat up for her, said:

'You've no idea how Madame is looking. She's ten years older since this morning. Holy mother! what a day this has been for her!'

'Wait till tomorrow,' said the oracular Valentine.

Later, when the women went up to bed in the attic, they saw a light under Hortense's door, and during the night Josephine, whose chamber was above Madame's, and who couldn't sleep (for sympathy, let us say), heard movements beneath her, which told that her mistress was even more wakeful than she.

High-Water Mark

～～～

FRANCIS BRET HARTE

WHEN the tide was out on the Dedlow Marsh its extended dreariness was patent. Its spongy, low-lying surface, sluggish, inky pools, and tortuous sloughs, twisting their slimy way, eel-like, toward the open bay, were all hard facts. So were the few green tussocks, with their scant blades, their amphibious flavour, and unpleasant dampness. And if you choose to indulge your fancy—although the flat monotony of the Dedlow Marsh was not inspiring—the wavy line of scattered drift gave an unpleasant consciousness of the spent waters, and made the dead certainty of the returning tide a gloomy reflection, which no present sunshine could dissipate. The greener meadow-land seemed oppressed with this idea, and made no positive attempt at vegetation until the work of reclamation should be complete. In the bitter fruit of the low cranberry-bushes one might fancy he detected a naturally sweet disposition curdled and soured by an injudicious course of too much regular cold water.

The vocal expression of the Dedlow Marsh was also melancholy and depressing. The sepulchral boom of the bittern, the shriek of the curlew, the scream of passing brent, the wrangling of quarrelsome teal, the sharp, querulous protest of the startled crane, and syllabled complaint of the 'killdeer' plover, were beyond the power of written expression. Nor was the aspect of these mournful fowls at all cheerful and inspiring. Certainly not the blue heron standing mid-leg deep in the water, obviously catching cold in a reckless disregard of wet feet and consequences; nor the mournful curlew, the dejected plover, or the low-spirited snipe, who saw fit to join him in his suicidal contemplation; nor the impassive kingfisher—an ornithological Marius—reviewing the desolate expanse; nor the black raven that went to and fro over the face of the marsh continually, but evidently couldn't make up his mind whether the waters had subsided, and felt low-spirited in the reflection that, after all this trouble, he wouldn't be able to give a definite answer. On the contrary, it was evident at a glance that the dreary expanse of Dedlow Marsh told unpleasantly on the birds, and that the season of migration was looked forward to with a feeling of relief and satisfaction by the full-grown, and of extravagant anticipation by the callow, brood. But if Dedlow Marsh was cheerless at the slack of the low tide, you should have seen it when the tide was strong and full; when the damp air blew chilly over the cold, glittering expanse, and came to the faces of those who looked seaward like another tide; when a steel-like glint marked the low hollows and the sinuous line of slough; when the great shell-incrusted trunks of fallen trees arose again, and went forth on their dreary, purposeless wanderings, drifting hither and thither, but getting no further toward any goal at the falling tide or the day's decline than the cursed Hebrew in the legend; when the glossy ducks swung silently, making neither ripple nor furrow on the shimmering surface; when the fog came in with the tide and shut out the blue above, even as the green below had been obliterated; when boatmen, last in that fog, paddling about in a hopeless way, started at what seemed the brushing of mermen's fingers on the boat's keel, or shrank from the tufts of grass spreading around like the floating hair of a corpse, and knew by these signs that they were lost upon Dedlow Marsh, and must make a night of it, and a gloomy one at that—then you might know something of Dedlow Marsh at high water.

Let me recall a story connected with this latter view which never failed to recur to my mind in my long gunning excursions upon Dedlow Marsh. Although the event was briefly recorded in the county paper, I had the story, in all its eloquent detail, from the lips of the principal actor. I cannot hope to catch the varying emphasis and peculiar colouring of feminine delineation, for my narrator was a woman; but I'll try to give at least its substance.

She lived midway of the great slough of Dedlow Marsh and a good-sized river, which debouched four miles beyond into an estuary formed by the Pacific Ocean, on the long sandy peninsula which constituted the south-western boundary of a noble bay. The house in which she lived was a small frame cabin raised from the marsh a few feet by stout piles, and was three miles distant from the settlements upon the river. Her husband was a logger—a profitable business in a county where the principal occupation was the manufacture of lumber.

It was the season of early spring, when her husband left on the ebb of a high tide, with a raft of logs for the usual transportation to the lower end of the bay. As she stood by the door of the little cabin when the voyagers departed she noticed a cold look in the south-eastern sky, and she remembered hearing her husband say to his companions that they must endeavour to complete their voyage before the coming of the south-westerly gale which he saw brewing. And that night it began to storm and blow harder than she had ever before experienced, and some great trees fell in the forest by the river, and the house rocked like her baby's cradle.

But, however the storm might roar about the little cabin, she knew that one she trusted had driven bolt and bar with his own strong hand, and that had he feared for her he would not have left her. This, and her domestic duties, and the care of her little sickly baby, helped to keep her mind from dwelling on the weather, except, of course, to hope that he was safely harboured with the logs at Utopia in the dreary distance. But she noticed that day, when she went out to feed the chickens and look after the cow, that the tide was up to the little fence of their garden-patch, and the roar of the surf on the south beach, though miles away, she could hear distinctly. And she began to think that she would like to have someone to talk with about matters, and she believed that if it had not been so far and so stormy, and the trail so impassable, she would have taken the baby and have gone over to Ryckman's, her nearest neighbour. But then, you see, he might have returned in the storm, all wet, with no one to see to him; and it was a long exposure for baby, who was croupy and ailing.

But that night, she never could tell why, she didn't feel like sleeping or even lying down. The storm had somewhat abated, but she still 'sat and sat', and even tried to read. I don't know whether it was a bible or some profane magazine that this poor woman read, but most probably the latter, for the words all ran together and made such sad nonsense that she was forced at last to put the book down and turn to that dearer volume which lay before her in the cradle, with its white initial leaf yet unsoiled, and try to look forward to its mysterious future. And, rocking the cradle, she thought of everything and every-body, but still was wide awake as ever.

It was nearly twelve o'clock when she at last laid down in her clothes. How long she slept she could not remember, but she awoke with a dreadful choking in her throat, and found herself standing, trembling all over, in the middle of the room, with her baby clasped to her breast, and she was 'saying something'. The baby cried and sobbed, and she walked up and down trying to hush it, when she heard a scratching at the door. She opened it fearfully, and was glad to see it was only old Pete, their dog, who crawled, dripping with water, into the room. She would like to have looked out, not in the faint hope of her husband's coming, but to see how things looked; but the wind shook the door so savagely that she could hardly hold it. Then she sat down a little while, and then she lay down again a little while. Lying close by the wall of the little cabin, she thought she heard once or twice something scrape slowly against the clapboards, like the scraping of branches. Then there was a little gurgling sound, 'like the baby made when it was swallowing'; then something went 'click-click' and 'cluck-cluck', so that she sat up in bed. When she did so she was attracted by something else that seemed creeping from the back door towards the centre of the room. It wasn't much wider than her little finger, but soon it swelled to the width of her hand, and began spreading all over the floor. It was water.

She ran to the front door and threw it wide open, and saw nothing but water. She ran to the back door and threw it open, and saw nothing but water. She ran to the side window, and, throwing that open, she saw nothing but water. Then she remembered hearing her husband once say that there was no danger in the tide, for that fell regularly, and people could calculate on it, and that he would rather live near the bay than the river, whose banks might overflow at any time. But was it the tide? So she ran again to the back door, and threw out a stick of wood. It drifted away towards the bay. She scooped up some of the water and put it eagerly to her lips. It was fresh and sweet. It was the river, and not the tide!

It was then—oh, God be praised for His goodness! she did neither faint nor fall; it was then—blessed be the Saviour, for it was His merciful hand that touched and strengthened her in this awful moment—that fear dropped from her like a garment, and her trembling ceased. It was then and thereafter that she never lost her self-command, through all the trials of that gloomy night.

She drew the bedstead towards the middle of the room, and placed a table upon it, and on that she put the cradle. The water on the floor was already over her ankles, and the house once or twice moved so perceptibly, and seemed to be racked so, that the closet doors all flew open. Then she heard the same rasping and thumping against the wall, and looking out, saw that a large uprooted tree, which had lain near the road at the upper end of the pasture, had floated down to the house. Luckily its long roots dragged in the soil and kept it from moving as rapidly as the current, for had it struck the house in its full career, even the strong nails and bolts in the piles could not have withstood the shock. The hound had leaped upon its knotty surface, and crouched near the roots shivering and whining. A ray of hope flashed across her mind. She drew a heavy blanket from the bed, and, wrapping it about the babe, waded in the deepening waters to the door. As the tree swung again, broadside on, making the little cabin creak and tremble, she leaped on to its trunk. By God's mercy she succeeded in obtaining a footing on its slippery surface, and, twining an arm about its roots, she held in the other her moaning child. Then something cracked near the front porch, and the whole front of the house she had just quitted fell forward—just as cattle fall on their knees before they lie down—and at the same moment the great redwood-tree swung round and drifted away with its living cargo into the black night.

For all the excitement and danger, for all her soothing of her crying babe, for all the whistling of the wind, for all the uncertainty of her situation, she still turned to look at the deserted and water-swept cabin. She remembered even then, and she wonders how foolish she was to think of it at that time, that she wished she had put on another dress and the baby's best clothes; and she kept praying that the house would be spared so that he, when he returned, would have something to come to, and it wouldn't be quite so desolate, and—how could he ever know what had become of her and baby? And at the thought she grew sick and faint. But she had something else to do besides worrying, for whenever the long roots of her ark struck an obstacle, the whole trunk made half a revolution, and twice dipped her in the black water. The hound, who kept distracting her by running up and down the tree and howling, at last fell off at one of these collisions. He swam for some time beside her, and she tried to get the poor beast upon the tree, but he 'acted silly' and wild, and at last she lost sight of him for ever. Then she and her baby were left alone. The light which had burned for a few minutes in the deserted cabin was quenched suddenly. She could not then tell whether she was drifting. The outline of the white dunes on the peninsula showed dimly ahead, and she judged the tree was moving in a line with the river. It must be about slack water, and she had probably reached the eddy formed by the confluence of the tide and the overflowing waters of the river.

Unless the tide fell soon, there was present danger of her drifting to its channel, and being carried out to sea or crushed in the floating drift. That peril averted, if she were carried out on the ebb toward the bay, she might hope to strike one of the wooded promontories of the peninsula, and rest till daylight. Sometimes she thought she heard voices and shouts from the river, and the bellowing of cattle and bleating of sheep. Then again it was only the ringing in her ears and throbbing of her heart. She found at about this time that she was so chilled and stiffened in her cramped position that she could scarcely move, and the baby cried so when she put it to her breast that she noticed the milk refused to flow; and she was so frightened at that, that she put her head under her shawl, and for the first time cried bitterly.

When she raised her head again, the boom of the surf was behind her, and she knew that her ark had again swung round. She dipped up the water to cool her parched throat, and found that it was salt as her tears. There was a relief, though, for by this sign she knew that she was drifting with the tide. It was then the wind went down, and the great and awful silence oppressed her. There was scarcely a ripple against the furrowed sides of the great trunk on which she rested, and around her all was black gloom and quiet. She spoke to the baby just to hear herself speak, and to know that she had not lost her voice. She thought then—it was queer, but she could not help thinking it—how awful must have been the night when the great ship swung over the Asiatic peak, and the sounds of creation were blotted out from the world. She thought, too, of mariners clinging to spars, and of poor women who were lashed to rafts, and beaten to death by the cruel sea. She tried to thank God that she was thus spared, and lifted her eyes from the baby, who had fallen into a fretful sleep. Suddenly, away to the southward, a great light lifted itself out of the gloom, and flashed and flickered, and flickered and flashed again. Her heart fluttered quickly against the baby's cold cheek. It was the lighthouse at the entrance of the bay. As she was yet wondering, the tree suddenly rolled a little, dragged a little, and then seemed to lie quiet and still. She put out her hand and the current gurgled against it. The tree was aground, and, by the position of the light and the noise of the surf, aground upon the Dedlow Marsh.

Had it not been for her baby, who was ailing and croupy, had it not been for the sudden drying up of that sensitive fountain, she would have felt safe and relieved. Perhaps it was this which tended to make all her impressions mournful and gloomy. As the tide rapidly fell, a great flock of black brent fluttered by her, screaming and crying. Then the plover flew up and piped mournfully, as they wheeled around the trunk, and at last fearlessly lit upon it like a grey cloud. Then the heron flew over and around her, shrieking and protesting, and at last dropped its gaunt legs only a few yards from her. But, strangest of all, a pretty white bird, larger than a dove—like a pelican, but not a pelican—circled around and around her. At last it lit upon a rootlet of the tree, quite over her shoulder. She put out her hand and stroked its beautiful white neck, and it never appeared to move. It stayed there so long that she thought she would lift up the baby to see it, and try to attract her attention. But when she did so, the child was so chilled and cold, and had such a blue look under the little lashes which it didn't raise at all, that she screamed aloud and the bird flew away, and she fainted.

Well, that was the worst of it, and perhaps it was not so much, after all, to any but herself. For when she recovered her senses it was bright sunlight, and dead low water. There was a confused noise of guttural voices about her, and an old squaw, singing an Indian 'hushaby', and rocking herself from side to side before a fire built on the Marsh, before which she, the recovered wife and mother, lay weak and weary. Her first thought was for her baby, and she was about to speak, when a young squaw, who must have been a mother herself, fathomed her thought and brought her the 'mowitch', pale but living, in such a queer little willow cradle all bound up, just like the squaw's own young one, that she laughed and cried together, and the young squaw and the old squaw showed their big white teeth and glinted their black eyes and said, 'Plenty get well, skeena mowitch', 'wagee man come plenty soon', and she could have kissed their brown faces in her joy. And then she found that they had been gathering berries on the marsh in their queer, comical baskets, and saw the skirt of her gown fluttering on the tree from afar, and the old squaw couldn't resist the temptation of procuring a new garment, and came down and discovered the 'wagee' woman and child. And of course she gave the garment to the old squaw, as you may imagine, and when *he* came at last and rushed up to her, looking about ten years older in his anxiety, she felt so faint again that they had to carry her to the canoe. For, you see, he knew nothing about the flood until he met the Indians at Utopia, and knew by the signs that the poor woman was his wife. And at the next high-tide he towed the tree away back home, although it wasn't worth the trouble, and built another house, using the old tree for the foundation and props, and called it after her, 'Mary's Ark!' But you may guess the next house was built above High-Water Mark. And that's all.

Not much, perhaps, considering the malevolent capacity of the Dedlow Marsh. But you must tramp over it at low water, or paddle over it at high tide, or get lost upon it once or twice in the fog, as I have, to understand properly Mary's adventure, or to appreciate duly the blessing of living beyond High-Water Mark.

The Open Boat

A Tale Intended to be after the Fact: Being the Experience of Four Men from the Sunk Steamer Commodore*

～～～

STEPHEN CRANE

I

N ONE of them knew the colour of the sky. Their eyes glanced level, and were fastened upon the waves that swept toward them. These waves were of the hue of slate, save for the tops, which were of foaming white, and all of the men knew the colours of the sea. The horizon narrowed and widened, and dipped and rose, and at all times its edge was jagged with waves that seemed thrust up in points like rocks.

Many a man ought to have a bathtub larger than the boat which here rode upon the sea. These waves were most wrongfully and barbarously abrupt and tall, and each froth-top was a problem in small-boat navigation.

The cook squatted in the bottom, and looked with both eyes at the six inches of gunwale which separated him from the ocean. His sleeves were rolled over his fat forearms, and the two flaps of his

---

* On 1 January 1897, Crane sailed as a war correspondent on the steamer *Commodore*, which was running arms from Florida to the Cuban rebels (in an effort to weaken the insurgents, Spain had established a blockade around the island). The steamer foundered (possibly as a result of sabotage), and Crane was cast adrift in an open boat with three or four crew members, including the captain, Edward Murphy, and the oiler, Billy Higgins.

unbuttoned vest dangled as he bent to bail out the boat. Often he said, 'Gawd! that was a narrow clip.' As he remarked it he invariably gazed eastward over the broken sea.

The oiler, steering with one of the two oars in the boat, sometimes raised himself suddenly to keep clear of water that swirled in over the stern. It was a thin little oar, and it seemed often ready to snap.

The correspondent, pulling at the other oar, watched the waves and wondered why he was there.

The injured captain, lying in the bow, was at this time buried in that profound dejection and indifference which comes, temporarily at least, to even the bravest and most enduring when, willy-nilly, the firm fails, the army loses, the ship goes down. The mind of the master of a vessel is rooted deep in the timbers of her, though he command for a day or a decade; and this captain had on him the stern impression of a scene in the greys of dawn of seven turned faces, and later a stump of a topmast with a white ball on it, that slashed to and fro at the waves, went low and lower, and down. Thereafter there was something strange in his voice. Although steady, it was deep with mourning, and of a quality beyond oration or tears.

'Keep'er a little more south, Billie,' said he.

'A little more south, sir,' said the oiler in the stern.

A seat in this boat was not unlike a seat upon a bucking broncho, and by the same token a broncho is not much smaller. The craft pranced and reared and plunged like an animal. As each wave came, and she rose for it, she seemed like a horse making at a fence outrageously high. The manner of her scramble over these walls of water is a mystic thing, and, moreover, at the top of them were ordinarily these problems in white water, the foam racing down from the summit of each wave requiring a new leap, and a leap from the air. Then, after scornfully bumping a crest, she would slide and race and splash down a long incline, and arrive bobbing and nodding in front of the next menace.

A singular disadvantage of the sea lies in the fact that after successfully surmounting one wave you discover that there is another behind it just as important and just as nervously anxious to do something effective in the way of swamping boats. In a ten-foot dinghy one can get an idea of the resources of the sea in the line of waves that is not probable to the average experience, which is never at sea in a dinghy. As each slaty wall of water approached, it shut all else from the view of the men in the boat, and it was not difficult to imagine that this particular wave was the final outburst of the ocean, the last effort of the grim water. There was a terrible grace in the move of the waves, and they came in silence, save for the snarling of the crests.

In the wan light the faces of the men must have been grey. Their eyes must have glinted in strange ways as they gazed steadily astern. Viewed from a balcony, the whole thing would doubtless have been weirdly picturesque. But the men in the boat had no time to see it, and if they had had leisure, there were other things to occupy their minds. The sun swung steadily up the sky, and they knew it was broad day because the colour of the sea changed from slate to emerald-green streaked with amber lights, and the foam was like tumbling snow. The process of the breaking day was unknown to them. They were aware only of this effect upon the colour of the waves that rolled toward them.

In disjointed sentences the cook and the correspondent argued as to the difference between a life-saving station and a house of refuge. The cook had said: 'There's a house of refuge just north of the Mosquito Inlet Light, and as soon as they see us they'll come off in their boat and pick us up.'

'As soon as who see us?' said the correspondent.

'The crew,' said the cook.

'Houses of refuge don't have crews,' said the correspondent. 'As I understand them, they are only places where clothes and grub are stored for the benefit of shipwrecked people. They don't carry crews.'

'Oh, yes, they do,' said the cook.

'No, they don't,' said the correspondent.

'Well, we're not there yet, anyhow,' said the oiler, in the stern.

'Well,' said the cook, 'perhaps it's not a house of refuge that I'm thinking of as being near Mosquito Inlet Light; perhaps it's a life-saving station.'

'We're not there yet,' said the oiler in the stern.

II

As the boat bounced from the top of each wave the wind tore through the hair of the hatless men, and as the craft plopped her stern down again the spray slashed past them. The crest of each of these waves was a hill, from the top of which the men surveyed for a moment a broad tumultuous expanse, shining and wind-riven. It was probably splendid, it was probably glorious, this play of the free sea, wild with lights of emerald and white and amber.

'Bully good thing it's an onshore wind,' said the cook. 'If not, where would we be? Wouldn't have a show.'

'That's right,' said the correspondent.

The busy oiler nodded his assent.

Then the captain, in the bow, chuckled in a way that expressed humour, contempt, tragedy, all in one. 'Do you think we've got much of a show now, boys?' said he.

Whereupon the three were silent, save for a trifle of hemming and hawing. To express any particular optimism at this time they felt to be childish and stupid, but they all doubtless possessed this sense of the situation in their minds. A young man thinks doggedly at such times. On the other hand, the ethics of their condition was decidedly against any open suggestion of hopelessness. So they were silent.

'Oh, well', said the captain, soothing his children, 'we'll get ashore all right.'

But there was that in his tone which made them think; so the oiler quoth, 'Yes! if this wind holds.'

The cook was bailing. 'Yes! if we don't catch hell in the surf.'

Canton-flannel gulls flew near and far. Sometimes they sat down on the sea, near patches of brown seaweed that rolled over the waves with a movement like carpets on a line in a gale. The birds sat comfortably in groups, and they were envied by some in the dinghy, for the wrath of the sea was no more to them than it was to a covey of prairie chickens a thousand miles inland. Often they came very close and stared at the men with black bead-like eyes. At these times they were uncanny and sinister in their unblinking scrutiny, and the men hooted angrily at them, telling them to be gone. One came, and evidently decided to alight on the top of the captain's head. The bird flew parallel to the boat and did not circle, but made short sidelong jumps in the air in chicken fashion. His black eyes were wistfully fixed upon the captain's head. 'Ugly brute,' said the oiler to the bird. 'You look as if you were made with a jack-knife.' The cook and the correspondent swore darkly at the creature. The captain naturally wished to knock it away with the end of the heavy painter, but he did not dare do it, because anything resembling an emphatic gesture would have capsized this freighted boat; and so, with his open hand, the captain gently and carefully waved the gull away. After it had been discouraged from the pursuit the captain breathed easier on account of his hair, and others breathed easier because the bird struck their minds at this time as being somehow gruesome and ominous.

In the meantime the oiler and the correspondent rowed. And also they rowed. They sat together in the same seat, and each rowed an oar. Then the oiler took both oars; then the correspondent took both oars; then the oiler; then the correspondent. They rowed and they rowed. The very ticklish part of the business was when the time came for the reclining one in the stern to take his turn at the oars. By the very last star of truth, it is easier to steal eggs from under a hen than it was to change seats in the dinghy. First the man in the stern slid his hand along the thwart and moved with care, as if he were of Sèvres. Then the man in the rowing-seat slid his hand along the other thwart. It was all done with the most extraordinary care. As the two sidled past each other, the whole party kept watchful eyes on the coming wave, and the captain cried: 'Look out, now! Steady, there!'

The brown mats of seaweed that appeared from time to time were like islands, bits of earth. They were travelling, apparently, neither one way nor the other. They were, to all intents, stationary. They informed the men in the boat that it was making progress slowly toward the land.

The captain, rearing cautiously in the bow after the dinghy soared on a great swell, said that he had seen the lighthouse at Mosquito Inlet. Presently the cook remarked that he had seen it. The correspondent was at the oars then, and for some reason he too wished to look at the lighthouse; but his back was toward the far shore, and the waves were important, and for some time he could not seize an opportunity to turn his head. But at last there came a wave more gentle than the others, and when at the crest of it he swiftly scoured the western horizon.

'See it?' said the captain.

'No,' said the correspondent, slowly; 'I didn't see anything.'

'Look again,' said the captain. He pointed. 'It's exactly in that direction.'

At the top of another wave the correspondent did as he was bid, and this time his eyes chanced on a small, still thing on the edge of the swaying horizon. It was precisely like the point of a pin. It took an anxious eye to find a lighthouse so tiny.

'Think we'll make it, Captain?'

'If this wind holds and the boat don't swamp, we can't do much else,' said the captain.

The little boat, lifted by each towering sea and splashed viciously by the crests, made progress that in the absence of seaweed was not apparent to those in her. She seemed just a wee thing wallowing, miraculously top up, at the mercy of five oceans. Occasionally a great spread of water, like white flames, swarmed into her.

'Bail her, cook,' said the captain, serenely.

'All right, Captain,' said the cheerful cook.

III

It would be difficult to describe the subtle brotherhood of men that was here established on the seas. No one said that it was so. No one mentioned it. But it dwelt in the boat, and each man felt it warm him. They were a captain, an oiler, a cook, and a correspondent, and they were friends—friends in a more curiously ironbound degree than may be common. The hurt captain, lying against the water jar in the bow, spoke always in a low voice and calmly; but he could never command a more ready and swiftly obedient crew than the motley three of the dinghy. It was more than a mere recognition of what was best for the common safety. There was surely in it a quality that was personal and heartfelt. And after this devotion to the commander of the boat, there was this comradeship, that the correspondent, for instance, who had been taught to be cynical of men, knew even at the time was the best experience of his life. But no one said that it was so. No one mentioned it.

'I wish we had a sail,' remarked the captain. 'We might try my overcoat on the end of an oar, and give you two boys a chance to rest.' So the cook and the correspondent held the mast and spread wide the overcoat; the oiler steered; and the little boat made good way with her new rig. Sometimes the oiler had to scull sharply to keep a sea from breaking into the boat, but otherwise sailing was a success.

Meanwhile the lighthouse had been growing slowly larger. It had now almost assumed colour, and appeared like a little grey shadow on the sky. The man at the oars could not be prevented from turning his head rather often to try for a glimpse of this little grey shadow.

At last, from the top of each wave, the men in the tossing boat could see land. Even as the lighthouse was an upright shadow on the sky, this land seemed but a long black shadow on the sea. It certainly was thinner than paper. 'We must be about opposite New Smyrna,' said the cook, who had coasted this shore often in schooners. 'Captain, by the way, I believe they abandoned that lifesaving station there about a year ago.'

'Did they?' said the captain.

The wind slowly died away. The cook and the correspondent were not now obliged to slave in order to hold high the oar. But the waves continued their old impetuous swooping at the dinghy, and the little craft, no longer under way, struggled woundily over them. The oiler or the correspondent took the oars again.

Shipwrecks are *apropos* of nothing. If men could only train for them and have them occur when the men had reached pink condition, there would be less drowning at sea. Of the four in the dinghy none had slept any time worth mentioning for two days and two nights previous to embarking in the dinghy, and in the excitement of clambering about the deck of a foundering ship they had also forgotten to eat heartily.

For these reasons, and for others, neither the oiler nor the correspondent was fond of rowing at this time. The correspondent wondered ingenuously how in the name of all that was sane could there be people who thought it amusing to row a boat. It was not an amusement; it was a diabolical punishment, and even a genius of mental aberrations could never conclude that it was anything but a horror to the muscles and a crime against the back. He mentioned to the boat in general how the amusement of rowing struck him, and the weary-faced oiler smiled in full sympathy. Previously to the foundering, by the way, the oiler had worked a double watch in the engine room of the ship.

'Take her easy now, boys,' said the captain. 'Don't spend yourselves. If we have to run a surf you'll need all your strength, because we'll sure have to swim for it. Take your time.'

Slowly the land arose from the sea. From a black line it became a line of black and a line of white—trees and sand. Finally the captain said that he could make out a house on the shore. 'That's the house of refuge, sure,' said the cook. 'They'll see us before long, and come out after us.'

The distant lighthouse reared high. 'The keeper ought to be able to make us out now, if he's looking through a glass,' said the captain. 'He'll notify the lifesaving people.'

'None of those other boats could have got ashore to give word of this wreck,' said the oiler, in a low voice, 'else the lifeboat would be out hunting us.'

Slowly and beautifully the land loomed out of the sea. The wind came again. It had veered from the northeast to the southeast. Finally a new sound struck the ears of the men in the boat. It was the low thunder of the surf on the shore. 'We'll never be able to make the lighthouse now,' said the captain. 'Swing her head a little more north, Billie.'

'A little more north, sir,' said the oiler.

Whereupon the little boat turned her nose once more down the wind, and all but the oarsman watched the shore grow. Under the influence of this expansion doubt and direful apprehension were leaving the minds of the men. The management of the boat was still most absorbing, but it could not prevent a quiet cheerfulness. In an hour, perhaps, they would be ashore.

Their backbones had become thoroughly used to balancing in the boat, and they now rode this wild colt of a dinghy like circus men. The correspondent thought that he had been drenched to the skin, but happening to feel in the top pocket of his coat, he found therein eight cigars. Four of them were soaked with seawater; four were perfectly scatheless. After a search, somebody produced three dry matches; and thereupon the four waifs rode impudently in their little boat and, with an assurance of an impending rescue shining in their eyes, puffed at the big cigars, and judged well and ill of all men. Everybody took a drink of water.

IV

'Cook,' remarked the captain, 'there don't seem to be any signs of life about your house of refuge.'

'No,' replied the cook. 'Funny they don't see us!'

A broad stretch of lowly coast lay before the eyes of the men. It was of low dunes topped with dark vegetation. The roar of the surf was plain, and sometimes they could see the white lip of a wave as it spun up the beach. A tiny house was blocked out black upon the sky. Southward, the slim lighthouse lifted its little grey length.

Tide, wind, and waves were swinging the dinghy northward. 'Funny they don't see us,' said the men.

The surf's roar was here dulled, but its tone was nevertheless thunderous and mighty. As the boat swam over the great rollers the men sat listening to this roar. 'We'll swamp sure,' said everybody.

It is fair to say here that there was not a lifesaving station within twenty miles in either direction; but the men did not know this fact, and in consequence they made dark and opprobrious remarks concerning the eyesight of the nation's lifesavers. Four scowling men sat in the dinghy and surpassed records in the invention of epithets.

'Funny they don't see us.'

The light-heartedness of a former time had completely faded. To their sharpened minds it was easy to conjure pictures of all kinds of incompetency and blindness and, indeed, cowardice. There was the shore of the populous land, and it was bitter and bitter to them that from it came no sign.

'Well,' said the captain, ultimately, 'I suppose we'll have to make a try for ourselves. If we stay out here too long, we'll none of us have strength left to swim after the boat swamps.'

And so the oiler, who was at the oars, turned the boat straight for the shore. There was a sudden tightening of muscles. There was some thinking.

'If we don't all get ashore,' said the captain—'if we don't all get ashore, I suppose you fellows know where to send news of my finish?'

They then briefly exchanged some addresses and admonitions. As for the reflections of the men, there was a great deal of rage in them. Perchance they might be formulated thus: 'If I am going to be drowned—if I am going to be drowned—if I am going to be drowned, why, in the name of the seven mad gods who rule the sea, was I allowed to come thus far and contemplate sand and trees? Was I brought here merely to have my nose dragged away as I was about to nibble the sacred cheese of life? It is preposterous. If this old ninny-woman, Fate, cannot do better than this, she should be deprived of the management of men's fortunes. She is an old hen who knows not her intention. If she has decided to drown me, why did she not do it in the beginning and save me all this trouble? The whole affair is absurd. . . . But no; she cannot mean to drown me. She dare not drown me. She cannot drown me. Not after all this work.' Afterward the man might have had an impulse to shake his fist at the clouds. 'Just you drown me, now, and then hear what I call you!'

The billows that came at this time were more formidable. They seemed always just about to break and roll over the little boat in a turmoil of foam. There was a preparatory and long growl in the speech of them. No mind unused to the sea would have concluded that the dinghy could ascend these sheer heights in time. The shore was still afar. The oiler was a wily surfman. 'Boys,' he said swiftly, 'she won't live three minutes more, and we're too far out to swim. Shall I take her to sea again, Captain?'

'Yes; go ahead!' said the captain.

This oiler, by a series of quick miracles and fast and steady oarsmanship, turned the boat in the middle of the surf and took her safely to sea again.

There was a considerable silence as the boat bumped over the furrowed sea to deeper water. Then somebody in gloom spoke: 'Well, anyhow, they must have seen us from the shore by now.'

The gulls went in slanting flight up the wind toward the grey, desolate east. A squall, marked by dingy clouds and clouds brick-red, like smoke from a burning building, appeared from the southeast.

'What do you think of those lifesaving people? Ain't they peaches?'

'Funny they haven't seen us.'

'Maybe they think we're out here for sport! Maybe they think we're fishin'. Maybe they think we're damned fools.'

It was a long afternoon. A changed tide tried to force them southward, but wind and wave said northward. Far ahead, where coastline, sea, and sky formed their mighty angle, there were little dots which seemed to indicate a city on the shore.

'St Augustine?'

The captain shook his head. 'Too near Mosquito Inlet.'

And the oiler rowed, and then the correspondent rowed; then the oiler rowed. It was a weary business. The human back can become the seat of more aches and pains than are registered in books for the composite anatomy of a regiment. It is a limited area, but it can become the theatre of innumerable muscular conflicts, tangles, wrenches, knots, and other comforts.

'Did you ever like to row, Billie?' asked the correspondent.

'No,' said the oiler; 'hang it!'

When one exchanged the rowing-seat for a place in the bottom of the boat, he suffered a bodily depression that caused him to be careless of everything save an obligation to wiggle one finger. There was cold seawater swashing to and fro in the boat, and he lay in it. His head, pillowed on a thwart, was within an inch of the swirl of a wave-crest, and sometimes a particularly obstreperous sea came inboard and drenched him once more. But these matters did not annoy him. It is almost certain that if the boat had capsized he would have tumbled comfortably out upon the ocean as if he felt sure that it was a great soft mattress.

'Look! There's a man on the shore!'

'Where?'

'There! See 'im? See 'im?'

'Yes, sure! He's walking along.'

'Now he's stopped. Look! He's facing us!'

'He's waving at us!'

'So he is! By thunder!'

'Ah, now we're all right! Now we're all right! There'll be a boat out here for us in half an hour.'

'He's going on. He's running. He's going up to that house there.'

The remote beach seemed lower than the sea, and it required a searching glance to discern the little black figure. The captain saw a floating stick, and they rowed to it. A bath towel was by some weird chance in the boat, and, tying this on the stick, the captain waved it. The oarsman did not dare turn his head, so he was obliged to ask questions.

'What's he doing now?'

'He's standing still again. He's looking, I think. . . . There he goes again—toward the house. . . . Now he's stopped again.'

'Is he waving at us?'

'No, not now; he was, though.'

'Look! There comes another man!'

'He's running.'

'Look at him go, would you!'

'Why, he's on a bicycle. Now he's met the other man. They're both waving at us. Look!'

'There comes something up the beach.'

'What the devil is that thing?'

'Why, it looks like a boat.'

'Why, certainly, it's a boat.'

'No; it's on wheels.'

'Yes, so it is. Well, that must be the lifeboat. They drag them along shore on a wagon.'

'That's the lifeboat, sure.'

'No, by God, it's—it's an omnibus.'

'I tell you it's a lifeboat.'

'It is not! It's an omnibus. I can see it plain. See? One of these big hotel omnibuses.'

'By thunder, you're right. It's an omnibus, sure as fate. What do you suppose they are doing with an omnibus? Maybe they are going around collecting the life-crew, hey?'

'That's it, likely. Look! There's a fellow waving a little black flag. He's standing on the steps of the omnibus. There come those other two fellows. Now they're all talking together. Look at the fellow with the flag. Maybe he ain't waving it!'

'That ain't a flag, is it? That's his coat. Why, certainly, that's his coat.'

'So it is; it's his coat. He's taken it off and is waving it around his head. But would you look at him swing it!'

'Oh, say, there isn't any lifesaving station there. That's just a winter-resort hotel omnibus that has brought over some of the boarders to see us drown.'

'What's that idiot with the coat mean? What's he signalling, any-how?'

'It looks as if he were trying to tell us to go north. There must be a lifesaving station up there.'

'No; he thinks we're fishing. Just giving us a merry hand. See? Ah, there, Willie!'

'Well, I wish I could make something out of those signals. What do you suppose he means?'

'He don't mean anything; he's just playing.'

'Well, if he'd just signal us to try the surf again, or to go to sea and wait, or go north, or go south, or go to hell, there would be some reason in it. But look at him! He just stands there and keeps his coat revolving like a wheel. The ass!'

'There come more people.'

'Now there's quite a mob. Look! Isn't that a boat?'

'Where? Oh, I see where you mean. No, that's no boat.'

'That fellow is still waving his coat.'

'He must think we like to see him do that. Why don't he quit it? It don't mean anything.'

'I don't know. I think he is trying to make us go north. It must be that there's a lifesaving station there somewhere.'

'Say, he ain't tired yet. Look at 'im wave!'

'Wonder how long he can keep that up. He's been revolving his coat ever since he caught sight of us. He's an idiot. Why aren't they getting men to bring a boat out? A fishing boat—one of those big yawls—could come out here all right. Why don't he do something?'

'Oh, it's all right now.'

'They'll have a boat out here for us in less than no time, now that they've seen us.'

A faint yellow tone came into the sky over the low land. The shadows on the sea slowly deepened. The wind bore coldness with it, and the men began to shiver.

'Holy smoke!' said one, allowing his voice to express his impious mood, 'if we keep on monkeying out here! If we've got to flounder out here all night!'

'Oh, we'll never have to stay here all night! Don't you worry. They've seen us now, and it won't be long before they'll come chasing out after us.'

The shore grew dusky. The man waving a coat blended gradually into this gloom, and it swallowed in the same manner the omnibus and the group of people. The spray, when it dashed uproariously over the side, made the voyagers shrink and swear like men who were being branded.

'I'd like to catch the chump who waved the coat. I feel like socking him one, just for luck.'

'Why? What did he do?'

'Oh, nothing, but then he seemed so damned cheerful.'

In the meantime the oiler rowed, and then the correspondent rowed, and then the oiler rowed. Grey-faced and bowed forward, they mechanically, turn by turn, plied the leaden oars. The form of the lighthouse had vanished from the southern horizon, but finally a pale star appeared, just lifting from the sea. The streaked saffron in the west passed before the all-merging darkness, and the sea to the east was black. The land had vanished, and was expressed only by the low and drear thunder of the surf.

'If I am going to be drowned—if I am going to be drowned—if I am going to be drowned, why, in the name of the seven mad gods who rule the sea, was I allowed to come thus far and contemplate sand and trees? Was I brought here merely to have my nose dragged away as I was about to nibble the sacred cheese of life?'

The patient captain, drooped over the water jar, was sometimes obliged to speak to the oarsman.

'Keep her head up! Keep her head up!'

'Keep her head up, sir.' The voices were weary and low.

This was surely a quiet evening. All save the oarsman lay heavily and listlessly in the boat's bottom. As for him, his eyes were just capable of noting the tall black waves that swept forward in a most sinister silence, save for an occasional subdued growl of a crest.

The cook's head was on a thwart, and he looked without interest at the water under his nose. He was deep in other scenes. Finally he spoke. 'Billie,' he murmured, dreamfully, 'what kind of pie do you like best?'

V

'Pie!' said the oiler and the correspondent, agitatedly. 'Don't talk about those things, blast you!'

'Well,' said the cook, 'I was just thinking about ham sandwiches, and . . .'

A night on the sea in an open boat is a long night. As darkness settled finally, the shine of the light, lifting from the sea in the south, changed to full gold. On the northern horizon a new light appeared, a small bluish gleam on the edge of the waters. These two lights were the furniture of the world. Otherwise there was nothing but waves.

Two men huddled in the stern, and distances were so magnificent in the dinghy that the rower was enabled to keep his feet partly warm by thrusting them under his companions. Their legs indeed extended far under the rowing-seat until they touched the feet of the captain forward. Sometimes, despite the efforts of the tired oarsman, a wave came piling into the boat, an icy wave of the night, and the chilling water soaked them anew. They would twist their bodies for a moment and groan, and sleep the dead sleep once more, while the water in the boat gurgled about them as the craft rocked.

The plan of the oiler and the correspondent was for one to row until he lost the ability, and then arouse the other from his sea-water couch in the bottom of the boat.

The oiler plied the oars until his head drooped forward and the overpowering sleep blinded him; and he rowed yet afterward. Then he touched a man in the bottom of the boat, and called his name. 'Will you spell me for a little while?' he said meekly.

'Sure, Billie,' said the correspondent, awaking and dragging himself to a sitting position. They exchanged places carefully, and the oiler, cuddling down in the seawater at the cook's side, seemed to go to sleep instantly.

The particular violence of the sea had ceased. The waves came without snarling. The obligation of the man at the oars was to keep the boat headed so that the tilt of the rollers would not capsize her, and to preserve her from filling when the crests rushed past. The black waves were silent and hard to be seen in the darkness. Often one was almost upon the boat before the oarsman was aware.

In a low voice the correspondent addressed the captain. He was not sure that the captain was awake, although this iron man seemed to be always awake. 'Captain, shall I keep her making for that light north, sir?'

The same steady voice answered him. 'Yes. Keep it about two points off the port bow.'

The cook had tied a lifebelt around himself in order to get even the warmth which this clumsy cork contrivance could donate, and he seemed almost stove-like when a rower, whose teeth invariably chattered wildly as soon as he ceased his labour, dropped down to sleep.

The correspondent, as he rowed, looked down at the two men sleeping underfoot. The cook's arm was around the oiler's shoulders, and, with their fragmentary clothing and haggard faces, they were the babes of the sea—a grotesque rendering of the old babes in the wood.

Later he must have grown stupid at his work, for suddenly there was a growling of water, and a crest came with a roar and a swash into the boat, and it was a wonder that it did not set the cook afloat in his lifebelt. The cook continued to sleep, but the oiler sat up, blinking his eyes and shaking with the new cold.

'Oh, I'm awful sorry, Billie,' said the correspondent, contritely.

'That's all right, old boy,' said the oiler, and lay down again and was asleep.

Presently it seemed that even the captain dozed, and the correspondent thought that he was the one man afloat on all the oceans. The wind had a voice as it came over the waves, and it was sadder than the end.

There was a long, loud swishing astern of the boat, and a gleaming trail of phosphorescence, like blue flame, was furrowed on the black waters. It might have been made by a monstrous knife.

Then there came a stillness, while the correspondent breathed with open mouth and looked at the sea.

Suddenly there was another swish and another long flash of bluish light, and this time it was alongside the boat, and might almost have been reached with an oar. The correspondent saw an enormous fin speed like a shadow through the water, hurling the crystalline spray and leaving the long glowing trail.

The correspondent looked over his shoulder at the captain. His face was hidden, and he seemed to be asleep. He looked at the babes of the sea. They certainly were asleep. So, being bereft of sympathy, he leaned a little way to one side and swore softly into the sea.

But the thing did not then leave the vicinity of the boat. Ahead or astern, on one side or the other, at intervals long or short, fled the long sparkling streak, and there was to be heard the *whirroo* of the dark fin. The speed and power of the thing was greatly to be admired. It cut the water like a gigantic and keen projectile.

The presence of this biding thing did not affect the man with the same horror that it would if he had been a picnicker. He simply looked at the sea dully and swore in an undertone.

Nevertheless, it is true that he did not wish to be alone with the thing. He wished one of his companions to awake by chance and keep him company with it. But the captain hung motionless over the water jar, and the oiler and the cook in the bottom of the boat were plunged in slumber.

VI

'If I am going to be drowned—if I am going to be drowned—if I am going to be drowned, why, in the name of the seven mad gods who rule the sea, was I allowed to come thus far and contemplate sand and trees?'

During this dismal night, it may be remarked that a man would conclude that it was really the intention of the seven mad gods to drown him, despite the abominable injustice of it. For it was certainly an abominable injustice to drown a man who had worked so hard, so hard. The man felt it would be a crime most unnatural. Other people had drowned at sea since galleys swarmed with painted sails, but still—

When it occurs to a man that nature does not regard him as important, and that she feels she would not maim the universe by disposing of him, he at first wishes to throw bricks at the temple, and he hates deeply the fact that there are no bricks and no temples. Any visible expression of nature would surely be pelleted with his jeers.

Then, if there be no tangible thing to hoot, he feels, perhaps, the desire to confront a personification and indulge in pleas, bowed to one knee, and with hands supplicant, saying, 'Yes, but I love myself.'

A high cold star on a winter's night is the word he feels that she says to him. Thereafter he knows the pathos of his situation.

The men in the dinghy had not discussed these matters, but each had, no doubt, reflected upon them in silence and according to his mind. There was seldom any expression upon their faces save the general one of complete weariness. Speech was devoted to the business of the boat.

To chime the notes of his emotion, a verse mysteriously entered the correspondent's head. He had even forgotten that he had forgotten this verse, but it suddenly was in his mind.

> A soldier of the Legion lay dying in Algiers;
> There was lack of woman's nursing, there was dearth of woman's tears;
> But a comrade stood beside him, and he took that comrade's hand,
> And he said, 'I never more shall see my own, my native land.'*

In his childhood the correspondent had been made acquainted with the fact that a soldier of the Legion lay dying in Algiers, but he had never regarded the fact as important. Myriads of his schoolfellows had

---

* From 'Bingen on the Rhine', a poem by Caroline Sheridan Norton (1808–77).

> A soldier of the Legion lay dying in Algiers,
> There was lack of woman's nursing, there was dearth of woman's tears;
> But a comrade stood beside him, while his life-blood ebbed away,
> And bent with pitying glances, to hear what he might say.
> The dying soldier faltered, as he took that comrade's hand,
> And he said, 'I nevermore shall see my own, my native land. . . .'

informed him of the soldier's plight, but the dinning had naturally ended by making him perfectly indifferent. He had never considered it his affair that a soldier of the Legion lay dying in Algiers, nor had it appeared to him as a matter for sorrow. It was less to him than the breaking of a pencil's point.

Now, however, it quaintly came to him as a human, living thing. It was no longer merely a picture of a few throes in the breast of a poet, meanwhile drinking tea and warming his feet at the grate; it was an actuality—stern, mournful, and fine.

The correspondent plainly saw the soldier. He lay on the sand with his feet out straight and still. While his pale left hand was upon his chest in an attempt to thwart the going of his life, the blood came between his fingers. In the far Algerian distance, a city of low square forms was set against a sky that was faint with the last sunset hues. The correspondent, plying the oars and dreaming of the slow and slower movements of the lips of the soldier, was moved by a profound and perfectly impersonal comprehension. He was sorry for the soldier of the Legion who lay dying in Algiers.

The thing which had followed the boat and waited had evidently grown bored at the delay. There was no longer to be heard the slash of the cut-water, and there was no longer the flame of the long trail. The light in the north still glimmered, but it was apparently no nearer to the boat. Sometimes the boom of the surf rang in the correspondent's ears, and he turned the craft seaward then and rowed harder. Southward, some one had evidently built a watch fire on the beach. It was too low and too far to be seen, but it made a shimmering, roseate reflection upon the bluff in back of it, and this could be discerned from the boat. The wind came stronger, and sometimes a wave suddenly raged out like a mountain cat, and there was to be seen the sheen and sparkle of a broken crest.

The captain, in the bow, moved on his water jar and sat erect. 'Pretty long night,' he observed to the correspondent. He looked at the shore. 'Those lifesaving people take their time.'

'Did you see that shark playing around?'

'Yes, I saw him. He was a big fellow, all right.'

'Wish I had known you were awake.'

Later the correspondent spoke into the bottom of the boat. 'Billie!' There was a slow and gradual disentanglement. 'Billie, will you spell me?'

'Sure,' said the oiler.

As soon as the correspondent touched the cold, comfortable sea-water in the bottom of the boat and had huddled close to the cook's lifebelt he was deep in sleep, despite the fact that his teeth played all the popular airs. This sleep was so good to him that it was but a moment before he heard a voice call his name in a tone that demonstrated the last stages of exhaustion. 'Will you spell me?'

'Sure, Billie.'

The light in the north had mysteriously vanished, but the correspondent took his course from the wide-awake captain.

Later in the night they took the boat further out to sea, and the captain directed the cook to take one oar at the stern and keep the boat facing the seas. He was to call out if he should hear the thunder of the surf. This plan enabled the oiler and the correspondent to get respite together. 'We'll give those boys a chance to get into shape again', said the captain. They curled down and, after a few preliminary chatterings and trembles, slept once more the dead sleep. Neither knew they had bequeathed to the cook the company of another shark, or perhaps the same shark.

As the boat caroused on the waves, spray occasionally bumped over the side and gave them a fresh soaking, but this had no power to break their repose. The ominous slash of the wind and the water affected them as it would have affected mummies.

'Boys,' said the cook, with the notes of every reluctance in his voice, 'she's drifted in pretty close. I guess one of you had better take her to sea again.' The correspondent, aroused, heard the crash of the toppled crests.

As he was rowing, the captain gave him some whiskey-and-water, and this steadied the chills out of him. 'If I ever get ashore and anybody shows me even a photograph of an oar—'

At last there was a short conversation.

'Billie! . . . Billie, will you spell me?'

'Sure,' said the oiler.

VII

When the correspondent again opened his eyes, the sea and the sky were each of the grey hue of the dawning. Later, carmine and gold was painted upon the waters. The morning appeared finally, in its splendour, with a sky of pure blue, and the sunlight flamed on the tips of the waves.

On the distant dunes were set many little black cottages, and a tall white windmill reared above them. No man, nor dog, nor bicycle appeared on the beach. The cottages might have formed a deserted village.

The voyagers scanned the shore. A conference was held in the boat. 'Well,' said the captain, 'if no help is coming, we might better try a run through the surf right away. If we stay out here much longer we will be too weak to do anything for ourselves at all.' The others silently acquiesced in this reasoning. The boat was headed for the beach. The correspondent wondered if none ever ascended the tall wind-tower, and if then they never looked seaward. This tower was a giant, standing with its back to the plight of the ants. It represented in a degree, to the correspondent, the serenity of nature amid the struggles of the individual—nature in the wind, and nature in the vision of men. She did not seem cruel to him then, nor beneficent, nor treacherous, nor wise. But she was indifferent, flatly indifferent. It is, perhaps, plausible that a man in this situation, impressed with the unconcern of the universe, should see the innumerable flaws of his life, and have them taste wickedly in his mind, and wish for another chance. A distinction between right and wrong seems absurdly clear to him, then, in this new ignorance of the grave-edge, and he understands that if he were given another opportunity he would mend his conduct and his words, and be better and brighter during an introduction or at a tea.

'Now, boys,' said the captain, 'she is going to swamp sure. All we can do is to work her in as far as possible, and then when she swamps, pile out and scramble for the beach. Keep cool now, and don't jump until she swamps sure.'

The oiler took the oars. Over his shoulders he scanned the surf. 'Captain,' he said, 'I think I'd better bring her about and keep her head-on to the seas and back her in.'

'All right, Billie,' said the captain. 'Back her in.' The oiler swung the boat then, and, seated in the stern, the cook and the correspondent were obliged to look over their shoulders to contemplate the lonely and indifferent shore.

The monstrous inshore rollers heaved the boat high until the men were again enabled to see the white sheets of water scudding up the slanted beach. 'We won't get in very close,' said the captain. Each time a man could wrest his attention from the rollers, he turned his glance toward the shore, and in the expression of the eyes during this contemplation there was a singular quality. The correspondent, observing the others, knew that they were not afraid, but the full meaning of their glances was shrouded.

As for himself, he was too tired to grapple fundamentally with the fact. He tried to coerce his mind into thinking of it, but the mind was dominated at this time by the muscles, and the muscles said they did not care. It merely occurred to him that if he should drown it would be a shame.

There were no hurried words, no pallor, no plain agitation. The men simply looked at the shore. 'Now, remember to get well clear of the boat when you jump,' said the captain.

Seaward the crest of a roller suddenly fell with a thunderous crash, and the long white comber came roaring down upon the boat.

'Steady now,' said the captain. The men were silent. They turned their eyes from the shore to the comber and waited. The boat slid up the incline, leaped at the furious top, bounced over it, and swung down the long back of the wave. Some water had been shipped, and the cook bailed it out.

But the next crest crashed also. The tumbling, boiling flood of white water caught the boat and whirled it almost perpendicular. Water swarmed in from all sides. The correspondent had his hands on the gunwale at this time, and when the water entered at that place he swiftly withdrew his fingers, as if he objected to wetting them.

The little boat, drunken with this weight of water, reeled and snuggled deeper into the sea.

'Bail her out, cook! Bail her out!' said the captain.

'All right, Captain,' said the cook.

'Now boys, the next one will do for us sure,' said the oiler. 'Mind to jump clear of the boat.'

The third wave moved forward, huge, furious, implacable. It fairly swallowed the dinghy, and almost simultaneously the men tumbled into the sea. A piece of lifebelt had lain in the bottom of the boat, and as the correspondent went overboard he held this to his chest with his left hand.

The January water was icy, and he reflected immediately that it was colder than he had expected to find it off the coast of Florida. This appeared to his dazed mind as a fact important enough to be noted at the time. The coldness of the water was sad; it was tragic. This fact was somehow mixed and confused with his opinion of his own situation, so that it seemed almost a proper reason for tears. The water was cold.

When he came to the surface he was conscious of little but the noisy water. Afterward he saw his companions in the sea. The oiler was ahead in the race. He was swimming strongly and rapidly. Off to the correspondent's left, the cook's great white and corked back bulged out of the water; and in the rear the captain was hanging with his one good hand to the keel of the overturned dinghy.

There is a certain immovable quality to a shore, and the correspondent wondered at it amid the confusion of the sea.

It seemed also very attractive; but the correspondent knew that it was a long journey, and he paddled leisurely. The piece of life preserver lay under him, and sometimes he whirled down the incline of a wave as if he were on a hand-sled.

But finally he arrived at a place in the sea where travel was beset with difficulty. He did not pause swimming to enquire what manner of current had caught him, but there his progress ceased. The shore was set before him like a bit of scenery on a stage, and he looked at it and understood with his eyes each detail of it.

As the cook passed, much further to the left, the captain was calling to him, 'Turn over on your back, cook! Turn over on your back and use the oar.'

'All right, sir.' The cook turned on his back, and, paddling with an oar, went ahead as if he were a canoe.

Presently the boat also passed to the left of the correspondent, with the captain clinging with one hand to the keel. He would have appeared like a man raising himself to look over a board fence if it were not for the extraordinary gymnastics of the boat. The correspondent marvelled that the captain could still hold to it.

They passed on nearer to shore—the oiler, the cook, the captain—and following them went the water jar, bouncing gaily over the seas.

The correspondent remained in the grip of this strange new enemy—a current. The shore, with its white slope of sand and its green bluff topped with little silent cottages, was spread like a picture before him. It was very near to him then, but he was impressed as one who, in a gallery, looks at a scene from Brittany or Algiers.

He thought: 'I am going to drown? Can it be possible? Can it be possible? Can it be possible?' Perhaps an individual must consider his own death to be the final phenomenon of nature.

But later a wave perhaps whirled him out of this small deadly current, for he found suddenly that he could again make progress toward the shore. Later still he was aware that the captain, clinging with one hand to the keel of the dinghy, had his face turned away from the shore and toward him, and was calling his name. 'Come to the boat! Come to the boat!'

In his struggle to reach the captain and the boat, he reflected that when one gets properly wearied drowning must really be a comfortable arrangement—a cessation of hostilities accompanied by a large degree of relief; and he was glad of it, for the main thing in his mind for some moments had been horror of the temporary agony. He did not wish to be hurt.

Presently he saw a man running along the shore. He was undressing with most remarkable speed. Coat, trousers, shirt, everything flew magically off him.

'Come to the boat!' called the captain.

'All right, Captain.' As the correspondent paddled, he saw the captain let himself down to bottom and leave the boat. Then the correspondent performed his one little marvel of the voyage. A large wave caught him and flung him with ease and supreme speed completely over the boat and far beyond it. It struck him even then as an event in gymnastics and a true miracle of the sea. An overturned boat in the surf is not a plaything to a swimming man.

The correspondent arrived in water that reached only to his waist, but his condition did not enable him to stand for more than a moment. Each wave knocked him into a heap, and the undertow pulled at him.

Then he saw the man who had been running and undressing, and undressing and running, come bounding into the water. He dragged ashore the cook, and then waded toward the captain; but the captain waved him away and sent him to the correspondent. He was naked— naked as a tree in winter; but a halo was about his head, and he shone like a saint. He gave a strong pull, and a long drag, and a bully heave at the correspondent's hand. The correspondent, schooled in the minor formulae, said, 'Thanks, old man.' But suddenly the man cried, 'What's that?' He pointed a swift finger. The correspondent said, 'Go.'

In the shallows, face downward, lay the oiler. His forehead touched sand that was periodically, between each wave, clear of the sea.

The correspondent did not know all that transpired afterward. When he achieved safe ground he fell, striking the sand with each particular part of his body. It was as if he had dropped from a roof, but the thud was grateful to him.

It seems that instantly the beach was populated with men with blankets, clothes, and flasks, and women with coffee-pots and all the remedies sacred to their minds. The welcome of the land to the men from the sea was warm and generous; but a still and dripping shape was carried slowly up the beach, and the land's welcome for it could only be the different and sinister hospitality of the grave.

When it came night, the white waves paced to and fro in the moonlight, and the wind brought the sound of the great sea's voice to the men on the shore, and they felt that they could then be interpreters.

Make Westing

~~~

JACK LONDON

'Whatever you do, make westing! make westing!'
Sailing directions for Cape Horn

For seven weeks the *Mary Rogers* had been between 50° south in the Atlantic and 50° south in the Pacific, which meant that for seven weeks she had been struggling to round Cape Horn. For seven weeks she had been either in dirt, or close to dirt, save once, and then, following upon six days of excessive dirt, which she had ridden out under the shelter of the redoubtable Terra del Fuego coast, she had almost gone ashore during a heavy swell in the dead calm that had suddenly fallen. For seven weeks she had wrestled with the Cape Horn greybeards, and in return been buffeted and smashed by them. She was a wooden ship, and her ceaseless straining had opened her seams, so that twice a day the watch took its turn at the pumps.

The *Mary Rogers* was strained, the crew was strained, and big Dan Cullen, master, was likewise strained. Perhaps he was strained most of all, for upon him rested the responsibility of that titanic struggle. He slept most of the time in his clothes, though he rarely slept. He haunted the deck at night, a great, burly, robust ghost, black with the sunburn of thirty years of sea and hairy as an orang-outang. He, in turn, was haunted by one thought of action, a sailing direction for the Horn: *Whatever you do, make westing! make westing!* It was an obsession. He thought of nothing else, except, at times, to blaspheme God for sending such bitter weather.

Make westing! He hugged the Horn, and a dozen times lay hove to with the iron Cape bearing east-by-north, or north-north-east, a score of miles away. And each time the eternal west wind smote him back and he made easting. He fought gale after gale, south to 64°, inside the Antarctic drift-ice, and pledged his immortal soul to the

Powers of darkness, for a bit of westing, for a slant to take him around. And he made easting. In despair, he had tried to make the passage through the Straits of Le Maire. Half-way through, the wind hauled to the north'ard of north-west, the glass dropped to 28.88, and he turned and ran before a gale of cyclonic fury, missing, by a hair's breadth, piling up the *Mary Rogers* on the black-toothed rocks. Twice he had made west to the Diego Ramirez Rocks, one of the times saved between two snow-squalls by sighting the gravestones of ships a quarter of a mile dead ahead.

Blow! Captain Dan Cullen instanced all his thirty years at sea to prove that never had it blown so before. The *Mary Rogers* was hove to at the time he gave the evidence, and, to clinch it, inside half an hour the *Mary Rogers* was hove down to the hatches. Her new main-topsail and brand new spencer were blown away like tissue paper; and five sails, furled and fast under double gaskets, were blown loose and stripped from the yards. And before morning the *Mary Rogers* was hove down twice again, and holes were knocked in her bulwarks to ease her decks from the weight of ocean that pressed her down.

On an average of once a week Captain Dan Cullen caught glimpses of the sun. Once, for ten minutes, the sun shone at midday, and ten minutes afterwards a new gale was piping up, both watches were shortening sail, and all was buried in the obscurity of a driving snow-squall. For a fortnight, once, Captain Dan Cullen was without a meridian or a chronometer sight. Rarely did he know his position within half of a degree, except when in sight of land; for sun and stars remained hidden behind the sky, and it was so gloomy that even at the best the horizons were poor for accurate observations. A grey gloom shrouded the world. The clouds were grey; the great driving seas were leaden grey; the smoking crests were a grey churning; even the occasional albatrosses were grey, while the snow-flurries were not white, but grey, under the sombre pall of the heavens.

Life on board the *Mary Rogers* was grey—grey and gloomy. The faces of the sailors were blue grey; they were afflicted with sea-cuts and sea-boils, and suffered exquisitely. They were shadows of men. For seven weeks, in the forecastle or on deck, they had not known what it was to be dry. They had forgotten what it was to sleep out a watch, and all watches it was, 'All hands on deck!' They caught the snatches of agonized sleep, and they slept in their oilskins ready for the everlasting call. So weak and worn were they that it took both watches to do the work of one. That was why both watches were on deck so much of the time. And no shadow of a man could shirk duty. Nothing less than a broken leg could enable a man to knock off work; and there were two such, who had been mauled and pulped by the seas that broke aboard.

One other man who was the shadow of a man was George Dorety. He was the only passenger on board, a friend of the firm, and he had elected to make the voyage for his health. But seven weeks off Cape Horn had not bettered his health. He gasped and panted in his bunk through the long, heaving nights; and when on deck he was so bundled up for warmth that he resembled a peripatetic old-clothes shop. At midday, eating at the cabin table in a gloom so deep that the swinging sea-lamps burned always, he looked as blue-grey as the sickest, saddest man for'ard. Nor did gazing across the table at Captain Dan Cullen have any cheering effect upon him. Captain Cullen chewed and scowled and kept silent. The scowls were for God, and with every chew he reiterated the sole thought of his existence, which was *make westing*. He was a big, hairy brute, and the sight of him was not stimulating to the other's appetite. He looked upon George Dorety as a Jonah, and told him so once each meal savagely transferring the scowl from God to the passenger and back again.

Nor did the mate prove a first aid to a languid appetite. Joshua Higgins by name, a seaman by profession and pull, but a pot-walloper by capacity, he was a loose-jointed, sniffling creature, heartless and selfish and cowardly, without a soul, in fear of his life of Dan Cullen, and a bully over the sailors, who knew that behind the mate was Captain Cullen, the law-giver and compeller, the driver and the destroyer, the incarnation of a dozen bucko mates. In that wild weather at the southern end of the earth, Joshua Higgins ceased washing. His grimy face usually robbed George Dorety of what little appetite he managed to accumulate. Ordinarily this lavatorial dereliction would have caught Captain Cullen's eye and vocabulary, but in the present his mind was filled with making westing, to the exclusion of all other things not contributory thereto. Whether the mate's face was clean or dirty had no bearing upon westing. Later on, when 50° south in the Pacific had been reached, Joshua Higgins would wash his face very abruptly. In the meantime, at the cabin table, where grey twilight alternated with lamplight while the lamps were being filled, George Dorety sat between the two men, one a tiger and the other a hyena, and wondered why God had made them. The second mate, Matthew Turner, was a true sailor and a man, but George Dorety did not have the solace of his company, for he ate by himself, solitary, when they had finished.

On Saturday morning, July 24, George Dorety awoke to a feeling of life and headlong movement. On deck he found the *Mary Rogers* running off before a howling south-easter. Nothing was set but the lower topsails and the foresail. It was all she could stand, yet she was making fourteen knots, as Mr Turner shouted in Dorety's ear when he came on deck. And it was all westing. She was going round the Horn at last . . . if the wind held. Mr Turner looked happy. The end of the struggle was in sight. But Captain Cullen did not look happy. He scowled at Dorety in passing. Captain Cullen did not want God to know that he was pleased with that wind. He had a conception of a malicious God, and believed in his secret soul that if God knew it was a desirable wind, God would promptly efface it and send a snorter from the west. So he walked softly before God, smothering his joy down under scowls and muttered curses, and, so, fooling God, for God was the only thing in the universe of which Dan Cullen was afraid.

All Saturday and Saturday night the *Mary Rogers* raced her westing. Persistently she logged her fourteen knots, so that by Sunday morning she had covered three hundred and fifty miles. If the wind held, she would make around. If it failed, and the snorter came from anywhere between south-west and north, back the *Mary Rogers* would be hurled and be no better off than she had been seven weeks before. And on Sunday morning the wind *was* failing. The big sea was going down and running smooth. Both watches were on deck setting sail after sail as fast as the ship could stand it. And now Captain Cullen went around brazenly before God, smoking a big cigar, smiling jubilantly, as if the failing wind delighted him, while down underneath he was raging against God for taking the life out of the blessed wind. *Make westing!* So he would, if God would only leave him alone. Secretly, he pledged himself anew to the Powers of Darkness, if they would let him make westing. He pledged himself so easily because he did not believe in the Powers of Darkness. He really believed only in God, though he did not know it. And in his inverted theology God was really the Prince of Darkness. Captain Cullen was a devil-worshipper, but he called the devil by another name, that was all.

At midday, after calling eight bells, Captain Cullen ordered the royals on. The men went aloft faster than they had gone in weeks. Not alone were they nimble because of the westing, but a benignant sun was shining down and limbering their stiff bodies. George Dorety stood aft, near Captain Cullen, less bundled in clothes than usual, soaking in the grateful warmth as he watched the scene. Swiftly and abruptly the incident occurred. There was a cry from the foreroyal-yard of 'Man overboard!' Somebody threw a lifebuoy over the side, and at the same instant the second mate's voice came aft, ringing and peremptory—

'Hard down your helm!'

The man at the wheel never moved a spoke. He knew better, for Captain Dan Cullen was standing alongside of him. He wanted to move a spoke, to move all the spokes, to grind the wheel down, hard down, for his comrade drowning in the sea. He glanced at Captain Dan Cullen, and Captain Dan Cullen gave no sign.

'Down! Hard down!' the second mate roared, as he sprang aft.

But he ceased springing and commanding, and stood still, when he saw Dan Cullen by the wheel. And big Dan Cullen puffed at his cigar and said nothing. Astern, and going astern fast, could be seen the sailor. He had caught the life-buoy and was clinging to it. Nobody spoke. Nobody moved. The men aloft clung to the royal yards and watched with terror-stricken faces. And the *Mary Rogers* raced on, making her westing. A long, silent minute passed.

'Who was it?' Captain Cullen demanded.

'Mops, sir,' eagerly answered the sailor at the wheel.

Mops topped a wave astern and disappeared temporarily in the trough. It was a large wave, but it was no greybeard. A small boat could live easily in such a sea, and in such a sea the *Mary Rogers* could easily come to. But she could not come to and make westing at the same time.

For the first time in all his years, George Dorety was seeing a real drama of life and death—a sordid little drama in which the scales balanced an unknown sailor named Mops against a few miles of longitude. At first he had watched the man astern, but now he watched big Dan Cullen, hairy and black, vested with power of life and death, smoking a cigar.

Captain Dan Cullen smoked another long, silent minute. Then he removed the cigar from his mouth. He glanced aloft at the spars of the *Mary Rogers*, and overside at the sea.

'Sheet home the royals!' he cried.

Fifteen minutes later they sat at table, in the cabin, with food served before them. On one side of George Dorety sat Dan Cullen, the tiger,

on the other side, Joshua Higgins, the hyena. Nobody spoke. On deck
the men were sheeting home the skysails. George Dorety could hear
their cries, while a persistent vision haunted him of a man called
Mops, alive and well, clinging to a lifebuoy miles astern in that lonely
ocean. He glanced at Captain Cullen, and experienced a feeling of
nausea, for the man was eating his food with relish, almost bolting it.

'Captain Cullen,' Dorety said, 'you are in command of this ship,
and it is not proper for me to comment now upon what you do. But
I wish to say one thing. There is a hereafter, and yours will be a hot
one.'

Captain Cullen did not even scowl. In his voice was regret as he
said——

'It was blowing a living gale. It was impossible to save the man.'

'He fell from the royal-yard,' Dorety cried hotly. 'You were setting
the royals at the time. Fifteen minutes afterwards you were setting the
skysails.'

'It was a living gale, wasn't it, Mr Higgins?' Captain Cullen said,
turning to the mate.

'If you'd brought her to, it'd have taken the sticks out of her' was
the mate's answer. 'You did the proper thing, Captain Cullen. The
man hadn't a ghost of a show.'

George Dorety made no answer, and to the meal's end no one
spoke. After that, Dorety had his meals served in his state-room.
Captain Cullen scowled at him no longer, though no speech was
exchanged between them, while the *Mary Rogers* sped north towards
warmer latitudes. At the end of the week, Dan Cullen cornered
Dorety on deck.

'What are you going to do when we get to 'Frisco?' he demanded
bluntly.

'I am going to swear out a warrant for your arrest,' Dorety answered
quietly. 'I am going to charge you with murder, and I am going to
see you hanged for it.'

'You're almighty sure of yourself,' Captain Cullen sneered, turning
on his heel.

A second week passed, and one morning found George Dorety
standing in the coach-house companionway at the for'ard end of the
long poop, taking his first gaze around the deck. The *Mary Rogers*
was reaching full-and-by, in a stiff breeze. Every sail was set and
drawing, including the staysails. Captain Cullen strolled for'ard along
the poop. He strolled carelessly, glancing at the passenger out of the

corner of his eye. Dorety was looking the other way, standing with head and shoulders outside the companionway, and only the back of his head was to be seen. Captain Cullen, with swift eye, embraced the mainstaysail-block and the head and estimated the distance. He glanced about him. Nobody was looking. Aft, Joshua Higgins, pacing up and down, had just turned his back and was going the other way. Captain Cullen bent over suddenly and cast the staysail-sheet off from its pin. The heavy block hurtled through the air, smashing Dorety's head like an egg-shell and hurtling on and back and forth as the staysail whipped and slatted in the wind. Joshua Higgins turned around to see what had carried away, and met the full blast of the vilest portion of Captain Cullen's profanity.

'I made the sheet fast myself', whimpered the mate in the first lull, 'with an extra turn to make sure. I remember it distinctly.'

'Made fast?' the Captain snarled back, for the benefit of the watch as it struggled to capture the flying sail before it tore to ribbons. 'You couldn't make your grandmother fast, you useless hell's scullion. If you made that sheet fast with an extra turn, why in hell didn't it stay fast? That's what I want to know. Why in hell didn't it stay fast?'

The mate whined inarticulately.

'Oh, shut up!' was the final word of Captain Cullen.

Half an hour later he was as surprised as any when the body of George Dorety was found inside the companionway on the floor. In the afternoon, alone in his room, he doctored up the log.

'Ordinary seaman, Karl Brun', he wrote, 'lost overboard from fore-royal-yard in a gale of wind. Was running at the time, and for the safety of the ship did not dare to come up the wind. Nor could a boat have lived in the sea that was running.'

On another page he wrote:

'Had often warned Mr Dorety about the danger he ran because of his carelessness on deck. I told him, once, that some day he would get his head knocked off by a block. A carelessly fastened mainstaysail sheet was the cause of the accident, which was deeply to be regretted because Mr Dorety was a favourite with all of us.'

Captain Dan Cullen read over his literary effort with admiration, blotted the page, and closed the log. He lighted a cigar and stared before him. He felt the *Mary Rogers* lift, and heel, and surge along, and knew that she was making nine knots. A smile of satisfaction slowly dawned on his black and hairy face. Well, anyway, he had made his westing and fooled God.

A Matter of Fact

RUDYARD KIPLING

And if ye doubt the tale I tell,
Steer through the South Pacific swell;
Go where the branching coral hives
Unending strife of endless lives,
Where, leagued about the 'wildered boat,
The rainbow jellies fill and float;
And, lilting where the laver lingers,
The starfish trips on all her fingers;
Where, 'neath his myriad spines ashock,
The sea-egg ripples down the rock;
An orange wonder dimly guessed,
From darkness where the cuttles rest,
Moored o'er the darker deeps that hide
The blind white Sea-snake and his bride
Who, drowsing, nose the long-lost ships
Let down through darkness to their lips.

The Palms.

ONCE a priest, always a priest; once a mason, always a mason; but once a journalist, always and for ever a journalist.

There were three of us, all newspaper men, the only passengers on a little tramp steamer that ran where her owners told her to go. She had once been in the Bilbao iron ore business, had been lent to the Spanish Government for service at Manilla; and was ending her days in the Cape Town coolie-trade, with occasional trips to Madagascar and even as far as England. We found her going to Southampton in ballast, and shipped in her because the fares were nominal. There was Keller, of an American paper, on his way back to the States from palace executions in Madagascar; there was a burly half-Dutchman, called Zuyland, who owned and edited a paper up country near Johannesburg; and there was myself, who had solemnly put away all journalism, vowing to forget that I had ever known the difference between an imprint and a stereo advertisement.

Ten minutes after Keller spoke to me, as the *Rathmines* cleared Cape Town, I had forgotten the aloofness I desired to feign, and was in heated discussion on the immorality of expanding telegrams beyond a certain fixed point. Then Zuyland came out of his cabin, and we were all at home instantly, because we were men of the same profession needing no introduction. We annexed the boat formally, broke open the passengers' bathroom door—on the Manilla lines the Dons do not wash—cleaned out the orange-peel and cigar-ends at the bottom of the bath, hired a Lascar to shave us throughout the voyage, and then asked each other's names.

Three ordinary men would have quarrelled through sheer boredom before they reached Southampton. We, by virtue of our craft, were anything but ordinary men. A large percentage of the tales of the world, the thirty-nine that cannot be told to ladies and the one that can, are common property coming of a common stock. We told them all, as a matter of form, with all their local and specific variants which are surprising. Then came, in the intervals of steady card-play, more personal histories of adventure and things seen and suffered: panics among white folk, when the blind terror ran from man to man on the Brooklyn Bridge, and the people crushed each other to death they knew not why; fires, and faces that opened and shut their mouths horribly at red-hot window frames; wrecks in frost and snow, reported from the sleet-sheathed rescue-tug at the risk of frostbite; long rides after diamond thieves; skirmishes on the veldt and in municipal committees with the Boers; glimpses of lazy tangled Cape politics and the mule-rule in the Transvaal; card-tales, horse-tales, woman-tales, by the score and the half hundred; till the first mate, who had seen more than us all put together, but lacked words to clothe his tales with, sat open-mouthed far into the dawn.

When the tales were done we picked up cards till a curious hand or a chance remark made one or other of us say, 'That reminds me of a man who—or a business which—' and the anecdotes would continue while the *Rathmines* kicked her way northward through the warm water.

In the morning of one specially warm night we three were sitting immediately in front of the wheel-house, where an old Swedish boatswain whom we called 'Frithiof the Dane' was at the wheel, pretending that he could not hear our stories. Once or twice Frithiof spun the spokes curiously, and Keller lifted his head from a long chair to ask, 'What is it? Can't you get any steerage-way on her?'

'There is a feel in the water', said Frithiof, 'that I cannot understand. I think that we run downhills or somethings. She steers bad this morning.'

Nobody seems to know the laws that govern the pulse of the big waters. Sometimes even a landsman can tell that the solid ocean is atilt, and that the ship is working herself up a long unseen slope; and sometimes the captain says, when neither full steam nor fair wind justifies the length of a day's run, that the ship is sagging downhill; but how these ups and downs come about has not yet been settled authoritatively.

'No, it is a following sea,' said Frithiof; 'and with a following sea you shall not get good steerage-way.'

The sea was as smooth as a duck-pond, except for a regular oily swell. As I looked over the side to see where it might be following us from, the sun rose in a perfectly clear sky and struck the water with its light so sharply that it seemed as though the sea should clang like a burnished gong. The wake of the screw and the little white streak cut by the log-line hanging over the stern were the only marks on the water as far as eye could reach.

Keller rolled out of his chair and went aft to get a pineapple from the ripening stock that was hung inside the after awning.

'Frithiof, the log-line has got tired of swimming. It's coming home,' he drawled.

'What?' said Frithiof, his voice jumping several octaves.

'Coming home', Keller repeated, leaning over the stern. I ran to his side and saw the log-line, which till then had been drawn tense over the stern railing, slacken, loop, and come up off the port quarter. Frithiof called up the speaking-tube to the bridge, and the bridge answered, 'Yes, nine knots.' Then Frithiof spoke again, and the answer was, 'What do you want of the skipper?' and Frithiof bellowed, 'Call him up.'

By this time Zuyland, Keller, and myself had caught something of Frithiof's excitement, for any emotion on shipboard is most contagious. The captain ran out of his cabin, spoke to Frithiof, looked at the log-line, jumped on the bridge, and in a minute we felt the steamer swing round as Frithiof turned her.

''Going back to Cape Town?' said Keller.

Frithiof did not answer, but tore away at the wheel. Then he beckoned us three to help, and we held the wheel down till the *Rathmines* answered it, and we found ourselves looking into the white of our own wake, with the still oily sea tearing past our bows, though we were not going more than half steam ahead.

The captain stretched out his arm from the bridge and shouted. A minute later I would have given a great deal to have shouted too, for one-half of the sea seemed to shoulder itself above the other half, and came on in the shape of a hill. There was neither crest, comb, nor curl-over to it; nothing but black water with little waves chasing each other about the flanks. I saw it stream past and on a level with the *Rathmines'* bow-plates before the steamer hove up her bulk to rise, and I argued that this would be the last of all earthly voyages for me. Then we lifted for ever and ever and ever, till I heard Keller saying in my ear, 'The bowels of the deep, good Lord!' and the *Rathmines* stood poised, her screw racing and drumming on the slope of a hollow that stretched downwards for a good half-mile.

We went down that hollow, nose under for the most part, and the air smelt wet and muddy, like that of an emptied aquarium. There was a second hill to climb; I saw that much: but the water came aboard and carried me aft till it jammed me against the wheel-house door, and before I could catch breath or clear my eyes again we were rolling to and fro in torn water, with the scuppers pouring like eaves in a thunderstorm.

'There were three waves,' said Keller; 'and the stokehold's flooded.'

The firemen were on deck waiting, apparently, to be drowned. The engineer came and dragged them below, and the crew, gasping, began to work the clumsy Board of Trade pump. That showed nothing serious, and when I understood that the *Rathmines* was really on the water, and not beneath it, I asked what had happened.

'The captain says it was a blow-up under the sea—a volcano,' said Keller.

'It hasn't warmed anything,' I said. I was feeling bitterly cold, and cold was almost unknown in those waters. I went below to change my clothes, and when I came up everything was wiped out in clinging white fog.

'Are there going to be any more surprises?' said Keller to the captain.

'I don't know. Be thankful you're alive, gentlemen. That's a tidal wave thrown up by a volcano. Probably the bottom of the sea has been lifted a few feet somewhere or other. I can't quite understand this cold spell. Our sea-thermometer says the surface water is 44°, and it should be 68° at least.'

'It's abominable,' said Keller, shivering. 'But hadn't you better attend to the fog-horn? It seems to me that I heard something.'

'Heard! Good heavens!' said the captain from the bridge, 'I should think you did.' He pulled the string of our fog-horn, which was a weak one. It sputtered and choked, because the stoke-hold was full of water and the fires were half-drowned, and at last gave out a moan. It was answered from the fog by one of the most appalling steam-sirens I have ever heard. Keller turned as white as I did, for the fog, the cold fog, was upon us, and any man may be forgiven for fearing a death he cannot see.

'Give her steam there!' said the captain to the engine-room. 'Steam for the whistle, if we have to go dead slow.'

We bellowed again, and the damp dripped off the awnings on to the deck as we listened for the reply. It seemed to be astern this time, but much nearer than before.

'The *Pembroke Castle* on us!' said Keller; and then, viciously, 'Well, thank God, we shall sink her too.'

'It's a side-wheel steamer,' I whispered. 'Can't you hear the paddles?'

This time we whistled and roared till the steam gave out, and the answer nearly deafened us. There was a sound of frantic threshing in the water, apparently about fifty yards away, and something shot past in the whiteness that looked as though it were grey and red.

'The *Pembroke Castle* bottom up,' said Keller, who, being a journalist, always sought for explanations. 'That's the colours of a Castle liner. We're in for a big thing.'

'The sea is bewitched,' said Frithiof from the wheel-house. 'There are *two* steamers!'

Another siren sounded on our bow, and the little steamer rolled in the wash of something that had passed unseen.

'We're evidently in the middle of a fleet,' said Keller quietly. 'If one doesn't run us down, the other will. Phew! What in creation is that?'

I sniffed, for there was a poisonous rank smell in the cold air—a smell that I had smelt before.

'If I was on land I should say that it was an alligator. It smells like musk,' I answered.

'Not ten thousand alligators could make that smell,' said Zuyland; 'I have smelt them.'

'Bewitched! Bewitched!' said Frithiof. 'The sea she is turned upside down, and we are walking along the bottom.'

Again the *Rathmines* rolled in the wash of some unseen ship, and a silver-grey wave broke over the bow, leaving on the deck a sheet of sediment—the grey broth that has its place in the fathomless deeps of the sea. A sprinkling of the wave fell on my face, and it was so cold that it stung as boiling water stings. The dead and most untouched deep water of the sea had been heaved to the top by the submarine volcano—the chill still water that kills all life and smells of desolation and emptiness. We did not need either the blinding fog or that indescribable smell of musk to make us unhappy—we were shivering with cold and wretchedness where we stood.

'The hot air on the cold water makes this fog,' said the captain; 'it ought to clear in a little time.'

'Whistle, oh! whistle, and let's get out of it,' said Keller.

The captain whistled again, and far and far astern the invisible twin steam-sirens answered us. Their blasting shriek grew louder, till at last it seemed to tear out of the fog just above our quarter, and I cowered while the *Rathmines* plunged bows under on a double swell that crossed.

'No more,' said Frithiof, 'it is not good any more. Let us get away, in the name of God.'

'Now if a torpedo-boat with a *City of Paris* siren went mad and broke her moorings and hired a friend to help her, it's just conceivable that we might be carried as we are now. Otherwise this thing is——'

The last words died on Keller's lips, his eyes began to start from his head, and his jaw fell. Some six or seven feet above the port bulwarks, framed in fog, and as utterly unsupported as the full moon, hung a Face. It was not human, and it certainly was not animal, for it did not belong to this earth as known to man. The mouth was open, revealing a ridiculously tiny tongue—as absurd as the tongue of an elephant; there were tense wrinkles of white skin at the angles of the drawn lips, white feelers like those of a barbel sprung from the lower jaw, and there was no sign of teeth within the mouth. But the horror of the face lay in the eyes, for those were sightless—white, in sockets as white as scraped bone, and blind. Yet for all this the face, wrinkled as the mask of a lion is drawn in Assyrian sculpture, was alive with rage and terror. One long white feeler touched our bulwarks. Then the face disappeared with the swiftness of a blindworm popping into its burrow, and the next thing that I remember is my own voice in my own ears, saying gravely to the mainmast, 'But the air-bladder ought to have been forced out of its mouth, you know.'

Keller came up to me, ashy white. He put his hand into his pocket, took a cigar, bit it, dropped it, thrust his shaking thumb into his mouth and mumbled, 'The giant gooseberry and the raining frogs! Gimme a light—gimme a light! Say, gimme a light.' A little bead of blood dropped from his thumb-joint.

I respected the motive, though the manifestation was absurd. 'Stop, you'll bite your thumb off', I said, and Keller laughed brokenly as he picked up his cigar. Only Zuyland, leaning over the port bulwarks, seemed self-possessed. He declared later that he was very sick.

'We've seen it,' he said, turning round. 'That is it.'

'What?' said Keller, chewing the unlighted cigar.

As he spoke the fog was blown into shreds, and we saw the sea, grey with mud, rolling on every side of us and empty of all life. Then in one spot it bubbled and became like the pot of ointment that the Bible speaks of. From that wide-ringed trouble a Thing came up—a grey and red Thing with a neck—a Thing that bellowed and writhed in pain. Frithiof drew in his breath and held it till the red letters of the ship's name, woven across his jersey, straggled and opened out as though they had been type badly set. Then he said with a little cluck in his throat, 'Ah me! It is blind. *Hur illa!* That thing is blind,' and a murmur of pity went through us all, for we could see that the thing on the water was blind and in pain. Something had gashed and cut the great sides cruelly and the blood was spurting out. The grey ooze of the undermost sea lay in the monstrous wrinkles of the back, and poured away in sluices. The blind white head flung back and battered the wounds, and the body in its torment rose clear of the red and grey waves till we saw a pair of quivering shoulders streaked with weed and rough with shells, but as white in the clear spaces as the hairless, maneless, blind, toothless head. Afterwards, came a dot on the horizon and the sound of a shrill scream, and it was as though a shuttle shot all across the sea in one breath, and a second head and neck tore through the levels, driving a whispering wall of water to right and left. The two Things met—the one untouched and the other in its death-throe—male and female, we said, the female coming to the male. She circled round him bellowing, and laid her neck across the curve of his great turtle-back, and he disappeared under water for an instant, but flung up again, grunting in agony while the blood ran. Once the entire head and neck shot clear of the water and stiffened, and I heard Keller saying, as though he was watching a street accident, 'Give him air. For God's sake, give him air.' Then the death-struggle began, with crampings and twistings and jerkings of the white bulk to and fro, till our little steamer rolled again, and each grey wave coated her plates with the grey slime. The sun was clear, there was no wind, and we watched, the whole crew, stokers and all, in wonder and pity, but chiefly pity. The Thing was so helpless, and, save for his mate, so alone. No human eye should have beheld him; it was monstrous and indecent to exhibit him there in trade waters between atlas degrees of latitude. He had been spewed up, mangled and dying, from his rest on the sea-floor, where he might have lived till the Judgement Day, and we saw the tides of his life go from him as an angry tide goes out across rocks in the teeth of a landward gale. His mate lay rocking on the water a little distance off, bellowing continually, and the smell of musk came down upon the ship making us cough.

At last the battle for life ended in a batter of coloured seas. We saw the writhing neck fall like a flail, the carcase turn sideways, showing the glint of a white belly and the inset of a gigantic hind leg or flipper. Then all sank, and sea boiled over it, while the mate swam round and round, darting her head in every direction. Though we might have feared that she would attack the steamer, no power on earth could have drawn any one of us from our places that hour. We watched, holding our breaths. The mate paused in her search; we could hear the wash beating along her sides; reared her neck as high as she could reach, blind and lonely in all that loneliness of the sea, and sent one desperate bellow booming across the swells as an oyster-shell skips across a pond. Then she made off to the westward, the sun shining on the white head and the wake behind it, till nothing was left to see but a little pin point of silver on the horizon. We stood on our course again; and the *Rathmines*, coated with the sea-sediment from bow to stern, looked like a ship made grey with terror.

'We must pool our notes,' was the first coherent remark from Keller. 'We're three trained journalists—we hold absolutely the biggest scoop on record. Start fair.'

I objected to this. Nothing is gained by collaboration in journalism when all deal with the same facts, so we went to work each according to his own lights. Keller triple-headed his account, talked about our 'gallant captain', and wound up with an allusion to American enterprise in that it was a citizen of Dayton, Ohio, that had seen the sea-serpent. This sort of thing would have discredited the Creation, much more a mere sea tale, but as a specimen of the picture-writing of a half-civilised people it was very interesting. Zuyland took a heavy column and a half, giving approximate lengths and breadths, and the whole list of the crew whom he had sworn on oath to testify to his facts. There was nothing fantastic or flamboyant in Zuyland. I wrote three-quarters of a leaded bourgeois column, roughly speaking, and refrained from putting any journalese into it for reasons that had begun to appear to me.

Keller was insolent with joy. He was going to cable from Southampton to the New York *World*, mail his account to America on the same day, paralyse London with his three columns of loosely knitted headlines, and generally efface the earth. 'You'll see how I work a big scoop when I get it', he said.

'Is this your first visit to England?' I asked.

'Yes', said he. 'You don't seem to appreciate the beauty of our scoop. It's pyramidal—the death of the sea-serpent! Good heavens alive, man, it's the biggest thing ever vouchsafed to a paper!'

'Curious to think that it will never appear in any paper, isn't it?' I said.

Zuyland was near me, and he nodded quickly.

'What do you mean?' said Keller. 'If you're enough of a Britisher to throw this thing away, I shan't. I thought you were a newspaperman.'

'I am. That's why I know. Don't be an ass, Keller. Remember, I'm seven hundred years your senior, and what your grandchildren may learn five hundred years hence, I learned from my grandfathers about five hundred years ago. You won't do it, because you can't.'

This conversation was held in open sea, where everything seems possible, some hundred miles from Southampton. We passed the Needles Light at dawn, and the lifting day showed the stucco villas on the green and the awful orderliness of England—line upon line, wall upon wall, solid stone dock and monolithic pier. We waited an hour in the Customs shed, and there was ample time for the effect to soak in.

'Now, Keller, you face the music. The *Havel* goes out today. Mail by her, and I'll take you to the telegraph-office,' I said.

I heard Keller gasp as the influence of the land closed about him, cowing him as they say Newmarket Heath cows a young horse unused to open courses.

'I want to retouch my stuff. Suppose we wait till we get to London?' he said.

Zuyland, by the way, had torn up his account and thrown it overboard that morning early. His reasons were my reasons.

In the train Keller began to revise his copy, and every time that he looked at the trim little fields, the red villas, and the embankments of the line, the blue pencil plunged remorselessly through the slips. He appeared to have dredged the dictionary for adjectives. I could think of none that he had not used. Yet he was a perfectly sound poker-player and never showed more cards than were sufficient to take the pool.

'Aren't you going to leave him a single bellow?' I asked sympathetically. 'Remember, everything goes in the States, from a trouser-button to a double-eagle.'

'That's just the curse of it', said Keller below his breath. 'We've played 'em for suckers so often that when it comes to the golden truth—I'd like to try this on a London paper. You have first call there, though.'

'Not in the least. I'm not touching the thing in our papers. I shall be happy to leave 'em all to you; but surely you'll cable it home?'

'No. Not if I can make the scoop here and see the Britishers sit up.'

'You won't do it with three columns of slushy headline, believe me. They don't sit up as quickly as some people.'

'I'm beginning to think that too. Does *nothing* make any difference in this country?' he said, looking out of the window. 'How old is that farmhouse?'

'New. It can't be more than two hundred years at the most.'

'Um. Fields, too?'

'That hedge there must have been clipped for about eighty years.'

'Labour cheap—eh?'

'Pretty much. Well, I suppose you'd like to try the *Times*, wouldn't you?'

'No', said Keller, looking at Winchester Cathedral. ''Might as well try to electrify a haystack. And to think that the *World* would take three columns and ask for more—with illustrations too! It's sickening.'

'But the *Times* might,' I began.

Keller flung his paper across the carriage, and it opened in its austere majesty of solid type—opened with the crackle of an encyclopedia.

'Might! You *might* work your way through the bow-plates of a cruiser. Look at that first page!'

'It strikes you that way, does it?' I said. 'Then I'd recommend you to try a light and frivolous journal.'

'With a thing like this of mine—of ours? It's sacred history!'

I showed him a paper which I conceived would be after his own heart, in that it was modelled on American lines.

'That's homey,' he said, 'but it's not the real thing. Now, I should like one of these fat old *Times* columns. Probably there'd be a bishop in the office, though.'

When we reached London Keller disappeared in the direction of the Strand. What his experiences may have been I cannot tell, but it seems that he invaded the office of an evening paper at 11.45 a.m. (I told him English editors were most idle at that hour), and mentioned my name as that of a witness to the truth of his story.

'I was nearly fired out,' he said furiously at lunch. 'As soon as I mentioned you, the old man said that I was to tell you that they didn't want any more of your practical jokes, and that you knew the hours to call if you had anything to sell, and that they'd see you condemned before they helped to puff one of your infernal yarns in advance. Say, what record do you hold for truth in this country, anyway?'

'A beauty. You ran up against it, that's all. Why don't you leave the English papers alone and cable to New York? Everything goes over there.'

'Can't you see that's just why?' he repeated.

'I saw it a long time ago. You don't intend to cable, then?'

'Yes, I do,' he answered, in the over-emphatic voice of one who does not know his own mind.

That afternoon I walked him abroad and about, over the streets that run between the pavements like channels of grooved and tongued lava, over the bridges that are made of enduring stone, through subways floored and sided with yard-thick concrete, between houses that are never rebuilt, and by river-steps hewn, to the eye, from the living rock. A black fog chased us into Westminister Abbey, and, standing there in the darkness, I could hear the wings of the dead centuries circling round the head of Litchfield A. Keller, journalist, of Dayton, Ohio, USA, whose mission it was to make the Britishers sit up.

He stumbled gasping into the thick gloom, and the roar of the traffic came to his bewildered ears.

'Let's go to the telegraph-office and cable', I said. 'Can't you hear the New York *World* crying for news of the great sea-serpent, blind, white, and smelling of musk, stricken to death by a submarine volcanio, and assisted by his loving wife to die in mid-ocean, as visualized by 'an American citizen, the breezy, newsy, brainy newspaper man of Dayton, Ohio? 'Rah for the Buckeye State. Step lively! Both gates! Szz! Boom! Aah!' Keller was a Princeton man, and he seemed to need encouragement.

'You've got me on your own ground,' said he, tugging at his overcoat pocket. He pulled out his copy, with the cable forms—for he had written out his telegram—and put them all into my hand, groaning, 'I pass. If I hadn't come to your cursed country—If I'd sent it off at Southampton—If I ever get you west of the Alleghannies, if——'

'Never mind, Keller. It isn't your fault. It's the fault of your country. If you had been seven hundred years older you'd have done what I am going to do.'

'What are you going to do?'

'Tell it as a lie.'

'Fiction?' This with the full-blooded disgust of a journalist for the illegitimate branch of the profession.

'You can call it that if you like. I shall call it a lie.'

And a lie it has become; for Truth is a naked lady, and if by accident she is drawn up from the bottom of the sea, it behoves a gentleman either to give her a print petticoat or to turn his face to the wall and vow that he did not see.

In The Abyss

H. G. WELLS

THE lieutenant stood in front of the steel sphere and gnawed a piece of pine splinter. 'What do you think of it, Steevens?' he asked.

'It's an idea,' said Steevens, in the tone of one who keeps an open mind.

'I believe it will smash—flat,' said the lieutenant.

'He seems to have calculated it all out pretty well,' said Steevens, still impartial.

'But think of the pressure,' said the lieutenant. 'At the surface of the water it's fourteen pounds to the inch, thirty feet down it's double that; sixty, treble; ninety, four times; nine hundred, forty times; five thousand, three hundred—that's a mile—it's two hundred and forty times fourteen pounds; that's—let's see—thirty hundredweight—a ton and a half, Steevens; *a ton and a half* to the square inch. And the ocean where he's going is five miles deep. That's seven and a half——'

'Sounds a lot,' said Steevens, 'but it's jolly thick steel.'

The lieutenant made no answer, but resumed his pine splinter. The object of their conversation was a huge ball of steel, having an exterior diameter of perhaps nine feet. It looked like the shot for some titanic piece of artillery. It was elaborately nested in a monstrous scaffolding built into the framework of the vessel, and the gigantic spars that were presently to sling it overboard gave the stern of the ship an appearance that had raised the curiosity of every decent sailor who had sighted it, from the Pool of London to the Tropic of Capricorn. In two places, one above the other, the steel gave place to a couple of circular windows of enormously thick glass, and one of these, set in a steel frame of great solidity, was now partially unscrewed. Both the men had seen the interior of this globe for the first time that morning. It

was elaborately padded with air cushions, with little studs sunk between bulging pillows to work the simple mechanism of the affair.
Everything was elaborately padded, even the Myers apparatus which
was to absorb carbonic acid and replace the oxygen inspired by its
tenant, when he had crept in by the glass manhole, and had been
screwed in. It was so elaborately padded that a man might have
been fired from a gun in it with perfect safety. And it had need to be,
for presently a man was to crawl in through that glass manhole, to
be screwed up tightly, and to be flung overboard, and to sink down—
down—down, for five miles, even as the lieutenant said. It had taken
the strongest hold of his imagination; it made him a bore at mess;
and he found Steevens, the new arrival aboard, a godsend to talk to
about it, over and over again.

'It's my opinion', said the lieutenant, 'that that glass will simply
bend in and bulge and smash, under a pressure of that sort. Daubrée
has made rocks run like water under big pressures—and, you mark
my words——'

'If the glass did break in,' said Steevens, 'what then?'

'The water would shoot in like a jet of iron. Have you ever felt a
straight jet of high pressure water? It would hit as hard as a bullet. It
would simply smash him and flatten him. It would tear down his
throat, and into his lungs; it would blow in his ears——'

'What a detailed imagination you have!' protested Steevens, who
saw things vividly.

'It's simple statement of the inevitable,' said the lieutenant.

'And the globe?'

'Would just give out a few little bubbles, and it would settle down
comfortably against the day of judgement, among the oozes and the
bottom clay—with poor Elstead spread over his own smashed cushions like butter over bread.'

He repeated this sentence as though he liked it very much. 'Like
butter over bread,' he said.

'Having a look at the jigger?' said a voice, and Elstead stood behind
them, spick and span in white, with a cigarette between his teeth, and
his eyes smiling out of the shadow of his ample hat-brim. 'What's
that about bread and butter, Weybridge? Grumbling as usual about
the insufficient pay of naval officers? It won't be more than a day now
before I start. We are to get the slings ready today. This clean sky
and gentle swell is just the kind of thing for swinging off a dozen tons
of lead and iron, isn't it?'

'It won't affect you much,' said Weybridge.

'No. Seventy or eighty feet down, and I shall be there in a dozen seconds, there's not a particle moving, though the wind shriek itself hoarse up above, and the water lifts half-way to the clouds. No. Down there.' He moved to the side of the ship and the other two followed him. All three leant forward on their elbows and stared down into the yellow-green water.

'*Peace*', said Elstead, finishing his thought aloud.

'Are you dead certain that clockwork will act?' asked Weybridge presently.

'It has worked thirty-five times,' said Elstead. 'It's bound to work.'

'But if it doesn't.'

'Why shouldn't it?'

'I wouldn't go down in that confounded thing,' said Weybridge, 'for twenty thousand pounds.'

'Cheerful chap you are,' said Elstead, and spat sociably at a bubble below.

'I don't understand yet how you mean to work the thing,' said Steevens.

'In the first place, I'm screwed into the sphere', said Elstead, 'and when I've turned the electric light off and on three times to show I'm cheerful, I'm swung out over the stern by that crane, with all those big lead sinkers slung below me. The top lead weight has a roller carrying a hundred fathoms of strong cord rolled up, and that's all that joins the sinkers to the sphere, except the slings that will be cut when the affair is dropped. We use cord rather than wire rope because it's easier to cut and more buoyant—necessary points, as you will see.

'Through each of these lead weights you notice there is a hole, and an iron rod will be run through that and will project six feet on the lower side. If that rod is rammed up from below, it knocks up a lever and sets the clockwork in motion at the side of the cylinder on which the cord winds.

'Very well. The whole affair is lowered gently into the water, and the slings are cut. The sphere floats—with the air in it, it's lighter than water—but the lead weights go down straight and the cord runs out. When the cord is all paid out, the sphere will go down too, pulled down by the cord.'

'But why the cord?' asked Steevens. 'Why not fasten the weights directly to the sphere?'

'Because of the smash down below. The whole affair will go rushing down, mile after mile, at a headlong pace at last. It would be knocked to pieces on the bottom if it wasn't for that cord. But the weights will hit the bottom, and directly they do, the buoyancy of the sphere will come into play. It will go on sinking slower and slower; come to a stop at last, and then begin to float upward again.

'That's where the clockwork comes in. Directly the weights smash against the sea bottom, the rod will be knocked through and will kick up the clockwork, and the cord will be rewound on the reel. I shall be lugged down to the sea bottom. There I shall stay for half an hour, with the electric light on, looking about me. Then the clockwork will release a spring knife, the cord will be cut, and up I shall rush again, like a soda-water bubble. The cord itself will help the flotation.'

'And if you should chance to hit a ship?' said Weybridge.

'I should come up at such a pace, I should go clean through it', said Elstead, 'like a cannon ball. You needn't worry about that.'

'And suppose some nimble crustacean should wriggle into your clockwork——'

'It would be a pressing sort of invitation for me to stop,' said Elstead, turning his back on the water and staring at the sphere.

They had swung Elstead overboard by eleven o'clock. The day was serenely bright and calm, with the horizon lost in haze. The electric glare in the little upper compartment beamed cheerfully three times. Then they let him down slowly to the surface of the water, and a sailor in the stern chains hung ready to cut the tackle that held the lead weights and the sphere together. The globe, which had looked so large on deck, looked the smallest thing conceivable under the stern of the ship. It rolled a little, and its two dark windows, which floated uppermost, seemed like eyes turned up in round wonderment at the people who crowded the rail. A voice wondered how Elstead liked the rolling. 'Are you ready?' sang out the commander. 'Ay, ay, sir!' 'Then let her go!'

The rope of the tackle tightened against the blade and was cut, and an eddy rolled over the globe in a grotesquely helpless fashion. Some one waved a handkerchief, some one else tried an ineffectual cheer, a middy was counting slowly, 'Eight, nine, ten!' Another roll, then with a jerk and a splash the thing righted itself.

It seemed to be stationary for a moment, to grow rapidly smaller, and then the water closed over it, and it became visible, enlarged by refraction and dimmer, below the surface. Before one could count three it had disappeared. There was a flicker of white light far down in the water, that diminished to a speck and vanished. Then there was nothing but a depth of water going down into blackness, through which a shark was swimming.

Then suddenly the screw of the cruiser began to rotate, the water was crickled, the shark disappeared in a wrinkled confusion, and a torrent of foam rushed across the crystalline clearness that had swallowed up Elstead. 'What's the idea?' said one A. B. to another.

'We're going to lay off about a couple of miles, 'fear he should hit us when he comes up,' said his mate.

The ship steamed slowly to her new position. Aboard her almost everyone who was unoccupied remained watching the breathing swell into which the sphere had sunk. For the next half-hour it is doubtful if a word was spoken that did not bear directly or indirectly on Elstead. The December sun was now high in the sky, and the heat very considerable.

'He'll be cold enough down there,' said Weybridge. 'They say that below a certain depth sea water's always just about freezing.'

'Where'll he come up?' asked Steevens. 'I've lost my bearings.'

'That's the spot,' said the commander, who prided himself on his omniscience. He extended a precise finger southeastward. 'And this, I reckon, is pretty nearly the moment', he said. 'He's been thirty-five minutes.'

'How long does it take to reach the bottom of the ocean?' asked Steevens.

'For a depth of five miles, and reckoning—as we did—an acceleration of two feet per second, both ways, is just about three-quarters of a minute.'

'Then he's overdue,' said Weybridge.

'Pretty nearly,' said the commander. 'I suppose it takes a few minutes for that cord of his to wind in.'

'I forgot that,' said Weybridge, evidently relieved.

And then began the suspense. A minute slowly dragged itself out, and no sphere shot out of the water. Another followed, and nothing broke the low oily swell. The sailors explained to one another that little point about the winding-in of the cord. The rigging was dotted with expectant faces. 'Come up, Elstead!' called one hairy-chested salt impatiently, and the others caught it up, and shouted as though they were waiting for the curtain of a theatre to rise.

The commander glanced irritably at them.

'Of course, if the acceleration is less than two', he said, 'he'll be all the longer. We aren't absolutely certain that was the proper figure. I'm no slavish believer in calculations.'

Steevens agreed concisely. No one on the quarter-deck spoke for a couple of minutes. Then Steevens's watchcase clicked.

When, twenty-one minutes after, the sun reached the zenith, they were still waiting for the globe to reappear, and not a man aboard had dared to whisper that hope was dead. It was Weybridge who first gave expression to that realization. He spoke while the sound of eight bells still hung in the air. 'I always distrusted that window,' he said quite suddenly to Steevens.

'Good God!' said Steevens; 'you don't think——?'

'Well!' said Weybridge, and left the rest to his imagination.

'I'm no great believer in calculations myself', said the commander dubiously, 'so that I'm not altogether hopeless yet.' And at midnight the gunboat was steaming slowly in a spiral round the spot where the globe had sunk, and the white beam of the electric light fled and halted and swept discontentedly onward again over the waste of phosphorescent waters under the little stars.

'If his window hasn't burst and smashed him', said Weybridge, 'then it's a cursed sight worse, for his clockwork has gone wrong, and he's alive now, five miles under our feet, down there in the cold and dark, anchored in that little bubble of his, where never a ray of light has shone or a human being lived, since the waters were gathered together. He's there without food, feeling hungry and thirsty and scared, wondering whether he'll starve or stifle. Which will it be? The Myers apparatus is running out, I suppose. How long do they last?'

'Good heavens!' he exclaimed; 'what little things we are! What daring little devils! Down there, miles and miles of water—all water, and all this empty water about us and this sky. Gulfs!' He threw his hands out, and as he did so, a little white streak swept noiselessly up the sky, travelled more slowly, stopped, became a motionless dot, as though a new star had fallen up into the sky. Then it went sliding back again and lost itself amidst the reflections of the stars and the white haze of the sea's phosphorescence.

At the sight he stopped, arm extended and mouth open. He shut his mouth, opened it again, and waved his arms with an impatient gesture. Then he turned, shouted 'El-stead ahoy!' to the first watch, and went at a run to Lindley and the searchlight. 'I saw him,' he said. 'Star-board there! His light's on, and he's just shot out of the water.

Bring the light round. We ought to see him drifting, when he lifts on the swell.'

But they never picked up the explorer until dawn. Then they almost ran him down. The crane was swung out and a boat's crew hooked the chain to the sphere. When they had shipped the sphere, they unscrewed the manhole and peered into the darkness of the interior (for the electric light chamber was intended to illuminate the water about the sphere, and was shut off entirely from its general cavity).

The air was very hot within the cavity, and the india rubber at the lip of the manhole was soft. There was no answer to their eager questions and no sound of movement within. Elstead seemed to be lying motionless, crumpled up in the bottom of the globe. The ship's doctor crawled in and lifted him out to the men outside. For a moment or so they did not know whether Elstead was alive or dead. His face, in the yellow light of the ship's lamps, glistened with perspiration. They carried him down to his own cabin.

He was not dead, they found, but in a state of absolute nervous collapse, and besides cruelly bruised. For some days he had to lie perfectly still. It was a week before he could tell his experiences.

Almost his first words were that he was going down again. The sphere would have to be altered, he said, in order to allow him to throw off the cord if need be, and that was all. He had had the most marvellous experience. 'You thought I should find nothing but ooze,' he said. 'You laughed at my explorations, and I've discovered a new world!' He told his story in disconnected fragments, and chiefly from the wrong end, so that it is impossible to re-tell it in his words. But what follows is the narrative of his experience.

It began atrociously, he said. Before the cord ran out, the thing kept rolling over. He felt like a frog in a football. He could see nothing but the crane and the sky overhead, with an occasional glimpse of the people on the ship's rail. He couldn't tell a bit which way the thing would roll next. Suddenly he would find his feet going up, and try to step, and over he went rolling, head over heels, and just anyhow, on the padding. Any other shape would have been more comfortable, but no other shape was to be relied upon under the huge pressure of the nethermost abyss.

Suddenly the swaying ceased; the globe righted, and when he had picked himself up, he saw the water all about him greeny-blue, with an attenuated light filtering down from above, and a shoal of little floating things went rushing up past him, as it seemed to him, towards the light. And even as he looked, it grew darker and darker, until the water above was as dark as the midnight sky, albeit of a greener shade, and the water below black. And little transparent things in the water developed a faint glint of luminosity, and shot past him in faint greenish streaks.

And the feeling of falling! It was just like the start of a lift, he said, only it kept on. One has to imagine what that means, that keeping on. It was then of all times that Elstead repented of his adventure. He saw the chances against him in an altogether new light. He thought of the big cuttlefish people knew to exist in the middle waters, the kind of things they find half digested in whales at times, or floating dead and rotten and half eaten by fish. Suppose one caught hold and wouldn't let go. And had the clockwork really been sufficiently tested? But whether he wanted to go on or to go back mattered not the slightest now.

In fifty seconds everything was as black as night outside, except where the beam from his light struck through the waters, and picked out every now and then some fish or scrap of sinking matter. They flashed by too fast for him to see what they were. Once he thinks he passed a shark. And then the sphere began to get hot by friction against the water. They had underestimated this, it seems.

The first thing he noticed was that he was perspiring, and then he heard a hissing growing louder under his feet, and saw a lot of little bubbles—very little bubbles they were—rushing upward like a fan through the water outside. Steam! He felt the window, and it was hot. He turned on the minute glow-lamp that lit his own cavity, looked at the padded watch by the studs, and saw he had been travelling now for two minutes. It came into his head that the window would crack through the conflict of temperatures, for he knew the bottom water is very near freezing.

Then suddenly the floor of the sphere seemed to press against his feet, the rush of bubbles outside grew slower and slower, and the hissing diminished. The sphere rolled a little. The window had not cracked, nothing had given, and he knew that the dangers of sinking, at any rate, were over.

In another minute or so he would be on the floor of the abyss. He thought, he said, of Steevens and Weybridge and the rest of them five miles overhead, higher to him than the very highest clouds that ever floated over land are to us, steaming slowly and staring down and wondering what had happened to him.

He peered out of the window. There were no more bubbles now, and the hissing had stopped. Outside there was a heavy blackness—as black as black velvet—except where the electric light pierced the empty water and showed the colour of it—a yellow-green. Then three things like shapes of fire swam into sight, following each other through the water. Whether they were little and near or big and far off he could not tell.

Each was outlined in a bluish light almost as bright as the lights of a fishing smack, a light which seemed to be smoking greatly, and all along the sides of them were specks of this, like the lighter portholes of a ship. Their phosphorescence seemed to go out as they came within the radiance of his lamp, and he saw then that they were little fish of some strange sort, with huge heads, vast eyes, and dwindling bodies and tails. Their eyes were turned towards him, and he judged they were following him down. He supposed they were attracted by his glare.

Presently others of the same sort joined them. As he went on down, he noticed that the water became of a pallid colour, and that little specks twinkled in his ray like motes in a sunbeam. This was probably due to the clouds of ooze and mud that the impact of his leaden sinkers had disturbed.

By the time he was drawn down to the lead weights he was in a dense fog of white that his electric light failed altogether to pierce for more than a few yards, and many minutes elapsed before the hanging sheets of sediment subsided to any extent. Then, lit by his light and by the transient phosphorescence of a distant shoal of fishes, he was able to see under the huge blackness of the super-incumbent water an undulating expanse of greyish-white ooze, broken here and there by tangled thickets of a growth of sea lilies, waving hungry tentacles in the air.

Further away were the graceful, translucent outlines of a group of gigantic sponges. About this floor there were scattered a number of bristling flattish tufts of rich purple and black, which he decided must be some sort of sea-urchin, and small, large-eyed or blind things having a curious resemblance, some to woodlice, and others to lobsters, crawled sluggishly across the track of the light and vanished into the obscurity again, leaving furrowed trails behind them.

Then suddenly the hovering swarm of little fishes veered about and came towards him as a flight of starlings might do. They passed over him like a phosphorescent snow, and then he saw behind them some larger creature advancing towards the sphere.

At first he could see it only dimly, a faintly moving figure remotely suggestive of a walking man, and then it came into the spray of light that the lamp shot out. As the glare struck it, it shut its eyes, dazzled. He stared in rigid astonishment.

It was a strange vertebrated animal. Its dark purple head was dimly suggestive of a chameleon, but it had such a high forehead and such a braincase as no reptile ever displayed before; the vertical pitch of its face gave it a most extraordinary resemblance to a human being.

Two large and protruding eyes projected from sockets in chameleon fashion, and it had a broad reptilian mouth with hoary lips beneath its little nostrils. In the position of the ears were two huge gill-covers, and out of these floated a branching tree of coralline filaments, almost like the tree-like gills that very young rays and sharks possess.

But the humanity of the face was not the most extraordinary thing about the creature. It was a biped; its almost globular body was poised on a tripod of two frog-like legs and a long thick tail, and its fore limbs, which grotesquely caricatured the human hand, much as a frog's do, carried a long shaft of bone, tipped with copper. The colour of the creature was variegated; its head, hands, and legs were purple; but its skin, which hung loosely upon it, even as clothes might do, was a phosphorescent grey. And it stood there blinded by the light.

At last this unknown creature of the abyss blinked its eyes open, and, shading them with its disengaged hand, opened its mouth and gave vent to a shouting noise, articulate almost as speech might be, that penetrated even the steel case and padded jacket of the sphere. How a shouting may be accomplished without lungs Elstead does not profess to explain. It then moved sideways out of the glare into the mystery of shadow that bordered it on either side, and Elstead felt rather than saw that it was coming towards him. Fancying the light had attracted it, he turned the switch that cut off the current. In another moment something soft dabbed upon the steel, and the globe swayed.

Then the shouting was repeated, and it seemed to him that a distant echo answered it. The dabbing recurred, and the globe swayed and ground against the spindle over which the wire was rolled. He stood in the blackness and peered out into the everlasting night of the abyss. And presently he saw, very faint and remote, other phosphorescent quasi-human forms hurrying towards him.

Hardly knowing what he did, he felt about in his swaying prison for the stud of the exterior electric light, and came by accident against his own small glow-lamp in its padded recess. The sphere twisted, and then threw him down; he heard shouts like shouts of surprise, and when he rose to his feet, he saw two pairs of stalked eyes peering into the lower window and reflecting his light.

In another moment hands were dabbing vigorously at his steel casing, and there was a sound, horrible enough in his position, of the metal protection of the clockwork being vigorously hammered. That, indeed, sent his heart into his mouth, for if these strange creatures succeeded in stopping that, his release would never occur. Scarcely had he thought as much when he felt the sphere sway violently, and the floor of it press hard against his feet. He turned off the small glow-lamp that lit the interior, and sent the ray of the large light in the separate compartment out into the water. The sea-floor and the man-like creatures had disappeared, and a couple of fish chasing each other dropped suddenly by the window.

He thought at once that these strange denizens of the deep sea had broken the rope, and that he had escaped. He drove up faster and faster, and then stopped with a jerk that sent him flying against the padded roof of his prison. For half a minute, perhaps, he was too astonished to think.

Then he felt that the sphere was spinning slowly, and rocking, and it seemed to him that it was also being drawn through the water. By crouching close to the window, he managed to make his weight effective and roll that part of the sphere downward, but he could see nothing save the pale ray of his light striking down ineffectively into the darkness. It occurred to him that he would see more if he turned the lamp off, and allowed his eyes to grow accustomed to the profound obscurity.

In this he was wise. After some minutes the velvety blackness became a translucent blackness, and then, far away, and as faint as the zodiacal light of an English summer evening, he saw shapes moving below. He judged these creatures had detached his cable, and were towing him along the sea bottom.

And then he saw something faint and remote across the undulations of the submarine plain, a broad horizon of pale luminosity that extended this way and that way as far as the range of his little window permitted him to see. To this he was being towed, as a balloon might be towed by men out of the open country into a town. He approached it very slowly, and very slowly the dim irradiation was gathered together into more definite shapes.

It was nearly five o'clock before he came over this luminous area, and by that time he could make out an arrangement suggestive of streets and houses grouped about a vast roofless erection that was grotesquely suggestive of a ruined abbey. It was spread out like a map below him. The houses were all roofless enclosures of walls, and their substance being, as he afterwards saw, of phosphorescent bones, gave the place an appearance as if it were built of drowned moonshine.

Among the inner caves of the place waving trees of crinoid stretched their tentacles, and tall, slender, glassy sponges shot like shining minarets and lilies of filmy light out of the general glow of the city. In the open spaces of the place he could see a stirring movement as of crowds of people, but he was too many fathoms above them to distinguish the individuals in those crowds.

Then slowly they pulled him down, and as they did so, the details of the place crept slowly upon his apprehension. He saw that the courses of the cloudy buildings were marked out with beaded lines of round objects, and then he perceived that at several points below him, in broad open spaces, were forms like the encrusted shapes of ships.

Slowly and surely he was drawn down, and the forms below him became brighter, clearer, more distinct. He was being pulled down, he perceived, towards the large building in the centre of the town, and he could catch a glimpse ever and again of the multitudinous forms that were lugging at his cord. He was astonished to see that the rigging of one of the ships, which formed such a prominent feature of the place, was crowded with a host of gesticulating figures regarding him, and then the walls of the great building rose about him silently, and hid the city from his eyes.

And such walls they were, of waterlogged wood, and twisted wirerope, and iron spars, and copper, and the bones and skulls of dead men. The skulls ran in zigzag lines and spirals and fantastic curves over the building; and in and out of their eye-sockets, and over the whole surface of the place, lurked and played a multitude of silvery little fishes.

Suddenly his ears were filled with a low shouting and a noise like the violent blowing of horns, and this gave place to a fantastic chant. Down the sphere sank, past the huge pointed windows, through which he saw vaguely a great number of these strange, ghostlike people regarding him, and at last he came to rest, as it seemed, on a kind of altar that stood in the centre of the place.

And now he was at such a level that he could see these strange people of the abyss plainly once more. To his astonishment, he perceived that they were prostrating themselves before him, all save one, dressed as it seemed in a robe of placoid scales, and crowned with a luminous diadem, who stood with his reptilian mouth opening and shutting, as though he led the chanting of the worshippers.

A curious impulse made Elstead turn on his small globe-lamp again, so that he became visible to these creatures of the abyss, albeit the glare made them disappear forthwith into night. At this sudden sight of him, the chanting gave place to a tumult of exultant shouts; and Elstead, being anxious to watch them, turned his light off again, and vanished from before their eyes. But for a time he was too blind to make out what they were doing, and when at last he could distinguish them, they were kneeling again. And thus they continued worshipping him, without rest or intermission, for a space of three hours.

Most circumstantial was Elstead's account of this astounding city and its people, these people of perpetual night, who have never seen sun or moon or stars, green vegetation, nor any living, air-breathing creatures, who know nothing of fire, nor any light but the phosphorescent light of living things.

Startling as is his story, it is yet more startling to find that scientific men, of such eminence as Adams and Jenkins, find nothing incredible in it. They tell me they see no reason why intelligent, water-breathing, vertebrated creatures, inured to a low temperature and enormous pressure, and of such a heavy structure, that neither alive nor dead would they float, might not live upon the bottom of the deep sea, and quite unsuspected by us, descendants like ourselves of the great Theriomorpha of the New Red Sandstone age.

We should be known to them, however, as strange meteoric creatures, wont to fall catastrophically dead out of the mysterious blackness of their watery sky. And not only we ourselves, but our ships, our metals, our appliances, would come raining down out of the night. Sometimes sinking things would smite down and crush them, as if it were the judgement of some unseen power above, and sometimes would come things of the utmost rarity or utility, or shapes of inspiring suggestion. One can understand, perhaps, something of their behaviour at the descent of a living man, if one thinks what a barbaric people might do, to whom an enhaloed, shining creature came suddenly out of the sky.

At one time or another Elstead probably told the officers of the *Ptarmigan* every detail of his strange twelve hours in the abyss. That he also intended to write them down is certain, but he never did, and so unhappily we have to piece together the discrepant fragments of his story from the reminiscences of Commander Simmons, Weybridge, Steevens, Lindley and the others.

We see the thing darkly in fragmentary glimpses—the huge ghostly building, the bowing, chanting people, with their dark chameleon-like heads and faintly luminous clothing, and Elstead, with his light turned on again, vainly trying to convey to their minds that the cord by which the sphere was held was to be severed. Minute after minute slipped away, and Elstead, looking at his watch, was horrified to find that he had oxygen only for four hours more. But the chant in his honour kept on as remorselessly as if it was the marching song of his approaching death.

The manner of his release he does not understand, but to judge by the end of cord that hung from the sphere, it had been cut through by rubbing against the edge of the altar. Abruptly the sphere rolled over, and he swept up, out of their world, as an ethereal creature clothed in a vacuum would sweep through our own atmosphere back to its native ether again. He must have torn out of their sight as a hydrogen bubble hastens upward from our air. A strange ascension it must have seemed to them.

The sphere rushed up with even greater velocity than, when weighted with the lead sinkers, it had rushed down. It became exceedingly hot. It drove up with the windows uppermost, and he remembers the torrent of bubbles frothing against the glass. Every moment he expected this to fly. Then suddenly something like a huge wheel seemed to be released in his head, the padded compartment began spinning about him, and he fainted. His next recollection was of his cabin, and of the doctor's voice.

But that is the substance of the extraordinary story that Elstead related in fragments to the officers of the *Ptarmigan*. He promised to write it all down at a later date. His mind was chiefly occupied with the improvement of his apparatus, which was effected at Rio.

It remains only to tell that on 2 February 1896, he made his second descent into the ocean abyss, with the improvements his first experience suggested. What happened we shall probably never know. He never returned. The *Ptarmigan* beat about over the point of his submersion, seeking him in vain for thirteen days. Then she returned to

Rio, and the news was telegraphed to his friends. So the matter remains for the present. But it is hardly probable that no further attempt will be made to verify his strange story of these hitherto unsuspected cities of the deep sea.

The Cruise of the Willing Mind

A. E. W. MASON

T HE cruise happened before the steam trawler ousted the smack from the North Sea. A few newspapers recorded it in half a dozen lines of small print which nobody read. But it became and—though nowadays the *Willing Mind* rots from month to month by the quay—remains staple talk at Gorleston alehouses on winter nights.

The crew consisted of Weeks, three fairly competent hands, and a baker's assistant, when the *Willing Mind* slipped out of Yarmouth. Alexander Duncan, the photographer from Derby, joined the smack afterwards under peculiar circumstances. Duncan was a timid person, but aware of his timidity. He was quite clear that his paramount business was to be a man; and he was equally clear that he was not successful in his paramount business. Meanwhile he pretended to be, hoping that on some miraculous day a sudden test would prove the straw man he was to have become real flesh and blood. A visit to a surgeon and the flick of a knife quite shattered that illusion. He went down to Yarmouth afterwards fairly disheartened. The test had been applied and he had failed.

Now Weeks was a particular friend of Duncan's. They had chummed together on Gorleston Quay some years before, perhaps because they were so dissimilar. Weeks had taught Duncan to sail a boat, and had once or twice taken him for a short trip on his smack; so that the first thing that Duncan did on his arrival at Yarmouth was to take the tram to Gorleston and to make enquiries.

A fisherman lounging against a winch replied to them—

'If Weeks is a friend o' yours I should get used to missin' 'im, as I tell his wife.'

There was at that time an ingenious system by which the skipper might buy his smack from the owner on the instalment plan—as people buy their furniture—only with a difference: for people sometimes get their furniture. The instalments had to be completed within a certain period. The skipper could do it—he could just do it; but he couldn't do it without running up one little bill here for stores, and another little bill there for sail mending. The owner worked in with the sailmaker, and just as the skipper was putting out to earn his last instalment he would find the bailiffs on board, his cruise would be delayed, he would be consequently behindhand with his instalment, and back would go the smack to the owner with a present of four-fifths of its price. Weeks had still to pay two hundred pounds, and had eight weeks to earn it in. The time was sufficient, although no more than sufficient. But he got the straight tip that his sailmaker would stop him; and getting together any sort of crew he could he slipped out at night with half his stores.

'Now the No'th Sea', concluded the fisherman, 'in November and December ain't a bobby's job.'

Duncan walked forward to the pierhead. He looked out at a grey tumbled sky shutting down on a grey tumbled sea. There were flecks of white cloud in the sky, flecks of white breakers on the sea, and it was all most dreary. He stood at the end of the jetty, and his great possibility came out of the grey to him. Weeks was shorthanded. Cribbed within a few feet of the smack's deck, there would be no chance for any man to shirk. Duncan acted on the impulse. He bought a fisherman's outfit at Gorleston, travelled up to London, got a passage the next morning on a Billingsgate fish-carrier, and that night went throbbing down the great water street of the Swin, past the green globes of the Mouse. The four flashes of the Outer Gabbard winked him goodbye away on the starboard, and at eleven o'clock the next night far out in the North Sea he saw the little city of lights swinging on the Dogger.

The *Willing Mind*'s boat came aboard the next morning, and Captain Weeks with it, who smiled grimly while Duncan explained how he had learnt that the smack was shorthanded.

'I can't put you ashore in Denmark,' said Weeks knowingly. 'There'll be seven weeks, it's true, for things to blow over; but I'll have to take you back to Yarmouth. And I can't afford a passenger. If you come, you come as a hand. I mean to own my smack at the end of this voyage.'

Duncan climbed after him into the boat. The *Willing Mind* had now six for her crew: Weeks; his son Willie, a lad of sixteen; Upton, the first hand; Deakin, the decky; Rall, the baker's assistant, and Alexander Duncan. And of these six four were almost competent. Deakin, it is true, was making only his second voyage; but Willie Weeks, though young, had begun early; and Upton, a man of forty, knew the banks and currents of the North Sea as well as Weeks.

'It's all right', said the skipper, 'if the weather holds.' And for a month the weather did hold, and the catches were good, and Duncan learned a great deal. He learnt how to keep a night watch from midnight till eight in the morning, and then stay on deck till noon: how to put his tiller up and down when his tiller was a wheel, and how to vary the order according as his skipper stood to windward or to lee; he learnt to box a compass and to steer by it; to gauge the leeway he was making by the angle of his wake and the black line in the compass; above all, he learnt to love the boat like a live thing, as a man loves his horse, and to want every scanty inch of brass on her to shine.

But it was not for this that Duncan had come down to the sea. He gazed out at night across the rippling starlit water and the smacks nestling upon it, and asked of his God: 'Is this all?' And his God answered him.

The beginning of it was the sudden looming of ships upon the horizon, very clear, till they looked like carved toys. The skipper got out his accounts and totted up his catches, and the prices they had fetched in Billingsgate Market. Then he went on deck and watched the sun set. There were no cloud-banks in the west, and he shook his head.

'It'll blow a bit from the east before morning,' he said, and he tapped on the barometer. Then he returned to his accounts and added them up again. After a little he looked up, and saw the first hand watching him with comprehension.

'Two or three really good hauls would do the trick,' suggested Weeks.

Upton nodded. 'If it was my boat I should chance it tomorrow before the weather blows up.'

Weeks drummed his fists on the table and agreed.

On the morrow the Admiral headed north for the Great Fisker Bank, and the fleet followed, with the exception of the *Willing Mind*. The *Willing Mind* lagged along in the rear without her topsails till about half-past two in the afternoon, when Captain Weeks became suddenly alert. He bore away till he was right before the wind, hoisted every scrap of sail he could carry; rigged out a spinnaker with his balloon foresail, and made a clean run for the coast of Denmark. Deakin explained the manœuvre to Duncan. 'The old man's goin' poachin'. He's after soles.'

'Keep a lookout, lads!' cried Weeks. 'It's not the Danish gunboat I'm afraid of: it's the fatherly English cruiser a-turning of us back.'

Darkness, however, found them unmolested. They crossed the three-mile limit at eight o'clock, and crept close in under the Danish headlands without a glimmer of light showing.

'I want all hands all night', said Weeks, 'and there's a couple of pounds for him as first sees the bogey-man.'

'Meaning the Danish gunboat,' explained Deakin.

The trawl was down before nine. The skipper stood by his lead, Upton took the wheel, and all night they trawled in the shallows, creeping silently beneath the dark headlands, bumping on the grounds, with a sharp eye forward and aft for the Danish gunboat. The wind veered round from the west. They hauled in at twelve and again at three and again at six, and they had just got their last catch on deck when Duncan saw by the first grey of the morning a dun-coloured trail of smoke hanging over a projecting knoll.

'There she is!' he cried.

'Yes, that's the gunboat', answered Weeks. 'She has waited too long. We can laugh at her with this wind.'

He put his smack about, and before the gunboat puffed round the headland, three miles away, was reaching northwards with his sails free. He rejoined the fleet that afternoon. 'Fifty-two boxes of soles!' said Weeks. 'And every one of them worth two pounds ten in Billingsgate Market. This smack's mine!' and he stamped on the deck in all the pride of ownership. 'We'll take a reef in,' he added. 'There's a no'th-easterly gale blowin' up and I don't know anything worse in the No'th Sea. The sea piles in upon you from Noofoundland, piles in till it strikes the banks. Then it breaks. You were right, Upton; we'll be lying hove-to in the morning.'

They were lying hove-to before the morning. Duncan, tossing about in his canvas cot, heard the skipper stamping overhead, and in an interval of the wind caught a snatch of song bawled out in a high voice. The song was not reassuring, for the two lines which Duncan caught ran as follows:

You never can tell when your death-bells are ringing,
You never can know when you're going to die.

Duncan tumbled on to the floor, fell about the cabin as he pulled
on his sea-boots and climbed up the companion. He clung to the
mizzen runners in a night of extraordinary blackness. To port and to
starboard the lights of the smacks rose on the crests and sank in the
troughs with such violence they had the air of being tossed up into
the sky and then extinguished in the water; while all round him there
flashed little points of white which suddenly lengthened out into a
horizontal line. There was one quite close to the quarter of the *Willing
Mind*. It stretched about the height of the main-gaff in a line of
white. The line suddenly descended towards him and became a sheet;
and then a voice bawled, 'Water! Jump! Down the companion! Jump!'
The line of white was a breaking wave.

There was a scamper of heavy boots, and a roar of water plunging
over the bulwarks, as though so many loads of wood had been
dropped on the deck. Duncan jumped for the cabin, Weeks and the
mate jumped the next second and the water sluiced down after them,
put out the fire, and washed them, choking and wrestling, about on
the cabin floor. Weeks was the first to disentangle himself, and he
turned fiercely to Duncan.

'What were you doing on deck? Upton and I keep the watch tonight. You stay below, and by God, I'll see you do it! I have fifty-two
boxes of soles to put aboard the fish cutter in the morning, and I'm
not going to lose lives before I do that! This smack's mine!'

Captain Weeks was transformed into a savage animal fighting for
his own. All night he and the mate stood on the deck and plunged
down the open companion with a torrent of water to hurry them. All
night Duncan lay in his bunk listening to the bellowing of the wind,
the great thuds of solid green wave on the deck, the horrid rush and
roaring of the seas as they broke loose to leeward from under the
smack's keel. And he listened to something more—the whimpering
of the baker's assistant in the next bunk. 'Three inches of deck!
What's the use of it! Lord ha' mercy on me, what's the use of it? No
more than an eggshell! We'll be broken in afore the morning, broken
in like a man's skull under a bludgeon. . . . I'm no sailor, I'm not; I'm
a baker. It isn't right I should die at sea!'

Duncan stopped his ears, and thought of the journey someone
would have to make to the fish cutter in the morning. There were
fifty-two boxes of soles to be put aboard.

He remembered the waves and the swirl of foam upon their crests and the wind. Two men would be needed to row the boat, and the boat must make three trips. The skipper and the first hand had been on deck all night. Accordingly he left them out of his reckoning. There remained four, or rather three, for the baker's assistant had ceased to count—Willie Weeks, Deakin, and himself, not a great number to choose from. He felt that he was within an ace of a panic, and not so far, after all, from that whimperer his neighbour. Two men to row the boat—two men! His hands clutched at the iron bar of his hammock; he closed his eyes tight; but the words were thundered out at him overhead in the whistle of the wind, and slashed at him by the water against the planks at his side. He found that his lips were framing excuses.

Nevertheless Duncan was on deck when the morning broke. It broke extraordinarily slowly, a niggardly filtering of grey sad light from the under edge of the sea. The bare topmasts of the smacks showed one after the other. Duncan watched each boat as it came into view with a keen suspense. This was a ketch, and that, and that other, for there was the peak of its reefed mainsail just visible, like a bird's wing, and at last he saw it—the thing he looked for, the steam fish cutter—lurching and rolling in the very middle of the fleet, whither she had crept up in the night. He stared at her; his belly was pinched with fear as a starveling's with hunger; and yet he was conscious that in a way he would have been disappointed if she had not been there.

'No other smack is shipping her fish,' quavered a voice at his elbow. It was the voice of the baker's assistant.

'But this smack is,' replied Weeks, and he set his mouth hard. 'And, what's more, my Willie is taking it aboard. Now, who'll go with Willie?'

'I will.'

Weeks swung round on Duncan and stared at him. Then he stared out to sea. Then he stared again at Duncan.

'You?'

'When I shipped as a hand on the *Willing Mind*, I took all a hand's risks.'

'And brought the willing mind,' said Weeks with a smile. 'Go, then! Some one must go. Get the boat tackle ready forward. Here, Willie, put your lifebelt on. You, too, Duncan, though God knows lifebelts won't be of no manner of use; but they'll save your insurance. Steady with the punt there! If it slips inboard off the rail there will be a broken back! And, Willie, don't get under the cutter's counter. She'll come atop of you and smash you like an egg. I'll drop you as close as I can to windward, and pick you up as close as I can to leeward.'

The boat was slung over into the water and loaded up with fish-boxes. Duncan and Willie Weeks took their places, and the boat slid away into a furrow. Duncan sat in the bow and rowed. Willie Weeks stood in the stern, facing him, and rowed and steered.

'Water!' said Willie every now and then and a wave curled over the bows and hit Duncan a stunning blow on the back.

'Row!' said Willie, and Duncan rowed and rowed. His hands were ice, he sat in water ice-cold, and his body perspired beneath his oilskins, but he rowed. Once, on the crest of a wave, he looked out and saw below them the deck of a smack, and the crew looking upwards at them as though they were a balloon. 'Row!' said Willie Weeks. Once, too, at the bottom of a slope down which they had bumped dizzily, Duncan again looked out, and saw the spar of a mainmast tossing high above his head just over the edge of a grey roller. 'Row!' said Weeks, and a moment later, 'Ship your oar!' and a rope caught him across the chest.

They were alongside the cutter.

Duncan made fast the rope.

'Push her off!' suddenly cried Willie, and grasped an oar. But he was too late. The cutter's bulwarks swung down towards him, disappeared under water, caught the punt fairly beneath the keel, and scooped it clean on to the deck, cargo and crew.

For a moment both men sat dazed upon their thwarts, stupidly staring. Then Willie exclaimed, 'And this is only the first trip!'

The two following trips however were made without accident, and Duncan found the work too arduous to allow him much thought of its danger.

'Fifty-two boxes at two pound ten,' Weeks cuckled as the boat was swung inboard. 'That's a hundred and four, and ten twos are twenty, and carry two, and ten fives are fifty, and two carried, and twenties into that make twenty-six. One hundred and thirty pounds—this smack's mine, every rope on her. I tell you what, Duncan: you've done me a good turn today, and I'll do you another. I'll land you at Helsund in Denmark, and you shall get clear away. All we can do now is to lie out this gale.'

Before the afternoon the air was dark with a swither of foam and spray blown off the waves in the thickness of a fog. The heavy bows of the smack beat into the seas with a thud and a hiss—the thud of a steam hammer, the hiss of molten iron plunged into water; the waves raced exultingly up to the bows from windward, and roared angrily away in a spume of foam from the ship's keel to lee; and the thrumming and screaming of the storm in the rigging exceeded all that Duncan had ever imagined. He clung to the stays appalled. This storm was surely the perfect expression of anger too persistent for mere fury. There seemed to be a definite aim of destruction, a deliberate attempt to wear the boat down, in the steady follow of wave upon wave, and in the steady volume of the wind.

Captain Weeks, too, had lost of a sudden all his exhilaration. He stood moodily by Duncan's side, his mind evidently labouring like his ship. He told Duncan stories to which Duncan would rather not have listened—the story of the man who slipped as he stepped from the deck into the punt, and, weighted by his boots, had sunk visibly down and down and down through the clearest calmest water without a struggle; the story of the punt which got its painter under its keel and drowned three men; the story of the full rigged ship which was driven across the seven-fathom part of the Dogger—the part that looks like a man's leg in the chart—and which was turned upside down through the back breaking.

'The skipper and the mate', said Weeks, 'got outside and clung to her bottom, and a steam cutter tried to get them off, but smashed them both with her iron counter instead. Look!' and he gloomily pointed his finger, 'I don't know why that breaker didn't hit us. I don't know what we should have done if it had. I can't think why it didn't hit us! Are you saved?'

Duncan was taken aback by the unexpected question, and answered vaguely—

'I hope so.'

'But you must know,' said Weeks, perplexed. The wind made a theological discussion difficult. Weeks curved his hand into a trumpet, and bawled into Duncan's ear: 'You are either saved or not saved! It's a thing one knows. You must know if you are saved, if you've felt the glow and illumination of it.' He suddenly broke off into a shout of triumph: 'But I got my fish on board the cutter. The *Willing Mind*'s the only boat that did.' Then he relapsed again into melancholy: 'But I'm troubled about the poachin'. The temptation was great but it wasn't right; and I'm not sure but what this storm ain't a judgement.'

He was silent for a little, and then cheered up. 'I tell you what. Since we're hove-to, we'll have a prayer meeting in the cabin tonight and smooth things over.'

The meeting was held after tea by the light of a smoking paraffin lamp with a broken chimney. The crew sat and smoked but with no thought of irreverence, the companion was open, so that the swish of the water and the man on deck alike joined in the hymns. Rall, the baker's assistant, who had once been a steady attendant at Revivalist meetings, led off with a Moody and Sankey hymn, and the crew followed, bawling at the top pitch of their lungs, with now and then some suggestion of a tune. The little stuffy smoke-laden cabin rang with the noise. It burst upwards through the companion way, loud and earnest and plaintive, and the winds caught it and carried it over the water, a thin and appealing cry. After the hymn Weeks prayed aloud, and extempore and most seriously. He prayed for each member of the crew by name, one by one, taking the opportunity to mention in detail each fault of which he had had to complain, and begging that the offender's chastisement might be light. Of Duncan he spoke in ambiguous terms.

'O Lord!' he prayed, and without any abatement of his sincerity, 'a strange gentleman, Mr Duncan, has come amongst us. O Lord! we do not know as much about Mr Duncan as You do, but still bless him, O Lord!' And so he came to himself: 'O Lord! this smack's mine, this little smack labouring in the No'th Sea is mine, Through my poachin' and Your lovin' kindness it's mine; and, O Lord, see that it don't cost me dear!' And the crew solemnly and fervently said 'Amen!'

But the smack was to cost him dear. For in the morning Duncan woke to find himself alone in the cabin. He thrust his head up the companion, and saw Weeks alone with a very grey face standing by the lashed wheel.

'Halloa!' said Duncan. 'Where's the binnacle?'

'Overboard,' said Weeks.

Duncan looked round the deck.

'Where's Willie and the crew?'

'Overboard,' said Weeks. 'All except Rall! He's below deck forward, and clean daft. Listen and you'll hear 'im. He's singing hymns for those in peril on the sea.'

Duncan stared in disbelief. The skipper's face drove the disbelief out of him.

'Why didn't you wake me?' he asked.

'What's the use? You want all the sleep you can get, because you an' me have got to sail my smack into Yarmouth. But I was minded to call you, lad,' he said, with a sort of cry leaping from his throat. 'The wave struck us at about twelve, and it's been mighty lonesome on deck since then with Willie callin' out of the sea. All night he's been callin' out of the welter of the sea. Funny that I haven't heard Upton or Deakin, but on'y Willie! All night until daybreak he called, first on one side of the smack and then on t'other. I don't think I'll tell his mother that. An' I don't see how I'm to put you on shore in Denmark, after all.'

What had happened Duncan put together from the curt utterances of Captain Weeks and the crazy lamentations of Rall. Weeks had roused all hands except Duncan to take the last reef in. They were forward by the mainmast at the time the wave struck them. Weeks himself was on the boom, threading the reefing rope through the eye of the sail. He shouted 'Water!' and the water roared on the deck, carrying the three men aft. Upton was washed over the taffrail. Weeks threw one end of the rope down, and Rall and Willie caught it and were swept overboard, dragging Weeks from the boom on to the deck and jamming him against the bulwarks.

The captain held on to the rope, setting his feet against the side. The smack lifted and dropped and tossed, and each movement wrenched his arms. He could not reach a cleat. Had he moved he would have been jerked overboard.

'I can't hold you both!' he cried, and then, setting his teeth and hardening his heart, he addressed his words to his son: 'Willie! I can't hold you both!' and immediately the weight upon the rope was less. With each drop of the stern the rope slackened, and Weeks gathered the slack in. He could now afford to move. He made the rope fast and hauled the one survivor on deck. He looked at him for a moment. 'Thank God, it's not my son!' he had the courage to say.

'And my heart's broke!' had gasped Rall. 'Fair broke.' And he had gone forward and sung hymns.

They saw little more of Rall. He came aft and fetched his meals away; but he was crazed and made a sort of kennel for himself forward, and the two men left on the smack had enough upon their hands to hinder them from waiting on him. The gale showed no sign of abatement; the fleet was scattered; no glimpse of the sun was visible at any time; and the binnacle was somewhere at the bottom of the sea.

'We may be making a bit of headway no'th, or a bit of leeway west,' said Weeks, 'or we may be doing a sternboard. All that I'm sure of is that you and me are one day going to open Gorleston Harbour. This smack's cost me too dear for me to lose her now. Lucky there's the tell-tale compass in the cabin to show us the wind hasn't shifted.'

All the energy of the man was concentrated upon this wrestle with the gale for the ownership of the *Willing Mind*; and he imparted his energy to his companion. They lived upon deck, wet and starved and perishing with the cold—the cold of December in the North Sea, when the spray cuts the face like a whipcord. They ate by snatches when they could, which was seldom; and they slept by snatches when they could, which was even less often. And at the end of the fourth day there came a blinding fall of snow and sleet, which drifted down the companion, sheeted the ropes with ice, and hung the yards with icicles, and which made every inch of brass a searing iron and every yard of the deck a danger to the foot.

It was when this storm began to fall that Weeks grasped Duncan fiercely by the shoulder.

'What is it you did on land?' he cried. 'Confess it, man! There may be some chance for us if you go down on your knees and confess it.'

Duncan turned as fiercely upon Weeks. Both men were overstrained with want of food and sleep.

'I'm not your Jonah—don't fancy it! I did nothing on land!'

'Then what did you come out for?'

'What did you? To fight and wrestle for your ship, eh? Well I came out to fight and wrestle for my immortal soul, and let it go at that!'

Weeks turned away, and as he turned, slipped on the frozen deck. A lurch of the smack sent him sliding into the rudder-chains, where to Duncan's despair he lay. Once he tried to rise, and fell back. Duncan hauled himself along the bulwarks to him.

'Hurt?'

'Leg broke. Get me down into the cabin. Lucky there's the tell-tale. We'll get the *Willing Mind* berthed by the quay, see if we don't.' That was still his one thought, his one belief.

Duncan hitched a rope round Weeks, underneath his arms, and lowered him as gently as he could down the companion.

'Lift me on to the table so that my head's just beneath the compass! Right! Now take a turn with the rope underneath the table, or I'll roll off. Push an oily under my head, and then go for'ard and see if you can find a fish-box. Take a look that the wheel's fast.'

It seemed to Duncan that the last chance was gone. There was just one inexperienced amateur to shift the sails and steer a seventy-ton ketch across the North Sea into Yarmouth Roads. He said nothing however of his despair to the indomitable man upon the table, and went forward in search of a fish-box. He split up the sides into rough splints and came aft with them.

'Thank 'ee, lad,' said Weeks. 'Just cut my boot away, and fix it up best you can.'

The tossing of the smack made the operation difficult and long. Weeks however never uttered a groan. Only Duncan once looked up and said—

'Halloa! You've hurt your face too. There's blood on your chin!'

'That's all right,' said Weeks with an effort. 'I reckon I've just bit through my lip.'

Duncan stopped his work.

'You've got a medicine chest, skipper, with some laudanum in it—?'

'Daren't!' replied Weeks. 'There's on'y you and me to work the ship. Fix up the job quick as you can, and I'll have a drink of Friar's Balsam afterwards. Seems to me the gale's blowing itself out, and if on'y the wind holds in the same quarter—' And thereupon he fainted.

Duncan bandaged up the leg, got Weeks round, gave him a drink of Friar's Balsam, set the teapot within his reach, and went on deck. The wind was certainly going down; the air was clearer of foam. He tallowed the lead and heaved it, and brought it down to Weeks. Weeks looked at the sand stuck on the tallow and tasted it, and seemed pleased.

'This gives me my longitude', said he, 'but not my latitude, worse luck. Still, we'll manage it. You'd better get our dinner now; any odd thing in the way of biscuits or a bit of cold fish will do, and then I think we'll be able to run.'

After dinner Duncan said: 'I'll put her about now.'

'No; wear her and let her jibe,' said Weeks, 'then you'll on'y have to ease your sheets.'

Duncan stood at the wheel, while Weeks, with the compass swinging above his head, shouted directions through the companion. They sailed the boat all that night with the wind on her quarter, and at daybreak Duncan brought her to and heaved his lead again. There was rough sand with blackish specks upon the tallow, and Weeks, when he saw it, forgot his broken leg.

'My word', he cried, 'we've hit the Fisker Bank! You'd best lash the wheel, get our breakfast, and take a spell of sleep on deck. Tie a string to your finger and pass it down to me, so that I can wake you up.'

Weeks waked him up at ten o'clock, and they ran south-west with a steady wind till six, when Weeks shouted—

'Take another cast with your lead.'

The sand upon the tallow was white like salt.

'Yes,' said Weeks; 'I thought we was hereabouts. We're on the edge of the Dogger, and we'll be in Yarmouth by the morning.' And all through the night the orders came thick and fast from the cabin. Weeks was on his own ground; he had no longer any need of the lead; he seemed no longer to need his eyes; he felt his way across the currents from the Dogger to the English coast; and at daybreak he shouted—

'Can you see land?'

'There's a mist.'

'Lie to, then, till the sun's up.'

Duncan lay the boat to for a couple of hours, till the mist was tinged with gold and the ball of the sun showed red on his starboard quarter. The mist sank, the brown sails of a smack thrust upwards through it; coastwards it shifted and thinned and thickened, as though cunningly to excite expectation as to what it hid. Again Weeks called out—

'See anything?'

'Yes,' said Duncan in a perplexed voice. 'I see something. Looks like a sort of medieval castle on a rock.'

A shout of laughter answered him.

'That's the Gorleston Hotel. The harbour mouth's just beneath. We've hit it fine,' and while he spoke the mist swept clear, and the long treeless esplanade of Yarmouth lay there a couple of miles from Duncan's eyes, glistening and gilded in the sun like a row of doll's houses.

'Haul in your sheets a bit,' said Weeks. 'Keep no'th of the hotel, for the tide'll set you up and we'll sail her in without dawdlin' behind a tug. Get your mainsail down as best you can before you make the entrance.'

Half an hour afterwards the smack sailed between the pierheads.

'Who are you?' cried the harbour master.

'The *Willing Mind.*'

'The *Willing Mind*'s reported lost with all hands.'

'Well, here's the *Willing Mind*,' said Duncan, 'and here's one of the hands.'

The irrepressible voice bawled up the companion to complete the sentence—

'And the owner's reposin' in his cabin.' But in a lower key he added words for his own ears: 'There's the old woman to meet. Lord! but the *Willing Mind* has cost me dear.'

The Terror of the Sea Caves

∾∾∾

CHARLES G. D. ROBERTS

I

It was in Singapore that big Jan Laurvik, the diver, heard about the lost pearls.

As he was passing the head of a mean-looking alley near the water-side, late one sweltering afternoon, he was halted by a sudden uproar of cries and curses. The noise came from a courtyard about twenty paces up the alley. It was a fight evidently, and Jan's blood responded with a sympathetic thrill. But the curses which he caught were all in Malay or Chinese, and he curbed his natural desire to rush in and help somebody. Though he knew both languages very well, he knew that he did not know, and never could know, the people who spoke those languages. Interference on the part of a stranger might be resented by both parties to the quarrel. He shrugged his great shoulders and walked on reluctantly.

Hardly three steps had he taken, however, when above the shrill cries a great voice shouted.

'Take that, you——' it began, in English. And at that it ended, with a kind of choking.

Jan Laurvik wheeled round in a flash and ran furiously for the door of the courtyard, which stood half open. He was a Norwegian, but English was as a native tongue to him; and amid the jumble of races in the East he counted all of European speech his brothers. An Englishman was being killed in there. The quarrel was clearly his.

Six feet two in height, swift, and of huge strength, with yellow hair, so light as to be almost white, waving thickly over a face that was sunburnt to a high red, his blue eyes flaming with the delight of battle, Jan burst in upon the mob of fighters. Several bodies lay on the floor.

One dark-faced, low-browed fellow, a Lascar apparently, with his back to the wall and a bloody kreese in his hand, was putting up a savage fight against five or six assailants, who seemed to be Chinamen and Malays. The body of the Englishman whose voice Jan had heard lay in an ugly heap against the wall, its head far back and almost severed.

Jan's practised eye took in everything at a glance. The heavy stick he carried was, for a mêlée like this, a better weapon than knife or gun. With a great bellowing roar he sprang upon the knot of fighters.

The result was almost instantaneous. The two nearest rascals went down at his first two strokes. At the sound of that huge roar of his all had turned their eyes; and the man at bay, seizing his opportunity, had cut down two more of his foes with lightning slashes of his blade. The remaining two, scattering and ducking, had leaped for the door like rabbits. Jan wheeled, and sprang after them. But they were too quick for him. As he reached the head of the alley they darted into a narrow doorway across the street which led into a regular warren of low structures. Knowing it would be madness to follow, Jan turned back to the courtyard, curious to find out what it had all been about.

The silence was now startling. As he entered, there was no sound but the painful breathing of the Lascar, whom he found sitting with his back against the wall, close beside the body of the Englishman. He was desperately slashed. His eyes were half closed; and Jan saw that there was little chance of his recovery. Besides that of the Englishman, there were six bodies lying on the floor, all apparently quite lifeless. Jan saw that the place was a kind of drinking den. The proprietor, a brutal-looking Chinaman, lay dead beside his jugs and bottles. Jan reached for a jug of familiar appearance, poured out a cup of arrack, and held it to the lips of the dying Lascar. At the first gulp of the potent spirit his eyes opened again. He swallowed it all eagerly, then straightened himself up, held out his hand in European fashion to Jan, and thanked him in Malayan.

'Who's that?' enquired Jan in the same tongue, pointing to the dead white man.

Grief and rage convulsed the fierce face of the wounded Lascar.

'He was my friend,' he answered. 'The sons of filthy mothers, they killed him!'

'Too bad!' said Jan sympathetically. 'But you gave a pretty good account of yourselves, you two. I like a man that can fight like you were fighting when I came in. What can I do for you?'

'I'm dead, pretty soon now!' said the fellow indifferently. And from the blood that was soaking down his shirt and spreading on the floor about him, Jan saw that the words were true. Anxious, however, to do something to show his goodwill, he pulled out his big red handkerchief, and knelt to bandage a gaping slash straight across the man's left forearm, from which the bright arterial blood was jumping hotly. As he bent, the fellow's eyes lifted and looked over his shoulder.

'Look out!' he screamed. Before the words were fairly out of his mouth Jan had thrown himself violently to one side and sprung to his feet. He was just in time. The knife of one of the Chinamen whom he had supposed to be dead was sticking in the wall beside the Lascar's arm.

Jan stared at the bodies—all, apparently, lifeless.

'That's the one did it,' cried the Lascar excitedly, pointing to one whom Jan had struck on the head with his stick. 'Put your knife into the son of a dog!'

But that was not the big Norseman's way. He wanted to assure himself. He went and bent over the limp-looking, sprawling shape, to examine it. As he did so, the slant eyes opened upon his with a flash of such maniacal hate that he started back. He was just in time to save his eyes, for the Chinaman had clutched at them like lightning with his long nails.

Startled and furious at this novel attack, Jan reached for his knife. But before he could get his hand on it the Chinaman had leaped into the air like a wild-cat, wound arms and legs about his body, and was struggling like a mad beast to set teeth into his throat. The attack was so miraculously swift, so disconcerting in its beast-like ferocity, that Jan felt a strange qualm that was almost akin to panic. Then a black rage swelled his muscles; and tearing the creature from him he dashed him down upon the floor, on the back of his neck, with a violence which left no need of pursuing the question further. Not till he had examined each of the bodies carefully, and tried them with his knife, did he turn again to the wounded Lascar leaning against the wall.

'Thank you, my friend!' he said simply.

'You're a good fighting man. You're—like him', answered the Lascar feebly, nodding toward the dead Englishman. 'Give me more arrack. I will tell you something. Hurry, for I go soon.'

Jan brought him the liquor, and he gulped it. Then from a pouch within his knotted silk waistband he hurriedly produced a bit of paper which he unfolded with trembling fingers. Jan saw that it was a rough map sketched with India ink and marked with Malayan characters. The Lascar peered about him with fierce eyes already growing dim.

'Are you sure they are all gone?' he demanded.

'Certain!' answered Jan, highly interested.

'They'll try their best to kill you,' went on the dying man. 'Don't let them. If you let them get the pearls, I'll come back and haunt you.'

'I won't let them kill me, and I won't let them get the pearls, if that's what it is that's made all the trouble. Don't worry about that,' responded Jan confidently, reaching out his great hand for the paper, which was evidently so precious that men were giving up their lives for it.

The man handed it over with a groping gesture, though his savage black eyes were wide open.

'That'll show you where the wreck of the junk lies, in seven or eight fathoms of water, close inshore. The pearls are in the deck-house. *He* kept them. The steamer was on a reef, going to pieces, and we came up just as the boats were putting off. We sunk them all, and got the pearls. And next night, in a storm, the junk was carried on to the rocks by a current we didn't know about. Only five of us got ashore— for the sharks were around, and the "killers", that night. *Him* and me, we were the only ones knew enough to make that map.'

Here the dying pirate—for such he had declared himself—sank forward with his face upon his knees. But with a mighty effort he sat up again and fixed Jan Laurvik with terrible eyes.

'Don't let the sons of a dog get them, or I will come back and choke you in your sleep,' he gasped, suddenly pointing a lean finger straight at the Norseman's face. Then his black eyes opened wide, a strange red light blazed up in them for an instant and faded. With a sigh he toppled over, dead, his head resting on the dead Englishman's feet.

II

Jan Laurvik looked down upon the slack form with a sort of grim indulgence. 'He was game, and he loved his comrade, though he *was* but a bloodthirsty pirate!' he muttered to himself.

With the paper folded small and hidden in his great palm, he glanced again from the door to see if any of the routed scoundrels were coming back. Satisfied on this point, he once more investigated

the dead bodies on the floor, to assure himself that all were as dead
as they appeared. Then he set himself to examine the precious paper,
which held out to his imagination all sorts of fascinating possibilities.
He knew that the swift boats carrying the proceeds of the pearl
fisheries were always eagerly watched by the piratical junks infesting
those waters, but carried an armament which secured them from all
interference. In case of wreck, however, the pirates' opportunity
would come. Jan knew that the story he had just heard was no
improbable one.

The map proved to be rough, but very intelligible. It indicated a
stretch of the eastern coast of Java, which Jan recognized; but the spot
where the junk had gone down was one to which passing ships always
gave a wide berth. It was a place of treacherous anchorage, of abrupt,
forbidding, uninhabited shore, and of violent currents that shifted
erratically. So much the better, thought Jan, for his investigations, if
only the pirate junk should prove to have been considerate enough to
sink in water not too deep for a diver to work in. There would be so
much the less danger of interruption.

Jan was on the point of hurrying away from the gruesome scene,
which might at any moment become a scene of excitement and an-
noying investigation, when a new idea flashed into his mind. It was
over this precious paper that all the trouble had been. The scoundrels
who had fled would undoubtedly return as soon as they dared and
would search for it. Finding it gone, they would conclude that he had
it; and they would be hot on his trail. He had no fancy for the sleepless
vigilance that this would entail upon him. He had no fancy for the
heavy armed expedition which it would force him to organize for
the pearl hunt. He saw his airy palaces toppling ignominiously to
earth. He saw that all he was likely to get was a slit throat.

As he glanced about him for a way out of his dilemma his eyes fell
on a bottle of India ink containing the fine-tipped brush with which
these Orientals did their writing. His resourcefulness awoke to this
chance. The moments were becoming very pearls themselves for
preciousness, but seizing the brush, he made a workable copy of the
map on the back of a letter which he had in his pocket. Then he made
a minute and very careful correction in the original, in such a manner
as to indicate that the position of the wreck was in a deep fiord some
fifty miles east of where it actually was. This done to his critical
satisfaction, he returned the map to its hiding-place in the dead
pirate's belt, and made all haste away. Not till he was back in the

European quarter did he feel himself secure. Once among his fellow whites, where he was a man of known standing and reputed to be the best diver in the Archipelago, he knew that he would run no risk of being connected with a drinking brawl of Lascars and pirates. As for the dead Englishman, he knew the odds were that the Singapore police would know all about him.

Jan Laurvik had a little capital. But he needed a trusty partner with more. To his experienced wits his other needs were clear. There would have to be a very seaworthy little steamer, powerfully engined for service on that stormy coast, and armed to defend herself against prowling pirate junks. This small and fit craft would have to be manned by a crew equally fit, and at the same time as small as possible, for the reason that in a venture of this sort everyone concerned would of necessity come in for a share of the winnings. Moreover, the fewer there were to know, the fewer the chances of the secret leaking out; and Jan was even more in dread of the Dutch Government getting wind of it than he was of the pirates picking up his trail.

Up to a certain point, he had no difficulty in verifying the dead pirate's story. He had heard of the wreck of the Dutch steamer *Viecht* on a reef off the Celebes, and of the massacre of all the crew and passengers, except one small boatload, by pirates. This had happened about eight months ago. Discreet enquiry developed the fact that the *Viecht* had carried about $300,000 worth of pearls. The evidence was sufficiently convincing, and the prize was sufficiently alluring to make it worth his while to risk the adventure.

It was with a certain amount of Northern deliberation that Jan Laurvik thought these points all out, and made up his mind what to do. Then he acted promptly. First he cabled to Calcutta, to one Captain Jerry Parsons, to join him in Singapore without fail by the very next steamer. Then he set himself unobtrusively to the task of finding the craft he wanted and looking up the equipment for her.

Captain Jerry Parsons was a New Englander, from Portland, Maine. He had been whaler, gold-hunter, filibuster, copra-trader, general-in-chief to a small Central American republic, and sheep-farmer in the Australian bush. At present he was conducting a more or less regular trade in precious stones among the lesser Indian potentates. He loved gain much, but he loved adventure more.

When he received the cable from his good friend Jan Laurvik, he knew that both were beckoning to him. With light-hearted zest he betook himself to the steamship offices, found a P. and O. boat sailing

on the morrow, and booked his passage. Throughout the journey he
amused himself with trying to guess what Jan Laurvik was after; and,
as it happened, almost the only thing he failed to think of was pearls.

When Captain Jerry reached Singapore Jan Laurvik told him the
story of the dead pirate's map.

'Let's see the map!' said he, chewing hard on the butt of his unlighted Manila.

Jan passed his copy over. The New Englander inspected it carefully,
in silence, for several minutes.

''Tain't much of a map!' said he at length disparagingly. 'You think
the varmint was straight?'

'In his way, yes,' answered Jan with conviction. 'He had it in him
to be straight in his way to a friend, which wouldn't hinder him cuttin'
the throats of a thousand chaps he didn't take an interest in.'

'When shall we start?' asked Captain Jerry. Now that his mind was
quite made up he took out his matchbox and carefully lighted his
cheroot.

The big Norseman's face lighted up with pleasure, and he reached
out his hand. The grip was all, in the way of a bargain, that was
needed between them.

'Why, tomorrow night!' he answered.

'Well,' said the New Englander, 'I'll draw some cash in the morning.'

The boat which Jan had hired was a fast and sturdy sea-going tug,
serviceable, but not designed for comfort. Jan had retained her engineer, a shrewd and close-mouthed Scotsman. Her sailing-master would
be Captain Jerry. For crew he had chosen a wiry little Welshman and
two lank leather-skinned Yankees. To these four, for whose honesty
and loyalty he trusted to his own insight as a reader of men, he
explained, partially, the nature of the undertaking, and agreed to give
them, over and above their wages, a substantial percentage of whatever treasure he might succeed in recovering. He had made his selection wisely, and every man of the four laid hold of the opportunity
with ardour.

The tug was swift enough to elude any of the junks infesting those
waters, but the danger was that she might be taken by surprise at her
anchorage while Laurvik was under water. He fitted her, therefore, with
a Maxim gun on the roof of the deck-house, and armed the crew
with repeating Winchesters.

Thus equipped, he felt ready for any perils that might confront
him above the surface of the water. As to what might lurk below

he felt somewhat less confident, as these he should have to face alone, and he remembered the ominous warning of his pirate friend, about the sharks and the 'killers'. For sharks Jan Laurvik had comparatively small concern; but for the 'killers', those swift and implacable little whales who fear no living thing, he entertained the highest respect.

On the evening of the day after Captain Jerry's arrival, the tug *Sarawak* steamed quietly out of the harbour. As this was a customary thing for her to do, it excited no particular comment among the frequenters of the waterside. By the pirates' spies, who abounded in the city, it was not considered an event worth noting.

The journey, across the Straits, and down the treacherous Javan Sea, was so prosperous that Jan Laurvik, his blood steeped in Norse superstition, began to feel uneasy. The sea was like a millpond all the way, and they were sighted by no one likely to interfere or ask questions. Jan distrusted Fortune when she seemed to smile too blandly. But Captain Jerry comforted him with the assurance that there'd be trouble enough ahead; and strangely enough, this singular variety of comfort quite relieved Jan's depression.

The unusual calm made it easy to hold close inshore, when they reached that portion of the coast where they must keep watch for the landmarks indicated on the pirate's map. Every reef and surface-ledge boiled ceaselessly in the smooth swell, and by that clear green sea they were saved the trouble of tedious soundings. When they came exactly abreast of a low headland which they had been watching for some time, it suddenly opened out into the semblance of a two-humped camel crouching sidewise to the sea, exactly as it was represented in Jan's map. Just beyond was a narrow bay, and across the middle of its mouth, with a dangerous passage on either side, stretched the reef on which the pirate junk had gone down. At this hour of low water the reef was showing its teeth and snarling with surf. At high tide it would be hidden, and a perfect snare of ships. According to the map, the wreck lay in some eight fathoms of water, midway of the outer crescent of the reef. Behind the reef, where the latter might serve them as a partial shelter from the sweep of the seas if a north-easter should blow up, they found tolerable anchorage for the tug. For the preliminary soundings, and for the diving operations, of course, Jan planned to use the launch. And, in order to take utmost advantage of the phenomenal calm, which seemed determined to smooth away every obstacle for the adventurers, Jan got instantly to work. Within a half-hour of the *Sarawak*'s anchoring he had the launch outside the reef with all his diving apparatus aboard, with Captain Jerry to manage the air-pump, and the Scottish engineer to run the motor.

III

Along the outer face of the reef, at a depth varying from eight to twelve fathoms, ran an irregular rocky shelf which dipped gradually seaward for several hundred yards, then dropped sheer to the ocean depths. In the warm water along this shelf swarmed a teeming life, of gay-coloured gigantic weeds, and of strange fish that outdid the brightest weeds in brilliancy and unexpectedness of hue. Where the tropic sunlight filtered dimly down through the beryl tide it sank into a marvellous garden whose flowers, for the most part, were living and moving forms, some monstrous, many terrifying, and almost all as grotesque in shape as they were radiant in colour. But in that insufficient, glimmering light, which was rather, to a human eye, a vaguely translucent, greenish darkness, these colours were almost blotted out. It took eyes adapted to the depth and gloom to differentiate them clearly.

In the great deeps, also, beyond the edge of the shelf, thronged life in swimming, crawling, or moveless forms, of every imagined and many unimagined shapes, from creatures so tiny that a whole colony could dwell at ease in the eye of a cambric needle, to the titanic squid, or cuttlefish, with oval body fifty feet in length and arms like writhing constrictors reaching twenty or thirty feet further. It was a life of noiseless but terrific activity, of unrelenting and incessant death, in a darkness streaked fitfully with phosphorescent gleams from the bodies of the darting, writhing, or pouncing creatures that slew and were slain in the stupendous silence.

Down to these dwellers in the profound had come some mysterious message or exciting influence, no man knows what, from the prolonged calm on the surface. It affected individuals among various species, in such a way that they moved upward, into a twilight where they were aliens and intruders. Among those so stung with unrest were several of the gigantic, pallid cuttles. Far offshore, one of these monsters came up and sprawled upon the surface in the unfriendly sun, his dreadful arms curling and uncurling like snakes, till a great sperm-whale, of scarcely more than his own size, came by and fell upon him ravenously, and devoured him.

Another of the restless monsters, however, kept his restlessness within the bounds of discretion. Slowly rising, a vast and spectral horror as he came up into the green light, he reached the rim of the ledge. The growing light had already made him uneasy, and he wanted no more of it. Here on the ledge, where food, though novel in character, was unlimited in supply, was variety enough to content him. Gorging himself as he went with everything that swam within reach of his darting tentacles, he moved over the rocky floor till he came to the wreck of the junk.

To his huge unwinking eyes of crystal black, which caught every tiniest ray of light in their smooth, appalling deeps, the wreck looked strange enough to attract his attention at once. It was quite unlike any rock-form which he had ever seen. Rather cautiously he advanced a giant tentacle to investigate it. But at the touch of the unfamiliar and alien substance the tentacle recoiled in aversion. The pale monster backed away. But the wreck made no attempt to pounce upon him. It seemed to have no fight in it. Possibly, on closer investigation, it might prove to be good to eat; and he was hungry. In fact, he was always hungry, for the irresistible corrosives in his great stomach—and he was nearly all stomach—were so swift in their action that whatever he swallowed was digested almost in the swallowing. Since coming upon the ledge he had clutched and devoured two small basking sharks, from six to eight feet long, and a sawfish fully ten feet long, who had not been on their guard against the approach of such a peril. Besides these substantial victims, countless small fry, of every kind, had been drawn deftly to the insatiable vortex of his maw. Nevertheless, his appetite was again crying out. He tried the wreck again, first carefully, then boldly, till the writhing tentacles, with their sensitive tips and suckers, had enveloped it from stem to stern and searched it inside and out. A few lurking fish and molluscs were snatched from the dark interior by those insinuating and inexorable feelers; and a toothsome harvest of anchored crustaceans was gathered from the hidden surfaces along beside the keel. But of the bodies of the pirates that had gone down in the sudden foundering there was nothing left but bones, which the myriad scavengers of the sea had polished to the barren smoothness of ivory.

While the pallid monster was occupied in the investigation of the wreck those two great bulging black mirrors of his eyes were sleeplessly alert to everything that passed above or about them. Once a swordfish, about seven feet long, sailed carelessly though swiftly some ten feet overhead. Up darted a livid tentacle, and fixed upon it with the deadly sucking discs. In vain the splendid and ferocious fish lashed out in the effort to wrench itself free. In vain it strove to plunge downward and pierce the puffy monster with its sword. In a second or two more tentacles were wrapped about it. Then, all force crushed out of it, it was dragged down and crammed into the conqueror's horrible mouth.

While its mouth was yet working with the satisfaction of this meal, the monster saw a graceful but massive black shape, nearly half as long as himself, swimming slowly between his eyes and the shining surface. At the sight a shudder of fear passed over him. Every waving tentacle shrank back and lay moveless, as if suddenly paralysed, and he flattened himself down as best he could beside the dark hulk of the wreck. Well he knew that dark shape was a whale—and a whale was the one being he knew of which he had cause to fear. Against those rending jaws his cable-like tentacles and tearing beak were of no avail, his unarmoured body utterly defenceless.

The whale, however—not a sperm, but one of a much smaller, though more savage species, the 'killer'—did not catch sight of the giant cuttlefish cringing below him. Intent on other game, he passed swiftly on. His presence, however, had for the moment destroyed the monster's appetite. Instead of continuing his search for food, he wanted a hiding-place. He could no longer be at ease for a moment there in the open.

Just behind the wreck the rock wall rose abruptly to the surface of the reef. Its base was hollowed into a series of low caves, where masses of softer rock had been eaten out from beneath a slanting stratum of more enduring material. The more spacious of these caves was immediately behind the wreck. It was exactly what the monster craved. He backed into it with alacrity, completely filling it with his spectral and swollen body. In the doorway the convex inky lenses of his eyes kept watch, moveless and all-seeing. And his ten pale-spotted tentacles, each thicker at the base than a man's thigh, lay outspread and hidden among the seaweeds, waiting for such victims as might come within reach of their lightning snap and coil.

The monster had no more than got himself fairly installed in his new quarters, when into the range of his awful eyes came a singular figure, descending slowly through the glimmering green directly over the wreck. It was not so long as the swordfish he had lately swallowed, but it was thick and massive-looking; and it was blunt at the ends, unlike any fish he had ever seen. Its eyes were enormous, round and bulging. From its head and from one of its curious round, thick fins, extended two slender antennæ straight up towards the surface, and so long that their extremities were beyond the monster's vision. It was indeed a strange-looking creature, but he felt sure that it would be very good to eat. In their concealment among the many-coloured seaweeds his tentacles thrilled with expectancy, and he waited, like some stupendous nightmare of a spider, to spring the moment the prey came within reach.

It chanced, however, that just as the strange creature, descending without any movement of its fins, did come within reach, there also appeared again, in the distance, the black form of the 'killer' whale, swimming far overhead. The monster changed his plans instantly. His interest in the newcomer died out. He became intent on nothing but keeping himself inconspicuous. The newcomer, unconscious of the terror lying in wait so near him and of the dark form patrolling the upper green, alighted upon the wreck and groped his way lumberingly into the cabin, dragging those two slim antennæ behind him.

IV

When Jan Laurvik, in his up-to-date and well-tested diving-suit, went down through the green twilight of the sea, he was doing what it was his profession to do, and he had few misgivings. He had confidence in his equipment, in his skill, and in his mate at the rope and the air-pump, Captain Jerry. For defence against any obtrusive shark or sawfish he carried a heavy, long-bladed, two-edged knife, by far the most effective weapon in deep water. This knife he wore in a sheath at his waist, with a cord attached to the handle so that it could not get away from him. He carried also a tiny electric battery supplying a strong lamp on the front of his head-piece just above his eyes.

From his long experience in sounding and in locating wrecks, Jan Laurvik had acquired an accuracy that seemed almost like divination. His soundings, in this instance, had been particularly thorough, because he did not wish to waste any more time than necessary at the depth in which he would have to work. He was not surprised, therefore, when he found himself descending upon the wreck of a junk. Moreover, as it was not an old wreck, he concluded that it was the junk which he was looking for. The wreck had settled almost on an even keel; and as he was familiar with craft of her type, he had no difficulty in finding his way about.

It was in the narrow, closet-like structure which served as the junk's cabin that the pirate had said the pearls would be found. The door was open. Turning on his light, which struggled with the water and diffused a ghostly glow, he found himself confronted by a hideous little joss of red-and-gilt lacquer. He knew it was lacquer, and of the best, for nothing else, except gold itself, would have withstood the months of soaking in sea-water. Jan grinned to himself, there within his rubber and copper shell, at this evidence of pirate piety. Then it occurred to him that a man like the pirate captain would probably have turned his piety to practical use. What better guardian of the treasure than a god? Dragging the gaudy deity from his altar, he found the altar hollow. In that secure receptacle lay a series of packages done up with careful precision in wrappings of oiled silk. He knew the style of wrapping very well. For all his coolness, his heart fell to thumping painfully at the sight of this vast wealth beneath his hand. Then he realized that the pressure of the water, and of the compressed air in his helmet, was beginning to tell upon him. In fierce but orderly haste he corded the packages about his middle and turned to leave the cabin. He would make another trip for the lacquer god, and for such other articles of value or *vertu* as the junk might contain.

Jan turned to leave the cabin. But in the doorway he started back with a shudder of dread and loathing. A slender, twisting thing, whitish in colour and minutely speckled with livid spots, reached in, and fastened upon his arm with soft-looking suckers which held like death.

Jan knew instantly what the pale, writhing thing was. Out flashed his knife. With a swift stroke he slashed off the detaining tip, where it had a thickness of perhaps two inches. The raw stump shrank back, like a severed worm, and Jan, leaping clear of the doorway, signalled furiously to be hauled up. But at the same instant two more of the curling white things came reaching over the bulwarks and fastened upon him—one upon his right arm, hampering him so that he was almost helpless, and the other upon his left leg just above the knee. He felt his signal promptly answered by a powerful tug on the rope. But he was anchored to the wreck as if he had grown to it.

Never before had Jan Laurvik felt the clutch of fear at his heart as he did at this moment. But not for an instant, in the horror, did he lose his presence of mind. He knew that in a pulling match with the giant devil-fish of the deeps his comrades in the boat far overhead would be nowhere. He had made a mistake in leaving the cabin. Frantically he signalled with his left hand, to 'slack away' on the rope; and at the same time, though hampered by the grip on his right arm, he managed to slash off the end of the feeler that had fixed upon his leg. On the instant, whipping the knife over to his left, he cut his right arm clear, and sprang back into the doorway.

Jan's idea was that by keeping just inside the cabin door he could defend himself from being surrounded by the assault of the writhing things. He knew that in the open he would speedily be enfolded and crushed, and engulfed between the jaws of the monstrous squid. But in the narrow doorway the swift play of his blade would have some chance. He gained the doorway. He got fairly inside it, indeed. But as he entered he was horrified to see the thick stump, whose tip he had shorn off, dart in with him and fix itself, by its bigger and more irresistible suckers, upon the middle of his breast. With a shiver he sliced off the fatal discs, in one long sweep of his blade; then turned like a flash to sever a pallid tip which had fastened upon his helmet.

Jan was now thankful enough that he had got himself into the narrow doorway. Seemingly undisturbed by the slashings and slicings which some of them had received, the whole ten squirming horrors now darted at the doorway. Jan's knife swooped this way and that; but as fast as he severed one clutch two more would make good. The cut tentacles grew to be the more terrifying, because their suckers were so big; and they themselves were so thick and hard to cut. Presently no fewer than three of the diabolical things laid their loathsome hold upon his right leg, below the knee, and began to haul it out through the door. Jan slashed at them madly, but not altogether effectually; for at this moment another tentacle had laid grip upon his arm below the elbow. He had just time to shift the knife again to his left and catch the jamb of the door, when he felt his helmet almost jerked from his head. This grip he dared not interfere with, lest he should cut, at the same time, the air-tube that fed his lungs, and drown like a rat in a hole. All he could do was to hold on to the door-jamb, and carve away savagely at the tentacles which were within reach. If he could get free of those, he calculated that he could then reach the one which had fastened to his headpiece by throwing himself over on his back and so bringing it within range of his vision and his knife. At this moment, however, just as the pressure upon his neck was becoming intolerable, he felt his head suddenly released. One of the great sucking discs had crushed in the glass of the electric lamp and fastened upon the live wire. The sensation it experienced was evidently not pleasant, for it let go promptly, and secured a new hold upon Jan's left arm.

This hold left him almost helpless, because he could no longer wield the knife freely with either hand. He felt himself slowly being pulled out of the doorway by his right leg. Throwing himself partly backward, and partly behind the door, he gained a firmer brace and at the same time brought his knife again into better play. He would fight to the very last gasp, but he felt that the odds had now gone overwhelmingly against him. The fear of death itself was not heavy upon him. He had faced it too often, and too coolly, for that. But at the manner of this death that confronted him his very soul sickened with loathing. As he thought of it, his horror was not lessened by the sight which now greeted his view. A colossal, swollen, leprous-looking bulk, pallid and spotted, was mounting over the bulwark. Two great oval lenses of clear blackness, set close together, were in the front of the bulk, just over the spot where the tentacles started. These gigantic, appalling, expressionless eyes were fixed upon him. The monster was coming aboard to see what kind of creature it was that was giving him so much trouble.

Jan saw that the end of the fight was very near. The thought, however, did not unnerve him. Rather, it put new fire into his nerves and muscles. By a tremendous wrench he succeeded in reaching with the knife the tentacle that bound his right arm. This freedom was like a new lease of life to him. He made swift play with his blade, so savagely that he was able to drag himself back almost completely into the cabin before the writhing horrors again closed upon him. But meanwhile the monster's gigantic body had gained the deck. Those two awful eyes were slowly drawing nearer; and below them he saw the viscid mouth opening and shutting in anticipation.

At this a kind of madness began to surge up in Jan Laurvik's overtaxed brain. His veins seemed to surge with fresh power, as if there were nothing too tremendous for him to accomplish. He was on the very point of stopping his resistance, plunging straight in among the arms, and burying his big blade in those unspeakable eyes.

It would be a satisfaction, at least, to force them to change their expression. And then, well, something might happen!

Before he could put this desperate scheme into execution, however, something did happen. Jan was aware of a sudden darkness overhead. The monster was evidently aware of it too, for every one of the twisting horrors suddenly shrank away, leaving Jan to lean up against the doorway, free. The next moment, a huge black shape descended perpendicularly upon the fleshy mountain of the monster's back, and a rush of water drove Jan backward into the cabin.

As the electric lamp had gone out when the glass was broken, Jan could see but dimly the awful battle of giants now going on before him. So excited was he that he forgot his own new peril. The danger was now that in the struggle one or other of the battling bulks might well crush the cabin flat, or entangle the air-tube and life-line. In either case Jan's finish would be swift; but in comparison with the loathsome death from which he had just been so miraculously saved, such an end seemed no very dreadful thing. He was altogether absorbed in watching the prowess of his avenging rescuer.

Skilled in deep-sea lore as he was, he knew the dark fury which had swooped down upon the devil-fish. It was a 'killer' whale, or grampus, the most redoubtable and implacable fighter of all the kindreds of the sea. Jan saw its wide jaws shear off three mighty tentacles at once, close at the base. The others writhed up hideously and fastened upon him, but under the surging of his resistless muscles their tissues tore apart like snapped cables. Huge masses of the monster's ghastly flesh were bitten off, and thrown aside. Then, gaining a grip that took in the monster's head and the roots of the tentacles, the killer shook his prey as a bulldog might shake a fat sheep. The tentacles straightened out sharply. Jan saw that the fight was over; and that it was high time for him to remove from that too strenuous neighbourhood. He gave the signal vehemently, and was drawn up, without attracting his dangerous rescuer's notice. When Captain Jerry hauled him in over the boatside, he fell in an unconscious heap.

When Jan came to himself he was in his bunk on the *Sarawak*. It was an utter physical and nervous exhaustion that had overcome him. His swoon had passed into a heavy sleep, and when he awoke he sat up with a start. Captain Jerry was at his side, bursting with suppressed curiosity; and the Scottish engineer was standing by the bunk.

'Waal, partner, you've delivered the goods all right!' drawled Captain Jerry. 'They're the stuff, not a doubt of it. But kind o' seemed to us up here you were having high jinks of one kind or another down there. What was it?'

'It was hell!' responded Jan with a shudder. Then he took hold of Captain Jerry's hand, and felt it, as if to make sure it was real, or as if he needed the feel of honest human flesh again to bring him to his senses.

'Ugh!' he went on, swinging out of the bunk. 'Let me get out into the sunlight again! Let me see the sky again! I'll tell you all about it by-an'-by, Jerry. But wait. Were all the packages on me, all right?'

'I reckon!' responded Captain Jerry. 'There was six of 'em tied on to you. I reckon they're worth the three hundred an' fifty thousand all right!'

'Well, let's get away from this place quick as we can get steam up again!' said Jan. 'There's more swag down there, I guess—lots of it. But I wouldn't go down again, nor send another man down, for all the millions we've all of us ever heard tell of. Mr McWha, how soon can we be moving?'

'Ten meenutes, more or less!' replied the Scotsman.

'All right! When we're outside of this accursed bay, an' round the "Camel" yonder, I'll tell you what it's like down there under that shiny green.'

False Colours

≈≈≈

W. W. JACOBS

'Of course, there is a deal of bullying done at sea at times,' said the night-watchman thoughtfully.

'The men call it bullying an' the officers call it discipline, but it's the same thing under another name. Still it's fair in a way. It gets passed on from one to another. Everybody aboard a'most has got somebody to bully, except, perhaps, the boy; he 'as the worst of it, unless he can manage to get the ship's cat by itself occasionally.

'I don't think sailormen mind being bullied. I never 'eard of its putting one off 'is feed yet, and that's the main thing, arter all's said and done.

'Fust officers are often worse than skippers. In the fust place, they know they ain't skippers, an' that alone is enough to put 'em in a bad temper, especially if they've 'ad their certifikit a good many years and can't get a vacancy.

'I remember, a good many years ago now, I was lying at Calcutta one time in the *Peewit*, as fine a barque as you'd wish to see, an' we 'ad a fust mate there as was a disgrace to 'is sects. A nasty, bullying, violent man, who used to call the hands names as they didn't know the meanings of and what was no use looking in the dictionary for.

'There was one chap aboard, Bill Cousins, as he used to make a partickler mark of. Bill 'ad the misfortin to 'ave red 'air, and the way the mate used to throw that in 'is face was disgraceful. Fortunately for us all, the skipper was a very decent sort of man, so that the mate was only at 'is worst when he wasn't by.

'We was sitting in the fo'c'sle at tea one arternoon, when Bill Cousins came down, an' we see at once 'e'd 'ad a turn with the mate. He sat all by hisself for some time simmering, an' then he broke out, "One o' these days I'll swing for 'im; mark my words."

' "Don't be a fool, Bill," ses Joe Smith.

' "If I could on'y mark 'im," ses Bill, catching his breath. "Just mark 'im fair an' square. If I could on'y 'ave 'im alone for ten minutes, with nobody standing by to see fair play. But, o' course, if I 'it 'im it's mutiny."

' "You couldn't do it if it wasn't, Bill," ses Joe Smith again.

' "He walks about the town as though the place belongs to 'im," said Ted Hill. "Most of us is satisfied to shove the niggers out o' the way, but he ups fist and 'its 'em if they comes within a yard of 'im."

' "Why don't they 'it 'im back?" ses Bill. "I would if I was them."

'Joe Smith grunted. "Well, why don't you?" he asked.

' "Cos I ain't a nigger," ses Bill.

' "Well, but you might be," ses Joe, very earnest. "Black your face an' 'ands an' legs, and dress up in them cotton things, and go ashore and get in 'is way."

' "If you will, I will, Bill," ses a chap called Bob Pullin.

'Well, they talked it over and over, and at last Joe, who seemed to take a great interest in it, went ashore and got the duds for 'em. They was a tight fit for Bill, Hindoos not being as wide as they might be, but Joe said if 'e didn't bend about he'd be all right, and Pullin, who was a smaller man, said his was fust class.

'After they were dressed, the next question was wot to use to colour them with; coal was too scratchy, an' ink Bill didn't like. Then Ted Hill burnt a cork and started on Bill's nose with it afore it was cool, an' Bill didn't like that.

' "Look 'ere," ses the carpenter, "nothin' seems to please you, Bill— it's my opinion you're backing out of it."

' "You're a liar," ses Bill.

' "Well, I've got some stuff in a can as might be boiled-down Hindoo for all you could tell the difference," ses the carpenter; 'and if you'll keep that ugly mouth of yours shut, I'll paint you myself.'

'Well, Bill was a bit flattered, the carpenter being a very superior sort of a man, and quite an artist in 'is way, an' Bill sat down an' let 'im do 'im with some stuff out of a can that made 'im look like a Hindoo what 'ad been polished. Then Bob Pullin was done too, an' when they'd got their turbins on, the change in their appearance was wonderful.

' "Feels a bit stiff," ses Bill, working 'is mouth.

' "That'll wear off," ses the carpenter; "it wouldn't be you if you didn't 'ave a grumble, Bill."

' ' "And mind and don't spare 'im, Bill," ses Joe. "There's two of you, an' if you only do wot's expected of you, the mate ought to 'ave a easy time abed this v'y'ge."

' "Let the mate start fust," ses Ted Hill. "He's sure to start on you if you only get in 'is way. Lord, I'd like to see his face when you start on *'im!*"

'Well, the two of 'em went ashore arter dark with the best wishes o' all on board, an' the rest of us sat down in the fo'c'sle spekerlating as to what sort o' time the mate was goin' to 'ave. He went ashore all right, because Ted Hill see 'im go, an' he noticed with partickler pleasure as 'ow he was dressed very careful.

'It must ha' been near eleven o'clock. I was sitting with Smith on the port side o' the galley, when we heard a 'ubbub approaching the ship. It was the mate just coming aboard. He was without 'is 'at; 'is necktie was twisted round 'is ear, and 'is shirt and 'is collar was all torn to shreds. The second and third officers ran up to him to see what was the matter, and while he was telling them, up comes the skipper.

' "You don't mean to tell me, Mr Fingall," ses the skipper in surprise, "that you've been knocked about like that by them mild and meek Hindoos?"

' "Hindoos, sir?" roared the mate. "Cert'nly not, sir. I've been assaulted like this by five German sailor-men. And I licked 'em all."

' "I'm glad to hear that," ses the skipper; and the second and third pats the mate on the back—just like you pat a dog you don't know.

' "Big fellows they was," ses he, "an' they give me some trouble. Look at my eye!"

'The second officer struck a match and looked at it, and it cert'nly was a beauty.

' "I hope you reported this at the police-station?" ses the skipper.

' "No, sir," ses the mate, holding up 'is 'ead. "I don't want no p'lice to protect me. Five's a large number, but I drove 'em off, and I don't think they'll meddle with any British fust-officers again."

' "You'd better turn in," ses the second, leading him off by the arm.

'The mate limped off with him, and as soon as the coast was clear we put our 'eads together and tried to make out how it was that Bill Cousins and Bob 'ad changed themselves into five German sailor-men.

' "It's the mate's pride," ses the carpenter. "He didn't like being knocked about by Hindoos."

'We thought it was that, but we had to wait nearly another hour afore the two came aboard, to make sure. There was a difference in the way they came aboard, too, from that of the mate. *They* didn't make no noise, and the fust thing we knew of their coming aboard was seeing a bare black foot waving feebly at the top of the fo'c'sle ladder feelin' for the step below.

'That was Bob. He came down without a word, and then we see 'e was holding another black foot and guiding it to where it should go. That was Bill, an' of all the 'orrid limp-looking blacks that you ever see, Bill was the worst when he got below. He just sat on a locker all of a heap and held 'is 'ead, which was swollen up, in 'is hands. Bob went and sat beside 'im, and there they sat, for all the world like two wax figgers instead o' human beings.

' "Well, you done it, Bill," ses Joe, after waiting a long time for them to speak. "Tell us all about it."

' "Nothin' to tell", ses Bill, very surly. "We knocked 'im about."

' "And he knocked us about," ses Bob, with a groan. "I'm sore all over, and as for my feet——"

' "Wot's the matter with them?" ses Joe.

' "Trod on," ses Bob, very short. "If my bare feet was trod on once they was a dozen times. I've never 'ad such a doing in all my life. He fought like a devil. I thought he'd ha' murdered Bill."

' "I wish 'e 'ad," ses Bill, with a groan; "my face is bruised and cut about cruel. I can't bear to touch it."

' "Do you mean to say the two of you couldn't settle 'im?" ses Joe, staring.

' "I mean to say we got a hiding," ses Bill. "We got close to him fust start off and got our feet trod on. Arter that it was like fighting a windmill, with sledge-hammers for sails."

'He gave a groan and turned over in his bunk, and when we asked him some more about it, he swore at us. They both seemed quite done up, and at last they dropped off to sleep just as they was, without even stopping to wash the black off or to undress themselves.

'I was awoke rather early in the morning by the sounds of somebody talking to themselves, and a little splashing of water. It seemed to go on a long while, and at last I leaned out of my bunk and see Bill bending over a bucket and washing himself and using bad langwidge.

' "Wot's the matter, Bill?" ses Joe, yawning and sitting up in bed.

' "My skin's that tender, I can hardly touch it," ses Bill, bending down and rinsing 'is face. "Is it all orf?"

' "Orf?" ses Joe; "no, o' course it ain't. Why don't you use some soap?"

' "*Soap*," answers Bill, mad-like; "why, I've used more soap than I've used for six months in the ordinary way."

' "That's no good," ses Joe; "give yourself a good wash."

'Bill put down the soap then very careful, and went over to 'im and told him all the dreadful things he'd do to him when he got strong agin, and then Bob Pullin got out of his bunk an' 'ad a try on *his* face. Him an' Bill kept washing, and then taking each other to the light and trying to believe it was coming off until they got sick of it, and then Bill, 'e up with his foot and capsized the bucket, and walked up and down the fo'c'sle raving.

' "Well, the carpenter put it on," ses a voice, "make 'im take it orf."

'You wouldn't believe the job we 'ad to wake that man up. He wasn't fairly woke till he was hauled out of 'is bunk an' set down opposite them two pore black fellers an' told to make 'em white again.

' "I don't believe as there's anything will touch it," he says at last. "I forgot all about that."

' "Do you mean to say," bawls Bill, "that we've got to be black all the rest of our life?"

' "Cert'nly not," ses the carpenter indignantly, "it'll wear off in time; shaving every morning 'll 'elp it, I should say."

' "I'll get my razor now," ses Bill in a awful voice; "don't let 'im go, Bob. I'll 'ack 'is head orf."

'He actually went off an' got his razor, but, o' course, we jumped out of our bunks and got between 'em and told him plainly that it was not to be, and then we set 'em down and tried everything we could think of, from butter and linseed oil to cold tea leaves used as a poultice, and all it did was to make 'em shinier an' shinier.

' "It's no good, I tell you," ses the carpenter, "it's the most lasting black I know. If I told you how much that stuff is a can, you wouldn't believe me."

' "Well, you're in it," ses Bill, his voice all of a tremble; "you done it so as we could knock the mate about. Whatever's done to us 'll be done to you too."

' "I don't think turps 'll touch it," ses the carpenter, getting up, "but we'll 'ave a try."

'He went and fetched the can and poured some out on a bit o' rag and told Bill to dab his face with it. Bill give a dab, and the next moment he rushed over with a scream and buried his head in a shirt wot Simmons was wearing at the time and began to wipe his face with it. Then he left the flustered Simmons an' shoved another chap away from the bucket, and buried his face in it and kicked and carried on like a madman. Then 'e jumped into his bunk again and buried 'is face in the clothes and rocked hisself and moaned as if he was dying.

' "Don't you use it, Bob," he ses at last.

' "'Tain't likely," ses Bob. "It's a good thing you tried it fust, Bill."

' "'Ave they tried holy-stone?" ses a voice from a bunk.

' "No, they ain't," ses Bob snappishly, "and what's more, they ain't goin' to."

'Both o' their tempers was so bad that we let the subject drop while we was at breakfast. The orkard persition of affairs could no longer be disregarded. Fust one chap threw out a 'int and then another, gradually getting a little stronger and stronger, until Bill turned round in a uncomfortable way and requested of us to leave off talking with our mouths full and speak up like Englishmen wot we meant.

' "You see, it's this way, Bill," ses Joe, soft-like. "As soon as the mate sees you there'll be trouble for all of us."

' "For all of us," repeats Bill, nodding.

' "Whereas," ses Joe, looking round for support, "if we gets up a little collection for you and you should find it convenient to desart——"

' "'Ear, 'ear," ses a lot o' voices. "Bravo, Joe."

' "Oh, desart is it?" ses Bill; "an' where are we goin' to desart to?"

' "Well, that we leave to you," ses Joe; "there's many a ship short-'anded as would be glad to pick up sich a couple of prime sailormen as you an' Bob."

' "Ah, an' wot about our black faces?" ses Bill, still in the same sneering, ungrateful sort o' voice.

' "That can be got over," ses Joe.

' "'Ow?" ses Bill and Bob together.

' "Ship as nigger cooks," ses Joe, slapping his knee and looking round triumphant.

'It's no good trying to do some people a kindness. Joe was perfectly sincere, and nobody could say but wot it wasn't a good idea, but o' course, Mr Bill Cousins must consider hisself insulted, and I can only suppose that the trouble he'd gone through 'ad affected his brain. Likewise Bob Pullin's. Anyway, that's the only excuse I can make for 'em. To cut a long story short, nobody 'ad any more breakfast, and no time to do anything until them two men was scrouged up in a corner an' 'eld there unable to move.

' "I'd never 'ave done 'em," ses the carpenter, arter it was all over, "if I'd known they was goin' to carry on like this. They wanted to be done."

' "The mate 'll half murder 'em," ses Ted Hill.

' "He'll 'ave 'em sent to jail, that's wot he'll do," ses Smith. "It's a serious matter to go ashore and commit assault and battery on the mate."

' "You're all in it," ses the voice o' Bill from the floor. "I'm going to make a clean breast of it. Joe Smith put us up to it, the carpenter blacked us, and the others encouraged us."

' "Joe got the clothes for us," ses Bob. "I know the place he got 'em from, too."

'The ingratitude o' these two men was sich that at first we decided to have no more to do with them, but better feelings prevailed, and we held a sort o' meeting to consider what was best to be done. An' everything that was suggested one o' them two voices from the floor found fault with and wouldn't 'ave, and at last we 'ad to go up on deck with nothing decided upon, except to swear 'ard and fast as we knew nothing about it.

' "The only advice we can give you," ses Joe, looking back at 'em, "is to stay down 'ere as long as you can."

'A'most the fust person we see on deck was the mate, an' a pretty sight he was. He'd got a bandage round 'is left eye, and a black ring round the other. His nose was swelled and his lip cut, but the other officers were making sich a fuss over 'im, that I think he rather gloried in it than otherwise.

' "Where's them other two 'ands?" he ses by-and-by, glaring out of 'is black eye.

' "Down below, sir, I b'lieve," ses the carpenter, all of a tremble.

' "Go an' send 'em up," ses the mate to Smith.

' "Yessir," ses Joe, without moving.

' "Well, go on then," roars the mate.

' "They ain't over and above well, sir, this morning," ses Joe.

' "Send 'em up, confound you," ses the mate, limping towards 'im.

'Well, Joe give 'is shoulders a 'elpless sort o' shrug and walked forward and bawled down the fo'c'sle.

' "They're coming, sir," he ses, walking back to the mate just as the skipper came out of 'is cabin.

'We all went on with our work as 'ard as we knew 'ow. The skipper was talking to the mate about 'is injuries, and saying unkind things about Germans, when he give a sort of a shout and staggered back staring. We just looked round, and there was them two blackamoors coming slowly towards us.

' "Good heavens, Mr Fingall," ses the old man. "What's this?"

'I never see sich a look on any man's face as I saw on the mate's then. Three times 'e opened 'is mouth to speak, and shut it agin without saying anything. The veins on 'is forehead swelled up tremendous and 'is cheeks was all blown out purple.

' "That's Bill Cousins' hair," ses the skipper to himself. "It's Bill Cousins' hair. It's Bill Cous——"

'Bob walked up to him, and Bill lagging a little way behind, and then he stops just in front of 'im and fetches up a sort o' little smile.

' "Don't you make those faces at me, sir," roars the skipper. "What do you mean by it? What have you been doing to yourselves?"

' "Nothin', sir," ses Bill 'umbly; "it was done to us."

'The carpenter, who was just going to cooper up a cask which had started a bit, shook like a leaf and gave Bill a look that would ha' melted a stone.

' "Who did it?" ses the skipper.

' "We've been the victims of a cruel outrage, sir," ses Bill, doing all 'e could to avoid the mate's eye, which wouldn't be avoided.

' "So I should think," ses the skipper. "You've been knocked about, too."

' "Yessir," ses Bill, very respectful; "me and Bob was ashore last night, sir, just for a quiet look round, when we was set on to by five furriners."

' "*What?*" ses the skipper; and I won't repeat what the mate said.

' "We fought 'em as long as we could, sir," ses Bill, "then we was both knocked senseless, and when we came to ourselves we was messed up like this 'ere."

' "What sort o' men were they?" asked the skipper, getting excited.

' "Sailormen, sir," ses Bob, putting in his spoke. "Dutchies or Germans, or something o' that sort."

' "Was there one tall man, with a fair beard?" ses the skipper, getting more and more excited.

' "Yessir," ses Bill, in a surprised sort o' voice.

' "Same gang", ses the skipper. "Same gang as knocked Mr Fingall about, you may depend upon it. Mr Fingall, it's a mercy for you you didn't get your face blacked too."

'I thought the mate would ha' burst. I can't understand how any man could swell as he swelled without bursting.

' "I don't believe a word of it," he ses at last.

' "Why not?" ses the skipper sharply.

' "Well, I don't," ses the mate, his voice trembling with passion. "I 'ave my reasons."

' "I s'pose you don't think these two poor fellows went and blacked themselves for fun, do you?" ses the skipper.

'The mate couldn't answer.

' "And then went and knocked themselves about for more fun?" ses the skipper, very sarcastic.

'The mate didn't answer. He looked round helpless like, and see the third officer swopping glances with the second, and all the men looking sly and amused, and I think if ever a man saw 'e was done 'e did at that moment.

'He turned away and went below, and the skipper, arter reading us all a little lecture on getting into fights without reason, sent the two chaps below agin and told 'em to turn in and rest. He was so good to 'em all the way 'ome, and took sich a interest in seeing 'em change from black to brown, and from light brown to spotted lemon, that the mate daren't do nothing to them, but gave us their share of what he owed them as well as an extra dose of our own.'

The Secret Sharer

~~~

JOSEPH CONRAD

I

On my right hand there were lines of fishing-stakes resembling a mysterious system of half-submerged bamboo fences, incomprehensible in its division of the domain of tropical fishes, and crazy of aspect as if abandoned for ever by some nomad tribe of fishermen now gone to the other end of the ocean; for there was no sign of human habitation as far as the eye could reach. To the left a group of barren islets, suggesting ruins of stone walls, towers, and block-houses, had its foundations set in a blue sea that itself looked solid, so still and stable did it lie below my feet; even the track of light from the westering sun shone smoothly, without that animated glitter which tells of an imperceptible ripple. And when I turned my head to take a parting glance at the tug which had just left us anchored outside the bar, I saw the straight line of the flat shore joined to the stable sea, edge to edge, with a perfect and unmarked closeness, in one levelled floor half brown, half blue under the enormous dome of the sky. Corresponding in their insignificance to the islets of the sea, two small clumps of trees, one on each side of the only fault in the impeccable joint, marked the mouth of the river Meinam we had just left on the first preparatory stage of our homeward journey; and, far back on the inland level, a larger and loftier mass, the grove surrounding the great Paknam pagoda, was the only thing on which the eye could rest from the vain task of exploring the monotonous sweep of the horizon. Here and there gleams as of a few scattered pieces of silver marked the windings of the great river; and on the nearest of them, just within the bar, the tug steaming right into the land became lost to my sight, hull and funnel and masts, as though the impassive earth had swallowed her up without an effort, without a tremor. My eye followed the light cloud of her smoke, now here, now there, above the plain, according to the devious curves of the stream, but always fainter and further away, till I lost it at last behind the mitre-shaped hill of the great pagoda. And then I was left alone with my ship, anchored at the head of the Gulf of Siam.

She floated at the starting-point of a long journey, very still in an immense stillness, the shadows of her spars flung far to the eastward by the setting sun. At that moment I was alone on her decks. There was not a sound in her—and around us nothing moved, nothing lived, not a canoe on the water, not a bird in the air, not a cloud in the sky. In this breathless pause at the threshold of a long passage we seemed to be measuring our fitness for a long and arduous enterprise, the appointed task of both our existences to be carried out, far from all human eyes, with only sky and sea for spectators and for judges.

There must have been some glare in the air to interfere with one's sight, because it was only just before the sun left us that my roaming eyes made out beyond the highest ridge of the principal islet of the group something which did away with the solemnity of perfect solitude. The tide of darkness flowed on swiftly; and with tropical suddenness a swarm of stars came out above the shadowy earth, while I lingered yet, my hand resting lightly on my ship's rail as if on the shoulder of a trusted friend. But, with all that multitude of celestial bodies staring down at one, the comfort of quiet communion with her was gone for good. And there were also disturbing sounds by this time—voices, footsteps forward; the steward flitted along the main-deck, a busily ministering spirit; a hand-bell tinkled urgently under the poop-deck. . . .

I found my two officers waiting for me near the supper table, in the lighted cuddy. We sat down at once, and as I helped the chief mate, I said:

'Are you aware that there is a ship anchored inside the islands? I saw her mastheads above the ridge as the sun went down.'

He raised sharply his simple face, overcharged by a terrible growth of whisker, and emitted his usual ejaculations: 'Bless my soul, sir! You don't say so!'

My second mate was a round-cheeked, silent young man, grave beyond his years, I thought; but as our eyes happened to meet I detected a slight quiver on his lips. I looked down at once. It was not my part to encourage sneering on board my ship. It must be said, too, that I knew very little of my officers. In consequence of certain events of no particular significance, except to myself, I had been appointed to the command only a fortnight before. Neither did I know much of the hands forward. All these people had been together for eighteen months or so, and my position was that of the only stranger on board. I mention this because it has some bearing on what is to follow. But what I felt most was my being a stranger to the ship; and if all the truth must be told, I was somewhat of a stranger to myself. The youngest man on board (barring the second mate), and untried as yet by a position of the fullest responsibility, I was willing to take the adequacy of the others for granted. They had simply to be equal to their tasks; but I wondered how far I should turn out faithful to that ideal conception of one's own personality every man sets up for himself secretly.

Meantime the chief mate, with an almost visible effect of collaboration on the part of his round eyes and frightful whiskers, was trying to evolve a theory of the anchored ship. His dominant trait was to take all things into earnest consideration. He was of a painstaking turn of mind. As he used to say, he 'liked to account to himself' for practically everything that came in his way, down to a miserable scorpion he had found in his cabin a week before. The why and the wherefore of that scorpion—how it got on board and came to select his room rather than the pantry (which was a dark place and more what a scorpion would be partial to), and how on earth it managed to drown itself in the inkwell of his writing-desk—had exercised him infinitely. The ship within the islands was much more easily accounted for; and just as we were about to rise from table he made his pronouncement. She was, he doubted not, a ship from home lately arrived. Probably she drew too much water to cross the bar except at the top of spring tides. Therefore she went into that natural harbour to wait for a few days in preference to remaining in an open roadstead.

'That's so,' confirmed the second mate, suddenly, in his slightly hoarse voice. 'She draws over twenty feet. She's the Liverpool ship *Sephora* with a cargo of coal. Hundred and twenty-three days from Cardiff.'

We looked at him in surprise.

'The tugboat skipper told me when he came on board for your letters, sir,' explained the young man. 'He expects to take her up the river the day after tomorrow.'

After thus overwhelming us with the extent of his information he slipped out of the cabin. The mate observed regretfully that he 'could not account for that young fellow's whims'. What prevented him telling us all about it at once, he wanted to know.

I detained him as he was making a move. For the last two days the crew had had plenty of hard work, and the night before they had very little sleep. I felt painfully that I—a stranger—was doing something unusual when I directed him to let all hands turn in without setting an anchor-watch. I proposed to keep on deck myself till one o'clock or thereabouts. I would get the second mate to relieve me at that hour.

'He will turn out the cook and the steward at four', I concluded, 'and then give you a call. Of course at the slightest sign of any sort of wind we'll have the hands up and make a start at once.'

He concealed his astonishment. 'Very well, sir.' Outside the cuddy he put his head in the second mate's door to inform him of my unheard-of caprice to take a five hours' anchor-watch on myself. I heard the other raise his voice incredulously—'What? The Captain himself?' Then a few more murmurs, a door closed, then another. A few moments later I went on deck.

My strangeness, which had made me sleepless, had prompted that unconventional arrangement, as if I had expected in those solitary hours of the night to get on terms with the ship of which I knew nothing, manned by men of whom I knew very little more. Fast alongside a wharf, littered like any ship in port with a tangle of unrelated things, invaded by unrelated shore people, I had hardly seen her yet properly. Now, as she lay cleared for sea, the stretch of her main-deck seemed to me very fine under the stars. Very fine, very roomy for her size, and very inviting. I descended the poop and paced the waist, my mind picturing to myself the coming passage through the Malay Archipelago, down the Indian Ocean, and up the Atlantic. All its phases were familiar enough to me, every characteristic, all the alternatives which were likely to face me on the high seas—everything! . . . except the novel responsibility of command. But I took heart from the reasonable thought that the ship was like other ships, the men like other men, and that the sea was not likely to keep any special surprises expressly for my discomfiture.

Arrived at that comforting conclusion, I bethought myself of a cigar and went below to get it. All was still down there. Everybody at the after end of the ship was sleeping profoundly. I came out again on the quarter-deck, agreeably at ease in my sleeping-suit on that warm breathless night, barefooted, a glowing cigar in my teeth, and, going forward, I was met by the profound silence of the fore end of the ship. Only as I passed the door of the forecastle I heard a deep, quiet, trustful sigh of some sleeper inside. And suddenly I rejoiced in the great security of the sea as compared with the unrest of the land, in my choice of that untempted life presenting no disquieting problems, invested with an elementary moral beauty by the absolute straightforwardness of its appeal and by the singleness of its purpose.

The riding-light in the fore-rigging burned with a clear, untroubled, as if symbolic, flame, confident and bright in the mysterious shades of the night. Passing on my way aft along the other side of the ship, I observed that the rope side-ladder, put over, no doubt, for the master of the tug when he came to fetch away our letters, had not been hauled in as it should have been. I became annoyed at this, for exactitude in small matters is the very soul of discipline. Then I reflected that I had myself peremptorily dismissed my officers from duty, and by my own act had prevented the anchor-watch being formally set and things properly attended to. I asked myself whether it was wise ever to interfere with the established routine of duties even from the kindest of motives. My action might have made me appear eccentric. Goodness only knew how that absurdly whiskered mate would 'account' for my conduct, and what the whole ship thought of that informality of their new captain. I was vexed with myself.

Not from compunction certainly, but, as it were mechanically, I proceeded to get the ladder in myself. Now a side-ladder of that sort is a light affair and comes in easily, yet my vigorous tug, which should have brought it flying on board, merely recoiled upon my body in a totally unexpected jerk. What the devil! . . . I was so astounded by the immovableness of that ladder that I remained stock-still, trying to account for it to myself like that imbecile mate of mine. In the end, of course, I put my head over the rail.

The side of the ship made an opaque belt of shadow on the darkling glassy shimmer of the sea. But I saw at once something elongated and pale floating very close to the ladder. Before I could form a guess a faint flash of phosphorescent light, which seemed to issue suddenly from the naked body of a man, flickered in the sleeping water with the elusive, silent play of summer lightning in a night sky. With a gasp I saw revealed to my stare a pair of feet, the long legs, a broad livid back immersed right up to the neck in a greenish cadaverous glow. One hand, awash, clutched the bottom rung of the ladder. He was complete but for the head. A headless corpse! The cigar dropped out of my gaping mouth with a tiny plop and a short hiss quite audible in the absolute stillness of all things under heaven. At that I suppose he raised up his face, a dimly pale oval in the shadow of the ship's side. But even then I could only barely make out down there the shape of his black-haired head. However, it was enough for the horrid, frost-bound sensation which had gripped me about the chest to pass off. The moment of vain exclamations was past, too. I only climbed on the spare spar and leaned over the rail as far as I could, to bring my eyes nearer to that mystery floating alongside.

As he hung by the ladder, like a resting swimmer, the sea-lightning played about his limbs at every stir; and he appeared in it ghastly, silvery, fish-like. He remained as mute as a fish, too. He made no motion to get out of the water, either. It was inconceivable that he should not attempt to come on board, and strangely troubling to suspect that perhaps he did not want to. And my first words were prompted by just that troubled incertitude.

'What's the matter?' I asked in my ordinary tone, speaking down to the face upturned exactly under mine.

'Cramp,' it answered, no louder. Then slightly anxious, 'I say, no need to call anyone.'

'I was not going to,' I said.

'Are you alone on deck?'

'Yes.'

I had somehow the impression that he was on the point of letting go the ladder to swim away beyond my ken—mysterious as he came. But, for the moment, this being appearing as if he had risen from the bottom of the sea (it was certainly the nearest land to the ship) wanted only to know the time. I told him. And he, down there, tentatively:

'I suppose your captain's turned in?'

'I am sure he isn't,' I said.

He seemed to struggle with himself, for I heard something like the low, bitter murmur of doubt. 'What's the good?' His next words came out with a hesitating effort.

'Look here, my man. Could you call him out quietly?'

I thought the time had come to declare myself.

'*I* am the captain.'

I heard a 'By Jove!' whispered at the level of the water. The phosphorescence flashed in the swirl of the water all about his limbs, his other hand seized the ladder.

'My name's Leggatt.'

The voice was calm and resolute. A good voice. The self-possession of that man had somehow induced a corresponding state in myself. It was very quietly that I remarked:

'You must be a good swimmer.'

'Yes. I've been in the water practically since nine o'clock. The question for me now is whether I am to let go this ladder and go on swimming till I sink from exhaustion, or—to come on board here.'

I felt this was no mere formula of desperate speech, but a real alternative in the view of a strong soul. I should have gathered from this that he was young; indeed, it is only the young who are ever confronted by such clear issues. But at the time it was pure intuition on my part. A mysterious communication was established already between us two—in the face of that silent, darkened tropical sea. I was young, too; young enough to make no comment. The man in the water began suddenly to climb up the ladder, and I hastened away from the rail to fetch some clothes.

Before entering the cabin I stood still, listening in the lobby at the foot of the stairs. A faint snore came through the closed door of the chief mate's room. The second mate's door was on the hook, but the darkness in there was absolutely soundless. He, too, was young and could sleep like a stone. Remained the steward, but he was not likely to wake up before he was called. I got a sleeping-suit out of my room and, coming back on deck, saw the naked man from the sea sitting on the main-hatch, glimmering white in the darkness, his elbows on his knees and his head in his hands. In a moment he had concealed his damp body in a sleeping-suit of the same grey-stripe pattern as the one I was wearing and followed me like my double on the poop. Together we moved right aft, barefooted, silent.

'What is it?' I asked in a deadened voice, taking the lighted lamp out of the binnacle, and raising it to his face.

'An ugly business.'

He had rather regular features; a good mouth; light eyes under some-what heavy, dark eyebrows; a smooth, square forehead; no growth on his cheeks; a small, brown moustache, and a well-shaped, round chin. His expression was concentrated, meditative, under the inspecting light of the lamp I held up to his face; such as a man thinking hard in solitude might wear. My sleeping-suit was just right for his size. A well-knit young fellow of twenty-five at most. He caught his lower lip with the edge of white, even teeth.

'Yes,' I said, replacing the lamp in the binnacle. The warm, heavy tropical night closed upon his head again.

'There's a ship over there,' he murmured.

'Yes, I know. The *Sephora*. Did you know of us?'

'Hadn't the slightest idea. I am the mate of her——' He paused and corrected himself. 'I should say I *was*.'

'Aha! Something wrong?'

'Yes. Very wrong indeed. I've killed a man.'

'What do you mean? Just now?'

'No, on the passage. Weeks ago. Thirty-nine south. When I say a man——'

'Fit of temper,' I suggested, confidently.

The shadowy, dark head, like mine, seemed to nod imperceptibly above the ghostly grey of my sleeping-suit. It was, in the night, as though I had been faced by my own reflection in the depths of a sombre and immense mirror.

'A pretty thing to have to own up to for a Conway boy', murmured my double, distinctly.

'You're a Conway boy?'

'I am,' he said, as if startled. Then, slowly . . . 'Perhaps you too——'

It was so; but being a couple of years older I had left before he joined. After a quick interchange of dates a silence fell; and I thought suddenly of my absurd mate with his terrific whiskers and the 'Bless my soul—you don't say so' type of intellect. My double gave me an inkling of his thoughts by saying: 'My father's a parson in Norfolk. Do you see me before a judge and jury on that charge? For myself I can't see the necessity. There are fellows that an angel from heaven ——And I am not that. He was one of those creatures that are just simmering all the time with a silly sort of wickedness. Miserable devils that have no business to live at all. He wouldn't do his duty and wouldn't let anybody else do theirs. But what's the good of talking! You know well enough the sort of ill-conditioned snarling cur——'

He appealed to me as if our experiences had been as identical as our clothes. And I knew well enough the pestiferous danger of such a character where there are no means of legal repression. And I knew well enough also that my double there was no homicidal ruffian. I did not think of asking him for details, and he told me the story roughly in brusque, disconnected sentences. I needed no more. I saw it all going on as though I were myself inside that other sleeping-suit.

'It happened while we were setting a reefed foresail, at dusk. Reefed foresail! You understand the sort of weather. The only sail we had left to keep the ship running; so you may guess what it had been like for days. Anxious sort of job, that. He gave me some of his cursed insolence at the sheet. I tell you I was overdone with this terrific weather that seemed to have no end to it. Terrific, I tell you—and a deep ship. I believe the fellow himself was half crazed with funk. It was no time for gentlemanly reproof, so I turned round and felled him like an ox. He up and at me. We closed just as an awful sea made for the ship. All hands saw it coming and took to the rigging, but I had him by the throat, and went on shaking him like a rat, the men above us yelling, "Look out! look out!" Then a crash as if the sky had fallen on my head. They say that for over ten minutes hardly anything was to be seen of the ship—just the three masts and a bit of the forecastle head and of the poop all awash driving along in a smother of foam. It was a miracle that they found us, jammed together behind the forebits. It's clear that I meant business, because I was holding him by the throat still when they picked us up. He was black in the face. It was too much for them. It seems they rushed us aft together, gripped as we were, screaming "Murder!" like a lot of lunatics, and broke into the cuddy. And the ship running for her life, touch and go all the time, any minute her last in a sea fit to turn your hair grey only a-looking at it. I understand that the skipper, too, started raving like the rest of them. The man had been deprived of sleep for more than a week, and to have this sprung on him at the height of a furious gale nearly drove him out of his mind. I wonder they didn't fling me overboard after getting the carcass of their precious ship-mate out of my fingers. They had rather a job to separate us, I've been told. A sufficiently fierce story to make an old judge and a respectable jury sit up a bit. The first thing I heard when I came to myself was the maddening howling of that endless gale, and on that the voice of the old man. He was hanging on to my bunk, staring into my face out of his sou'wester.

' "Mr Leggatt, you have killed a man. You can act no longer as chief mate of this ship." '

His care to subdue his voice made it sound monotonous. He rested a hand on the end of the skylight to steady himself with, and all that time did not stir a limb, so far as I could see. 'Nice little tale for a quiet tea-party,' he concluded in the same tone.

One of my hands, too, rested on the end of the skylight; neither did I stir a limb, so far as I knew. We stood less than a foot from each other. It occurred to me that if old 'Bless my soul—you don't say so' were to put his head up the companion and catch sight of us, he would think he was seeing double, or imagine himself come upon a scene of weird witchcraft; the strange captain having a quiet confabulation by the wheel with his own grey ghost. I became very much concerned to prevent anything of the sort. I heard the other's soothing undertone.

'My father's a parson in Norfolk,' it said. Evidently he had forgotten he had told me this important fact before. Truly a nice little tale.

'You had better slip down into my stateroom now,' I said, moving off stealthily. My double followed my movements; our bare feet made no sound; I let him in, closed the door with care, and, after giving a call to the second mate, returned on deck for my relief.

'Not much sign of any wind yet,' I remarked when he approached.

'No, sir. Not much,' he assented, sleepily, in his hoarse voice, with just enough deference, no more, and barely suppressing a yawn.

'Well, that's all you have to look out for. You have got your orders.'

'Yes, sir.'

I paced a turn or two on the poop and saw him take up his position face forward with his elbow in the ratlines of the mizzen-rigging before I went below. The mate's faint snoring was still going on peacefully. The cuddy lamp was burning over the table on which stood a vase with flowers, a polite attention from the ship's provision merchant—the last flowers we should see for the next three months at the very least. Two bunches of bananas hung from the beam symmetrically, one on each side of the rudder-casing. Everything was as before in the ship—except that two of her captain's sleeping-suits were simultaneously in use, one motionless in the cuddy, the other keeping very still in the captain's stateroom.

It must be explained here that my cabin had the form of the capital letter L, the door being within the angle and opening into the short part of the letter. A couch was to the left, the bed-place to the right; my writing-desk and the chronometers' table faced the door. But anyone opening it, unless he stepped right inside, had no view of what I call the long (or vertical) part of the letter. It contained some lockers surmounted by a bookcase; and a few clothes, a thick jacket or two, caps, oilskin coat, and such like, hung on hooks. There was at the bottom of that part a door opening into my bathroom, which could be entered also directly from the saloon. But that way was never used.

The mysterious arrival had discovered the advantage of this particular shape. Entering my room, lighted strongly by a big bulkhead lamp swung on gimbals above my writing-desk, I did not see him anywhere till he stepped out quietly from behind the coats hung in the recessed part.

'I heard somebody moving about, and went in there at once,' he whispered.

I, too, spoke under my breath.

'Nobody is likely to come in here without knocking and getting permission.'

He nodded. His face was thin and the sunburn faded, as though he had been ill. And no wonder. He had been, I heard presently, kept under arrest in his cabin for nearly seven weeks. But there was nothing sickly in his eyes or in his expression. He was not a bit like me, really; yet, as we stood leaning over my bed-place, whispering side by side, with our dark heads together and our backs to the door, anybody bold enough to open it stealthily would have been treated to the uncanny sight of a double captain busy talking in whispers with his other self.

'But all this doesn't tell me how you came to hang on to our side-ladder,' I enquired, in the hardly audible murmurs we used, after he had told me something more of the proceedings on board the *Sephora* once the bad weather was over.

'When we sighted Java Head I had had time to think all those matters out several times over. I had six weeks of doing nothing else, and with only an hour or so every evening for a tramp on the quarter-deck.'

He whispered, his arms folded on the side of my bed-place, staring through the open port. And I could imagine perfectly the manner of this thinking out—a stubborn if not a steadfast operation; something of which I should have been perfectly incapable.

'I reckoned it would be dark before we closed with the land,' he continued, so low that I had to strain my hearing, near as we were to each other, shoulder touching shoulder almost. 'So I asked to speak to the old man. He always seemed very sick when he came to see me—as if he could not look me in the face. You know, that foresail saved the ship. She was too deep to have run long under bare poles. And it was I that managed to set it for him. Anyway, he came. When I had him in my cabin—he stood by the door looking at me as if I had the halter round my neck already—I asked him right away to leave my cabin door unlocked at night while the ship was going through Sunda Straits. There would be the Java coast within two or three miles, off Angier Point. I wanted nothing more. I've had a prize for swimming my second year in the Conway.'

'I can believe it,' I breathed out.

'God only knows why they locked me in every night. To see some of their faces you'd have thought they were afraid I'd go about at night strangling people. Am I a murdering brute? Do I look it? By Jove! if I had been he wouldn't have trusted himself like that into my room. You'll say I might have chucked him aside and bolted out, there and then—it was dark already. Well, no. And for the same reason I wouldn't think of trying to smash the door. There would have been a rush to stop me at the noise, and I did not mean to get into a confounded scrimmage. Somebody else might have got killed—for I would not have broken out only to get chucked back, and I did not want any more of that work. He refused, looking more sick than ever. He was afraid of the men, and also of that old second mate of his who had been sailing with him for years—a grey-headed old humbug; and his steward, too, had been with him devil knows how long— seventeen years or more—a dogmatic sort of loafer who hated me like poison, just because I was the chief mate. No chief mate ever made more than one voyage in the *Sephora*, you know. Those two old chaps ran the ship. Devil only knows what the skipper wasn't afraid of (all his nerve went to pieces altogether in that hellish spell of bad weather we had)—of what the law would do to him—of his wife, perhaps. Oh, yes! she's on board. Though I don't think she would have meddled. She would have been only too glad to have me out of the ship in any way. The "brand of Cain" business, don't you see. That's all right. I was ready enough to go off wandering on the face of the earth—and that was price enough to pay for an Abel of that sort. Anyhow, he wouldn't listen to me. "This thing must take its course. I represent the law here." He was shaking like a leaf. "So you won't?" "No!" "Then I hope you will be able to sleep on that," I said, and turned my back on him. "I wonder that *you* can," cries he, and locks the door.

'Well, after that, I coudn't. Not very well. That was three weeks ago. We have had a slow passage through the Java Sea; drifted about Carimata for ten days. When we anchored here they thought, I suppose, it was all right. The nearest land (and that's five miles) is the ship's destination; the consul would soon set about catching me; and there would have been no object in bolting to these islets there. I don't suppose there's a drop of water on them. I don't know how it was, but tonight that steward, after bringing me my supper, went out to let me eat it, and left the door unlocked. And I ate it—all there was, too. After I had finished I strolled out on the quarter-deck. I don't know that I meant to do anything. A breath of fresh air was all I wanted, I believe. Then a sudden temptation came over me. I kicked off my slippers and was in the water before I had made up my mind fairly. Somebody heard the splash and they raised an awful hullabaloo. "He's gone! Lower the boats! He's committed suicide! No, he's swimming." Certainly I was swimming. It's not so easy for a swimmer like me to commit suicide by drowning. I landed on the nearest islet before the boat left the ship's side. I heard them pulling about in the dark, hailing, and so on, but after a bit they gave up. Everything quieted down and the anchorage became as still as death. I sat down on a stone and began to think. I felt certain they would start searching for me at daylight. There was no place to hide on those stony things— and if there had been, what would have been the good? But now I was clear of that ship, I was not going back. So after a while I took off all my clothes, tied them up in a bundle with a stone inside, and dropped them in the deep water on the outer side of that islet. That was suicide enough for me. Let them think what they liked, but I didn't mean to drown myself. I meant to swim till I sank—but that's not the same thing. I struck out for another of these little islands, and it was from that one that I first saw your riding-light. Something to swim for. I went on easily, and on the way I came upon a flat rock a foot or two above water. In the daytime, I dare say, you might make it out with a glass from your poop. I scrambled up on it and rested myself for a bit. Then I made another start. That last spell must have been over a mile.'

His whisper was getting fainter and fainter, and all the time he stared straight out through the porthole, in which there was not even a star to be seen. I had not interrupted him. There was something that made comment impossible in his narrative, or perhaps in himself; a sort of feeling, a quality, which I can't find a name for. And when he ceased, all I found was a futile whisper: 'So you swam for our light?'

'Yes—straight for it. It was something to swim for. I couldn't see any stars low down because the coast was in the way, and I couldn't see the land, either. The water was like glass. One might have been swimming in a confounded thousand-feet deep cistern with no place for scrambling out anywhere; but what I didn't like was the notion of swimming round and round like a crazed bullock before I gave out;

and as I didn't mean to go back . . . No. Do you see me being hauled back, stark naked, off one of these little islands by the scruff of the neck and fighting like a wild beast? Somebody would have got killed for certain, and I did not want any of that. So I went on. Then your ladder——'

'Why didn't you hail the ship?' I asked, a little louder.

He touched my shoulder lightly. Lazy footsteps came right over our heads and stopped. The second mate had crossed from the other side of the poop and might have been hanging over the rail, for all we knew.

'He couldn't hear us talking—could he?' My double breathed into my very ear, anxiously.

His anxiety was an answer, a sufficient answer, to the question I had put to him. An answer containing all the difficulty of that situation. I closed the porthole quietly, to make sure. A louder word might have been overheard.

'Who's that?' he whispered then.

'My second mate. But I don't know much more of the fellow than you do.'

And I told him a little about myself. I had been appointed to take charge while I least expected anything of the sort, not quite a fortnight ago. I didn't know either the ship or the people. Hadn't had the time in port to look about me or size anybody up. And as to the crew, all they knew was that I was appointed to take the ship home. For the rest, I was almost as much of a stranger on board as himself, I said. And at the moment I felt it most acutely. I felt that it would take very little to make me a suspect person in the eyes of the ship's company.

He had turned about meantime; and we, the two strangers in the ship, faced each other in identical attitudes.

'Your ladder——' he murmured, after a silence. 'Who'd have thought of finding a ladder hanging over at night in a ship anchored out here! I felt just then a very unpleasant faintness. After the life I've been leading for nine weeks, anybody would have got out of condition. I wasn't capable of swimming round as far as your rudder-chains. And, lo and behold! there was a ladder to get hold of. After I gripped it I said to myself, "What's the good?" When I saw a man's head looking over I thought I would swim away presently and leave him shouting—in whatever language it was. I didn't mind being looked at. I—I liked it. And then you speaking to me so quietly—as if you had expected me—made me hold on a little longer. It had been a confounded lonely time—I don't mean while swimming. I was glad to talk a little to somebody that didn't belong to the *Sephora*. As to asking for the captain, that was a mere impulse. It could have been no use, with all the ship knowing about me and the other people pretty certain to be round here in the morning. I don't know—I wanted to be seen, to talk with somebody, before I went on. I don't know what I would have said . . . "Fine night, isn't it?" or something of the sort.'

'Do you think they will be round here presently?' I asked with some incredulity.

'Quite likely', he said, faintly.

He looked extremely haggard all of a sudden. His head rolled on his shoulders.

'H'm. We shall see then. Meantime get into that bed', I whispered. 'Want help? There.'

It was a rather high bed-place with a set of drawers underneath. This amazing swimmer really needed the lift I gave him by seizing his leg. He tumbled in, rolled over on his back, and flung one arm across his eyes. And then, with his face nearly hidden, he must have looked exactly as I used to look in that bed. I gazed upon my other self for a while before drawing across carefully the two green serge curtains which ran on a brass rod. I thought for a moment of pinning them together for greater safety, but I sat down on the couch, and once there I felt unwilling to rise and hunt for a pin. I would do it in a moment. I was extremely tired, in a peculiarly intimate way, by the strain of stealthiness, by the effort of whispering and the general secrecy of this excitement. It was three o'clock by now and I had been on my feet since nine, but I was not sleepy; I could not have gone to sleep. I sat there, fagged out, looking at the curtains, trying to clear my mind of the confused sensation of being in two places at once, and greatly bothered by an exasperating knocking in my head. It was a relief to discover suddenly that it was not in my head at all, but on the outside of the door. Before I could collect myself the words 'Come in' were out of my mouth, and the steward entered with a tray, bringing in my morning coffee. I had slept, after all, and I was so frightened that I shouted, 'This way! I am here, steward', as though he had been miles away. He put down the tray on the table next the couch and only then said, very quietly, 'I can see you are here, sir.' I felt him give me a keen look, but I dared not meet his eyes just then. He must have wondered why I had drawn the curtains of my bed before going to sleep on the couch. He went out, hooking the door open as usual.

I heard the crew washing decks above me. I knew I would have been told at once if there had been any wind. Calm, I thought, and I was doubly vexed. Indeed, I felt dual more than ever. The steward reappeared suddenly in the doorway. I jumped up from the couch so quickly that he gave a start.

'What do you want here?'

'Close your port, sir—they are washing decks.'

'It is closed', I said, reddening.

'Very well, sir.' But he did not move from the doorway and returned my stare in an extraordinary, equivocal manner for a time. Then his eyes wavered, all his expression changed, and in a voice unusually gentle, almost coaxingly:

'May I come in to take the empty cup away, sir?'

'Of course!' I turned my back on him while he popped in and out. Then I unhooked and closed the door and even pushed the bolt. This sort of thing could not go on very long. The cabin was as hot as an oven, too. I took a peep at my double, and discovered that he had not moved, his arm was still over his eyes; but his chest heaved; his hair was wet; his chin glistened with perspiration. I reached over him and opened the port.

'I must show myself on deck,' I reflected.

Of course, theoretically, I could do what I liked, with no one to say nay to me within the whole circle of the horizon; but to lock my cabin door and take the key away I did not dare. Directly I put my head out of the companion I saw the group of my two officers, the second mate barefooted, the chief mate in long india-rubber boots, near the break of the poop, and the steward half-way down the poop-ladder talking to them eagerly. He happened to catch sight of me and dived, the second ran down on the main-deck shouting some order or other, and the chief mate came to meet me, touching his cap.

There was a sort of curiosity in his eye that I did not like. I don't know whether the steward had told them that I was 'queer' only, or downright drunk, but I know the man meant to have a good look at me. I watched him coming with a smile which, as he got into point-blank range, took effect and froze his very whiskers. I did not give him time to open his lips.

'Square the yards by lifts and braces before the hands go to breakfast.'

It was the first particular order I had given on board that ship; and I stayed on deck to see it executed, too. I had felt the need of asserting myself without loss of time. That sneering young cub got taken down a peg or two on that occasion, and I also seized the opportunity of having a good look at the face of every foremast man as they filed past me to go to the after braces. At breakfast time, eating nothing myself, I presided with such frigid dignity that the two mates were only too glad to escape from the cabin as soon as decency permitted; and all the time the dual working of my mind distracted me almost to the point of insanity. I was constantly watching myself, my secret self, as dependent on my actions as my own personality, sleeping in that bed, behind that door which faced me as I sat at the head of the table. It was very much like being mad, only it was worse because one was aware of it.

I had to shake him for a solid minute, but when at last he opened his eyes it was in the full possession of his senses, with an enquiring look.

'All's well so far,' I whispered. 'Now you must vanish into the bathroom.'

He did so, as noiseless as a ghost, and then I rang for the steward, and facing him boldly, directed him to tidy up my stateroom while I was having my bath—'and be quick about it'. As my tone admitted of no excuses, he said, 'Yes, sir,' and ran off to fetch his dustpan and brushes. I took a bath and did most of my dressing, splashing, and whistling softly for the steward's edification, while the secret sharer of my life stood drawn up bolt upright in that little space, his face looking very sunken in daylight, his eyelids lowered under the stern, dark line of his eyebrows drawn together by a slight frown.

When I left him there to go back to my room the steward was finishing dusting. I sent for the mate and engaged him in some insignificant conversation. It was, as it were, trifling with the terrific character of his whiskers; but my object was to give him an opportunity for a good look at my cabin. And then I could at last shut, with a clear conscience, the door of my stateroom and get my double back into the recessed part. There was nothing else for it. He had to sit still on a small folding stool, half smothered by the heavy coats hanging there. We listened to the steward going into the bathroom out of the saloon, filling the water-bottles there, scrubbing the bath, setting things to rights, whisk, bang, clatter—out again into the saloon—turn the key—click. Such was my scheme for keeping my second self invisible. Nothing better could be contrived under the circumstances. And there we sat; I at my writing-desk ready to appear busy with some papers, he behind me out of sight of the door. It would not have been prudent to talk in daytime; and I could not have stood the excitement of that queer sense of whispering to myself. Now and then, glancing over my shoulder, I saw him far back there, sitting rigidly on the low stool, his bare feet close together, his arms folded, his head hanging on his breast—and perfectly still. Anybody would have taken him for me.

I was fascinated by it myself. Every moment I had to glance over my shoulder. I was looking at him when a voice outside the door said:

'Beg pardon, sir.'

'Well!' . . . I kept my eyes on him, and so when the voice outside the door announced, 'There's a ship's boat coming our way, sir,' I saw him give a start—the first movement he had made for hours. But he did not raise his bowed head.

'All right. Get the ladder over.'

I hesitated. Should I whisper something to him? But what? His immobility seemed to have been never disturbed. What could I tell him he did not know already? . . . Finally I went on deck.

II

THE skipper of the *Sephora* had a thin red whisker all round his face, and the sort of complexion that goes with hair of that colour; also the particular, rather smeary shade of blue in the eyes. He was not exactly a showy figure; his shoulders were high, his stature but middling—one leg slightly more bandy than the other. He shook hands, looking vaguely around. A spiritless tenacity was his main characteristic, I judged. I behaved with a politeness which seemed to disconcert him. Perhaps he was shy. He mumbled to me as if he were ashamed of what he was saying; gave his name (it was something like Archbold—but at this distance of years I hardly am sure), his ship's name, and a few other particulars of that sort, in the manner of a criminal making a reluctant and doleful confession. He had had terrible weather on the passage out—terrible—terrible—wife aboard, too.

By this time we were seated in the cabin and the steward brought in a tray with a bottle and glasses. 'Thanks! No.' Never took liquor.

Would have some water, though. He drank two tumblerfuls. Terrible thirsty work. Ever since daylight had been exploring the islands round his ship.

'What was that for—fun?' I asked, with an appearance of polite interest.

'No!' He sighed. 'Painful duty.'

As he persisted in his mumbling and I wanted my double to hear every word, I hit upon the notion of informing him that I regretted to say I was hard of hearing.

'Such a young man, too!' he nodded, keeping his smeary blue, unintelligent eyes fastened upon me. What was the cause of it—some disease? he enquired, without the least sympathy and as if he thought that, if so, I'd got no more than I deserved.

'Yes; disease,' I admitted in a cheerful tone which seemed to shock him. But my point was gained, because he had to raise his voice to give me his tale. It is not worthwhile to record that version. It was just over two months since all this had happened, and he had thought so much about it that he seemed completely muddled as to its bearings, but still immensely impressed.

'What would you think of such a thing happening on board your own ship? I've had the *Sephora* for these fifteen years. I am a well-known shipmaster.'

He was densely distressed—and perhaps I should have sympathized with him if I had been able to detach my mental vision from the unsuspected sharer of my cabin as though he were my second self. There he was on the other side of the bulkhead, four or five feet from us, no more, as we sat in the saloon. I looked politely at Captain Archbold (if that was his name), but it was the other I saw, in a grey sleeping-suit, seated on a low stool, his bare feet close together, his arms folded, and every word said between us falling into the ears of his dark head bowed on his chest.

'I have been at sea now, man and boy, for seven-and-thirty years, and I've never heard of such a thing happening in an English ship. And that it should be my ship. Wife on board, too.'

I was hardly listening to him.

'Don't you think', I said, 'that the heavy sea which, you told me, came aboard just then might have killed the man? I have seen the sheer weight of a sea kill a man very neatly, by simply breaking his neck.'

'Good God!' he uttered, impressively, fixing his smeary blue eyes on me. 'The sea! No man killed by the sea ever looked like that.'

He seemed positively scandalized at my suggestion. And as I gazed at him, certainly not prepared for anything original on his part, he advanced his head close to mine and thrust his tongue out at me so suddenly that I couldn't help starting back.

After scoring over my calmness in this graphic way he nodded wisely. If I had seen the sight, he assured me, I would never forget it as long as I lived. The weather was too bad to give the corpse a proper sea burial. So next day at dawn they took it up on the poop, covering its face with a bit of bunting; he read a short prayer, and then, just as it was, in its oilskins and long boots, they launched it amongst those mountainous seas that seemed ready every moment to swallow up the ship herself and the terrified lives on board of her.

'That reefed foresail saved you', I threw in.

'Under God—it did,' he exclaimed fervently. 'It was by a special mercy, I firmly believe, that it stood some of those hurricane squalls.'

'It was the setting of that sail which——' I began.

'God's own hand in it,' he interrupted me. 'Nothing less could have done it. I don't mind telling you that I hardly dared give the order. It seemed impossible that we could touch anything without losing it, and then our last hope would have been gone.'

The terror of that gale was on him yet. I let him go on for a bit, then said, casually—as if returning to a minor subject:

'You were very anxious to give up your mate to the shore people, I believe?'

He was. To the law. His obscure tenacity on that point had in it something incomprehensible and a little awful; something, as it were, mystical, quite apart from his anxiety that he should not be suspected of 'countenancing any doings of that sort'. Seven-and-thirty virtuous years at sea, of which over twenty of immaculate command, and the last fifteen in the *Sephora*, seemed to have laid him under some pitiless obligation.

'And you know', he went on, groping shamefacedly amongst his feelings, 'I did not engage that young fellow. His people had some interest with my owners. I was in a way forced to take him on. He looked very smart, very gentlemanly, and all that. But do you know—I never liked him, somehow. I am a plain man. You see, he wasn't exactly the sort for the chief mate of a ship like the *Sephora*.'

I had become so connected in thoughts and impressions with the secret sharer of my cabin that I felt as if I, personally, were being given to understand that I, too, was not the sort that would have done for the chief mate of a ship like the *Sephora*. I had no doubt of it in my mind.

'Not at all the style of man. You understand,' he insisted, superfluously, looking hard at me.

I smiled urbanely. He seemed at a loss for a while.

'I suppose I must report a suicide.'

'Beg pardon?'

'Sui-cide! That's what I'll have to write to my owners directly I get in.'

'Unless you manage to recover him before tomorrow', I assented, dispassionately. . . . 'I mean, alive.'

He mumbled something which I really did not catch, and I turned my ear to him in a puzzled manner. He fairly bawled:

'The land—I say, the mainland is at least seven miles off my anchorage.'

'About that.'

My lack of excitement, of curiosity, of surprise, of any sort of pronounced interest, began to arouse his distrust. But except for the felicitous pretence of deafness I had not tried to pretend anything. I had felt utterly incapable of playing the part of ignorance properly, and therefore was afraid to try. It is also certain that he had brought some ready-made suspicions with him, and that he viewed my politeness as a strange and unnatural phenomenon. And yet how else could I have received him? Not heartily! That was impossible for psychological reasons, which I need not state here. My only object was to keep off his enquiries. Surlily? Yes, but surliness might have provoked a point-blank question. From its novelty to him and from its nature, punctilious courtesy was the manner best calculated to restrain the man. But there was the danger of his breaking through my defence bluntly. I could not, I think, have met him by a direct lie, also for psychological (not moral) reasons. If he had only known how afraid I was of his putting my feeling of identity with the other to the test! But, strangely enough—(I thought of it only afterwards)—I believe that he was not a little disconcerted by the reverse side of that weird situation, by something in me that reminded him of the man he was seeking—suggested a mysterious similitude to the young fellow he had distrusted and disliked from the first.

However that might have been, the silence was not very prolonged. He took another oblique step.

'I reckon I had no more than a two-mile pull to your ship. Not a bit more.'

'And quite enough, too, in this awful heat,' I said.

Another pause full of mistrust followed. Necessity, they say, is mother of invention, but fear, too, is not barren of ingenious suggestions. And I was afraid he would ask me point-blank for news of my other self.

'Nice little saloon, isn't it?' I remarked, as if noticing for the first time the way his eyes roamed from one closed door to the other. 'And very well fitted out, too. Here, for instance', I continued, reaching over the back of my seat negligently and flinging the door open, 'is my bathroom.'

He made an eager movement, but hardly gave it a glance. I got up, shut the door of the bathroom, and invited him to have a look round, as if I were very proud of my accommodation. He had to rise and be shown round, but he went through the business without any raptures whatever.

'And now we'll have a look at my stateroom,' I declared, in a voice as loud as I dared to make it, crossing the cabin to the starboard side with purposely heavy steps.

He followed me in and gazed around. My intelligent double had vanished. I played my part.

'Very convenient—isn't it?'

'Very nice. Very comf. . .' He didn't finish and went out brusquely as if to escape from some unrighteous wiles of mine. But it was not to be. I had been too frightened not to feel vengeful; I felt I had him on the run, and I meant to keep him on the run. My polite insistence must have had something menacing in it, because he gave in suddenly. And I did not let him off a single item; mate's room, pantry, store-rooms, the very sail-locker which was also under the poop—he had to look into them all. When at last I showed him out on the quarter-deck he drew a long, spiritless sigh, and mumbled dismally that he must really be going back to his ship now. I desired my mate, who had joined us, to see to the captain's boat.

The man of whiskers gave a blast on the whistle which he used to wear hanging round his neck, and yelled, '*Sephora*'s away!' My double down there in my cabin must have heard, and certainly could not feel more relieved than I. Four fellows came running out from somewhere forward and went over the side, while my own men, appearing on deck too, lined the rail. I escorted my visitor to the gangway ceremoniously, and nearly overdid it. He was a tenacious beast. On the very ladder he lingered, and in that unique, guiltily conscientious manner of sticking to the point:

'I say . . . you . . . you don't think that——'

I covered his voice loudly:

'Certainly not. . . . I am delighted. Goodbye.'

I had an idea of what he meant to say, and just saved myself by the privilege of defective hearing. He was too shaken generally to insist, but my mate, close witness of that parting, looked mystified and his face took on a thoughtful cast. As I did not want to appear as if I wished to avoid all communication with my officers, he had the opportunity to address me.

'Seems a very nice man. His boat's crew told our chaps a very extraordinary story, if what I am told by the steward is true. I suppose you had it from the captain, sir?'

'Yes. I had a story from the captain.'

'A very horrible affair—isn't it, sir?'

'It is.'

'Beats all these tales we hear about murders in Yankee ships.'

'I don't think it beats them. I don't think it resembles them in the least.'

'Bless my soul—you don't say so! But of course I've no acquaintance whatever with American ships, not I, so I couldn't go against your knowledge. It's horrible enough for me. . . . But the queerest part is that those fellows seemed to have some idea the man was hidden aboard here. They had really. Did you ever hear of such a thing?'

'Preposterous—isn't it?'

We were walking to and fro athwart the quarter-deck. No one of the crew forward could be seen (the day was Sunday), and the mate pursued:

'There was some little dispute about it. Our chaps took offence. "As if we would harbour a thing like that", they said. "Wouldn't you like to look for him in our coal-hole?" Quite a tiff. But they made it up in the end. I suppose he did drown himself. Don't you, sir?'

'I don't suppose anything.'

'You have no doubt in the matter, sir?'

'None whatever.'

I left him suddenly. I felt I was producing a bad impression, but with my double down there it was most trying to be on deck. And it was almost as trying to be below. Altogether a nerve-trying situation.

But on the whole I felt less torn in two when I was with him. There was no one in the whole ship whom I dared take into my confidence. Since the hands had got to know his story, it would have been impossible to pass him off for anyone else, and an accidental discovery was to be dreaded now more than ever. . . .

The steward being engaged in laying the table for dinner, we could talk only with our eyes when I first went down. Later in the afternoon we had a cautious try at whispering. The Sunday quietness of the ship was against us; the stillness of air and water around her was against us; the elements, the men were against us—everything was against us in our secret partnership; time itself—for this could not go on forever. The very trust in Providence was, I suppose, denied to his guilt. Shall I confess that this thought cast me down very much? And as to the chapter of accidents which counts for so much in the book of success, I could only hope that it was closed. For what favourable accident could be expected?

'Did you hear everything?' were my first words as soon as we took up our position side by side, leaning over my bed-place.

He had. And the proof of it was his earnest whisper, 'The man told you he hardly dared to give the order.'

I understood the reference to be to that saving foresail.

'Yes. He was afraid of it being lost in the setting.'

'I assure you he never gave the order. He may think he did, but he never gave it. He stood there with me on the break of the poop after the maintopsail blew away, and whimpered about our last hope—positively whimpered about it and nothing else—and the night coming on! To hear one's skipper go on like that in such weather was enough to drive any fellow out of his mind. It worked me up into a sort of desperation. I just took it into my own hands and went away from him, boiling, and—But what's the use telling you? *You* know! . . . Do you think that if I had not been pretty fierce with them I should have got the men to do anything? Not it! The bo's'n perhaps? Perhaps! It wasn't a heavy sea—it was a sea gone mad! I suppose the end of the world will be something like that; and a man may have the heart to see it coming once and be done with it—but to have to face it day after day——I don't blame anybody. I was precious little better than the rest. Only—I was an officer of that old coal-wagon, anyhow——'

'I quite understand,' I conveyed that sincere assurance into his ear. He was out of breath with whispering; I could hear him pant slightly.

It was all very simple. The same strung-up force which had given twenty-four men a chance, at least, for their lives, had, in a sort of recoil, crushed an unworthy mutinous existence.

But I had no leisure to weigh the merits of the matter—footsteps in the saloon, a heavy knock. 'There's enough wind to get under way with, sir.' Here was the call of a new claim upon my thoughts and even upon my feelings.

'Turn the hands up,' I cried through the door. 'I'll be on deck directly.'

I was going out to make the acquaintance of my ship. Before I left the cabin our eyes met—the eyes of the only two strangers on board. I pointed to the recessed part where the little camp-stool awaited him and laid my finger on my lips. He made a gesture—somewhat vague—a little mysterious, accompanied by a faint smile, as if of regret.

This is not the place to enlarge upon the sensations of a man who feels for the first time a ship move under his feet to his own independent word. In my case they were not unalloyed. I was not wholly alone with my command; for there was that stranger in my cabin. Or rather, I was not completely and wholly with her. Part of me was absent. That mental feeling of being in two places at once affected me physically as if the mood of secrecy had penetrated my very soul. Before an hour had elapsed since the ship had begun to move, having occasion to ask the mate (he stood by my side) to take a compass bearing of the Pagoda, I caught myself reaching up to his ear in whispers. I say I caught myself, but enough had escaped to startle the man. I can't describe it otherwise than by saying that he shied. A grave, preoccupied manner, as though he were in possession of some perplexing intelligence, did not leave him henceforth. A little later I moved away from the rail to look at the compass with such a stealthy gait that the helmsman noticed it—and I could not help noticing the unusual roundness of his eyes. These are trifling instances, though it's to no commander's advantage to be suspected of ludicrous eccentricities. But I was also more seriously affected. There are to a seaman certain words, gestures, that should in given conditions come as naturally, as instinctively as the winking of a menaced eye. A certain order should spring on to his lips without thinking; a certain sign should get itself made, so to speak, without reflection. But all unconscious alertness had abandoned me. I had to make an effort of will to recall myself back (from the cabin) to the conditions of the moment. I felt that I was appearing an irresolute commander to those people who were watching me more or less critically.

And, besides, there were the scares. On the second day out, for instance, coming off the deck in the afternoon (I had straw slippers on my bare feet) I stopped at the open pantry door and spoke to the steward. He was doing something there with his back to me. At the sound of my voice he nearly jumped out of his skin, as the saying is, and incidentally broke a cup.

'What on earth's the matter with you?' I asked, astonished.

He was extremely confused. 'Beg your pardon, sir. I made sure you were in your cabin.'

'You see I wasn't.'

'No, sir. I could have sworn I had heard you moving in there not a moment ago. It's most extraordinary . . . very sorry, sir.'

I passed on with an inward shudder. I was so identified with my secret double that I did not even mention the fact in those scanty, fearful whispers we exchanged. I suppose he had made some slight noise of some kind or other. It would have been miraculous if he hadn't at one time or another. And yet, haggard as he appeared, he looked always perfectly self-controlled, more than calm—almost invulnerable. On my suggestion he remained almost entirely in the bathroom, which, upon the whole, was the safest place. There could be really no shadow of an excuse for anyone ever wanting to go in there, once the steward had done with it. It was a very tiny place. Sometimes he reclined on the floor, his legs bent, his head sustained on one elbow. At others I would find him on the camp-stool, sitting in his grey sleeping-suit and with his cropped dark hair like a patient, unmoved convict. At night I would smuggle him into my bed-place, and we would whisper together, with the regular footfalls of the officer of the watch passing and repassing over our heads. It was an infinitely miserable time. It was lucky that some tins of fine preserves were stowed in a locker in my stateroom; hard bread I could always get hold of; and so he lived on stewed chicken, paté de foie gras, asparagus, cooked oysters, sardines—on all sorts of abominable sham delicacies out of tins. My early morning coffee he always drank; and it was all I dared do for him in that respect.

Every day there was the horrible manœuvring to go through so that my room and then the bathroom should be done in the usual way. I came to hate the sight of the steward, to abhor the voice of that harmless man. I felt that it was he who would bring on the disaster of discovery. It hung like a sword over our heads.

The fourth day out, I think (we were then working down the east side of the Gulf of Siam, tack for tack, in light winds and smooth water)—the fourth day, I say, of this miserable juggling with the unavoidable, as we sat at our evening meal, that man, whose slightest movement I dreaded, after putting down the dishes ran up on deck busily. This could not be dangerous. Presently he came down again; and then it appeared that he had remembered a coat of mine which I had thrown over a rail to dry after having been wetted in a shower which had passed over the ship in the afternoon. Sitting stolidly at the head of the table I became terrified at the sight of the garment on his arm. Of course he made for my door. There was no time to lose.

'Steward,' I thundered. My nerves were so shaken that I could not govern my voice and conceal my agitation. This was the sort of thing that made my terrifically whiskered mate tap his forehead with his forefinger. I had detected him using that gesture while talking on deck with a confidential air to the carpenter. It was too far to hear a word, but I had no doubt that this pantomime could only refer to the strange new captain.

'Yes, sir,' the pale-faced steward turned resignedly to me. It was this maddening course of being shouted at, checked without rhyme or reason, arbitrarily chased out of my cabin, suddenly called into it, sent flying out of his pantry on incomprehensible errands, that accounted for the growing wretchedness of his expression.

'Where are you going with that coat?'

'To your room, sir.'

'Is there another shower coming?'

'I'm sure I don't know, sir. Shall I go up again and see, sir?'

'No! never mind.'

My object was attained, as of course my other self in there would have heard everything that passed. During this interlude my two officers never raised their eyes off their respective plates; but the lip of that confounded cub, the second mate, quivered visibly.

I expected the steward to hook my coat on and come out at once. He was very slow about it; but I dominated my nervousness sufficiently not to shout after him. Suddenly I became aware (it could be heard plainly enough) that the fellow for some reason or other was opening the door of the bathroom. It was the end. The place was literally not big enough to swing a cat in. My voice died in my throat and I went stony all over. I expected to hear a yell of surprise and terror, and made a movement, but had not the strength to get on my legs. Everything remained still. Had my second self taken the poor wretch by the throat? I don't know what I could have done next moment if I had not seen the steward come out of my room, close the door, and then stand quietly by the sideboard.

'Saved,' I thought. 'But, no! Lost! Gone! He was gone!'

I laid my knife and fork down and leaned back in my chair. My head swam. After a while, when sufficiently recovered to speak in a steady voice, I instructed my mate to put the ship round at eight o'clock himself.

'I won't come on deck,' I went on. 'I think I'll turn in, and unless the wind shifts I don't want to be disturbed before midnight. I feel a bit seedy.'

'You did look middling bad a little while ago,' the chief mate remarked without showing any great concern.

They both went out, and I stared at the steward clearing the table. There was nothing to be read on that wretched man's face. But why did he avoid my eyes I asked myself. Then I thought I should like to hear the sound of his voice.

'Steward!'

'Sir!' Startled as usual.

'Where did you hang up that coat?'

'In the bathroom, sir.' The usual anxious tone. 'It's not quite dry yet, sir.'

For some time longer I sat in the cuddy. Had my double vanished as he had come? But of his coming there was an explanation, whereas his disappearance would be inexplicable. . . . I went slowly into my dark room, shut the door, lighted the lamp, and for a time dared not turn round. When at last I did I saw him standing bolt-upright in the narrow recessed part. It would not be true to say I had a shock, but an irresistible doubt of his bodily existence flitted through my mind. Can it be, I asked myself, that he is not visible to other eyes than mine? It was like being haunted. Motionless, with a grave face, he raised his hands slightly at me in a gesture which meant clearly, 'Heavens! what a narrow escape!' Narrow indeed. I think I had come creeping quietly as near insanity as any man who has not actually gone over the border. That gesture restrained me, so to speak.

The mate with the terrific whiskers was now putting the ship on the other tack. In the moment of profound silence which follows upon the hands going to their stations I heard on the poop his raised voice: 'Hard alee!' and the distant shout of the order repeated on the maindeck. The sails, in that light breeze, made but a faint fluttering noise. It ceased. The ship was coming round slowly; I held my breath in the renewed stillness of expectation; one wouldn't have thought that there was a single living soul on her decks. A sudden brisk shout, 'Mainsail haul!' broke the spell, and in the noisy cries and rush overhead of the men running away with the main-brace we two, down in my cabin, came together in our usual position by the bed-place.

He did not wait for my question. 'I heard him fumbling here and just managed to squat myself down in the bath,' he whispered to me. 'The fellow only opened the door and put his arm in to hang the coat up. All the same——'

'I never thought of that,' I whispered back, even more appalled than before at the closeness of the shave, and marvelling at that something unyielding in his character which was carrying him through so finely. There was no agitation in his whisper. Whoever was being driven distracted, it was not he. He was sane. And the proof of his sanity was continued when he took up the whispering again.

'It would never do for me to come to life again.'

It was something that a ghost might have said. But what he was alluding to was his old captain's reluctant admission of the theory of suicide. It would obviously serve his turn—if I had understood at all the view which seemed to govern the unalterable purpose of his action.

'You must maroon me as soon as ever you can get amongst these islands off the Cambodje shore,' he went on.

'Maroon you! We are not living in a boy's adventure tale,' I protested. His scornful whispering took me up.

'We aren't indeed! There's nothing of a boy's tale in this. But there's nothing else for it. I want no more. You don't suppose I am afraid of what can be done to me? Prison or gallows or whatever they may please. But you don't see me coming back to explain such things to an old fellow in a wig and twelve respectable tradesmen, do you? What can they know whether I am guilty or not—or of *what* I am guilty, either? That's my affair. What does the Bible say? "Driven off the face of the earth." Very well. I am off the face of the earth now. As I came at night so I shall go.'

'Impossible!' I murmured. 'You can't.'

'Can't? . . . Not naked like a soul on the Day of Judgement. I shall freeze on to this sleeping-suit. The Last Day is not yet—and . . . you have understood thoroughly. Didn't you?'

I felt suddenly ashamed of myself. I may say truly that I under-stood—and my hesitation in letting that man swim away from my ship's side had been a mere sham sentiment, a sort of cowardice.

'It can't be done now till next night,' I breathed out. 'The ship is on the off-shore tack and the wind may fail us.'

'As long as I know that you understand,' he whispered. 'But of course you do. It's a great satisfaction to have got somebody to understand. You seem to have been there on purpose.' And in the same whisper, as if we two whenever we talked had to say things to each other which were not fit for the world to hear, he added, 'It's very wonderful.'

We remained side by side talking in our secret way—but sometimes silent or just exchanging a whispered word or two at long intervals. And as usual he stared through the port. A breath of wind came now and again into our faces. The ship might have been moored in dock, so gently and on an even keel she slipped through the water, that did not murmur even at our passage, shadowy and silent like a phantom sea.

At midnight I went on deck, and to my mate's great surprise put the ship round on the other tack. His terrible whiskers flitted round me in silent criticism. I certainly should not have done it if it had been only a question of getting out of that sleepy gulf as quickly as possible. I believe he told the second mate, who relieved him, that it was a great want of judgement. The other only yawned. That intoler-able cub shuffled about so sleepily and lolled against the rails in such a slack, improper fashion that I came down on him sharply.

'Aren't you properly awake yet?'

'Yes, sir! I am awake.'

'Well, then, be good enough to hold yourself as if you were. And keep a look-out. If there's any current we'll be closing with some islands before daylight.'

The east side of the gulf is fringed with islands, some solitary, others in groups. On the blue background of the high coast they seem to float on silvery patches of calm water, arid and grey, or dark green and rounded like clumps of evergreen bushes, with the larger ones, a mile or two long, showing the outlines of ridges, ribs of grey rock under the dank mantle of matted leafage. Unknown to trade, to travel, almost to geography, the manner of life they harbour is an unsolved secret. There must be villages—settlements of fishermen at least—on the largest of them, and some communication with the world is probably kept up by native craft. But all that forenoon, as we headed for them, fanned along by the faintest of breezes, I saw no sign of man or canoe in the field of the telescope I kept on pointing at the scattered group.

At noon I gave no orders for a change of course, and the mate's whiskers became much concerned and seemed to be offering themselves unduly to my notice. At last I said:

'I am going to stand right in. Quite in—as far as I can take her.'

The stare of extreme surprise imparted an air of ferocity also to his eyes, and he looked truly terrific for a moment.

'We're not doing well in the middle of the gulf,' I continued, casually. 'I am going to look for the land breezes tonight.'

'Bless my soul! Do you mean, sir, in the dark amongst the lot of all them islands and reefs and shoals?'

'Well—if there are any regular land breezes at all on this coast one must get close inshore to find them, mustn't one?'

'Bless my soul!' he exclaimed again under his breath. All that afternoon he wore a dreamy, contemplative appearance which in him was a mark of perplexity. After dinner I went into my stateroom as if I meant to take some rest. There we two bent our dark heads over a half-unrolled chart lying on my bed.

'There,' I said. 'It's got to be Koh-ring. I've been looking at it ever since sunrise. It has got two hills and a low point. It must be inhabited. And on the coast opposite there is what looks like the mouth of a biggish river—with some town, no doubt, not far up. It's the best chance for you that I can see.'

'Anything. Koh-ring let it be.'

He looked thoughtfully at the chart as if surveying chances and distances from a lofty height—and following with his eyes his own figure wandering on the blank land of Cochin-China, and then passing off that piece of paper clean out of sight into uncharted regions. And it was as if the ship had two captains to plan her course for her. I had been so worried and restless running up and down that I had not had the patience to dress that day. I had remained in my sleeping-suit, with straw slippers and a soft floppy hat. The closeness of the heat in the gulf had been most oppressive, and the crew were used to seeing me wandering in that airy attire.

'She will clear the south point as she heads now,' I whispered into his ear. 'Goodness only knows when, though, but certainly after dark. I'll edge her in to half a mile, as far as I may be able to judge in the dark——'

'Be careful,' he murmured, warningly—and I realized suddenly that all my future, the only future for which I was fit, would perhaps go irretrievably to pieces in any mishap to my first command.

I could not stop a moment longer in the room. I motioned him to get out of sight and made my way on the poop. That unplayful cub had the watch. I walked up and down for a while thinking things out, then beckoned him over.

'Send a couple of hands to open the two quarterdeck ports,' I said, mildly.

He actually had the impudence, or else so forgot himself in his wonder at such an incomprehensible order, as to repeat:

'Open the quarterdeck ports! What for, sir?'

'The only reason you need concern yourself about is because I tell you to do so. Have them open wide and fastened properly.'

He reddened and went off, but I believe made some jeering remark to the carpenter as to the sensible practice of ventilating a ship's quarterdeck. I know he popped into the mate's cabin to impart the fact to him because the whiskers came on deck, as it were by chance, and stole glances at me from below—for signs of lunacy or drunkenness, I suppose.

A little before supper, feeling more restless than ever, I rejoined, for a moment, my second self. And to find him sitting so quietly was surprising, like something against nature, inhuman.

I developed my plan in a hurried whisper.

'I shall stand in as close as I dare and then put her round. I will presently find means to smuggle you out of here into the sail-locker, which communicates with the lobby. But there is an opening, a sort of square for hauling the sails out, which gives straight on the quarterdeck and which is never closed in fine weather, so as to give air to the sails. When the ship's way is deadened in stays and all the hands are aft at the main-braces you will have a clear road to slip out and get overboard through the open quarterdeck port. I've had them both fastened up. Use a rope's end to lower yourself into the water so as to avoid a splash—you know. It could be heard and cause some beastly complication.'

He kept silent for a while, then whispered, 'I understand.'

'I won't be there to see you go,' I began with an effort. 'The rest . . . I only hope I have understood, too.'

'You have. From first to last'—and for the first time there seemed to be a faltering, something strained in his whisper. He caught hold of my arm, but the ringing of the supper bell made me start. He didn't, though; he only released his grip.

After supper I didn't come below again till well past eight o'clock. The faint, steady breeze was loaded with dew; and the wet, darkened sails held all there was of propelling power in it. The night, clear and starry, sparkled darkly, and the opaque, lightless patches shifting slowly against the low stars were the drifting islets. On the port bow there was a big one more distant and shadowily imposing by the great space of sky it eclipsed.

On opening the door I had a back view of my very own self looking at a chart. He had come out of the recess and was standing near the table.

'Quite dark enough,' I whispered.

He stepped back and leaned against my bed with a level, quiet glance. I sat on the couch. We had nothing to say to each other. Over our heads the officer of the watch moved here and there. Then I heard him move quickly. I knew what that meant. He was making for the companion; and presently his voice was outside my door.

'We are drawing in pretty fast, sir. Land looks rather close.'

'Very well,' I answered. 'I am coming on deck directly.'

I waited till he was gone out of the cuddy, then rose. My double moved too. The time had come to exchange our last whispers, for neither of us was ever to hear each other's natural voice.

'Look here!' I opened a drawer and took out three sovereigns. 'Take this anyhow. I've got six and I'd give you the lot, only I must keep a little money to buy some fruit and vegetables for the crew from native boats as we go through Sunda Straits.'

He shook his head.

'Take it,' I urged him, whispering desperately. 'No one can tell what——'

He smiled and slapped meaningly the only pocket of the sleeping-jacket. It was not safe, certainly. But I produced a large old silk handkerchief of mine, and tying the three pieces of gold in a corner, pressed it on him. He was touched, I suppose, because he took it at last and tied it quickly round his waist under the jacket, on his bare skin.

Our eyes met; several seconds elapsed, till, our glances still mingled, I extended my hand and turned the lamp out. Then I passed through the cuddy, leaving the door of my room wide open. . . . 'Steward!'

He was still lingering in the pantry in the greatness of his zeal, giving a rub-up to a plated cruet stand the last thing before going to bed. Being careful not to wake up the mate, whose room was opposite, I spoke in an undertone.

He looked round anxiously. 'Sir!'

'Can you get me a little hot water from the galley?'

'I am afraid, sir, the galley fire's been out for some time now.'

'Go and see.'

He flew up the stairs.

'Now,' I whispered, loudly, into the saloon—too loudly, perhaps, but I was afraid I couldn't make a sound. He was by my side in an instant—the double captain slipped past the stairs—through a tiny dark passage . . . a sliding door. We were in the sail-locker, scrambling on our knees over the sails. A sudden thought struck me. I saw myself wandering barefooted, bareheaded, the sun beating on my dark poll. I snatched off my floppy hat and tried hurriedly in the dark to ram it on my other self. He dodged and fended off silently. I wonder what he thought had come to me before he understood and suddenly desisted. Our hands met gropingly, lingered united in a steady, motionless clasp for a second. . . . No word was breathed by either of us when they separated.

I was standing quietly by the pantry door when the steward returned.

'Sorry, sir. Kettle barely warm. Shall I light the spirit-lamp?'

'Never mind.'

I came out on deck slowly. It was now a matter of conscience to shave the land as close as possible—for now he must go overboard whenever the ship was put in stays. Must! There could be no going back for him. After a moment I walked over to leeward and my heart flew into my mouth at the nearness of the land on the bow. Under any other circumstances I would not have held on a minute longer. The second mate had followed me anxiously.

I looked on till I felt I could command my voice.

'She may weather,' I said then in a quiet tone.

'Are you going to try that, sir?' he stammered out incredulously.

I took no notice of him and raised my tone just enough to be heard by the helmsman.

'Keep her good full.'

'Good full, sir.'

The wind fanned my cheek, the sails slept, the world was silent. The strain of watching the dark loom of the land grow bigger and denser was too much for me. I had shut my eyes—because the ship must go closer. She must! The stillness was intolerable. Were we standing still?

When I opened my eyes the second view started my heart with a thump. The black southern hill of Koh-ring seemed to hang right over the ship like a towering fragment of the everlasting night. On that enormous mass of blackness there was not a gleam to be seen, not a sound to be heard. It was gliding irresistibly towards us and yet seemed already within reach of the hand. I saw the vague figures of the watch grouped in the waist, gazing in awed silence.

'Are you going on, sir?' enquired an unsteady voice at my elbow.

I ignored it. I had to go on.

'Keep her full. Don't check her way. That won't do now,' I said, warningly.

'I can't see the sails very well,' the helmsman answered me, in strange, quavering tones.

Was she close enough? Already she was, I won't say in the shadow of the land, but in the very blackness of it, already swallowed up as it were, gone too close to be recalled, gone from me altogether.

'Give the mate a call,' I said to the young man who stood at my elbow as still as death. 'And turn all hands up.'

My tone had a borrowed loudness reverberated from the height of the land. Several voices cried out together: 'We are all on deck, sir.'

Then stillness again, with the great shadow gliding closer, towering higher, without light, without a sound. Such a hush had fallen on the ship that she might have been a bark of the dead floating in slowly under the very gate of Erebus.

'My God! Where are we?'

It was the mate moaning at my elbow. He was thunderstruck, and as it were deprived of the moral support of his whiskers. He clapped his hands and absolutely cried out, 'Lost!'

'Be quiet,' I said, sternly.

He lowered his tone, but I saw the shadowy gesture of his despair. 'What are we doing here?'

'Looking for the land wind.'

He made as if to tear his hair, and addressed me recklessly.

'She will never get out. You have done it, sir. I knew it'd end in something like this. She will never weather, and you are too close now to stay. She'll drift ashore before she's round. O my God!'

I caught his arm as he was raising it to batter his poor devoted head, and shook it violently.

'She's ashore already,' he wailed, trying to tear himself away.

'Is she? . . . Keep good full there!'

'Good full, sir,' cried the helmsman in a frightened, thin, childlike voice.

I hadn't let go the mate's arm and went on shaking it. 'Ready about, do you hear? You go forward'—shake—'and stop there'—shake—'and hold your noise'—shake—'and see these head-sheets properly over-hauled'—shake, shake—shake.

And all the time I dared not look towards the land lest my heart should fail me. I released my grip at last and he ran forward as if fleeing for dear life.

I wondered what my double there in the sail-locker thought of this commotion. He was able to hear everything—and perhaps he was able to understand why, on my conscience, it had to be thus close—no less. My first order 'Hard alee!' re-echoed ominously under the tower-ing shadow of Koh-ring as if I had shouted in a mountain gorge. And then I watched the land intently. In that smooth water and light wind it was impossible to feel the ship coming-to. No! I could not feel her. And my second self was making now ready to slip out and lower himself overboard. Perhaps he was gone already . . .?

The great black mass brooding over our very mastheads began to pivot away from the ship's side silently. And now I forgot the secret stranger ready to depart, and remembered only that I was a total stranger to the ship. I did not know her. Would she do it? How was she to be handled?

I swung the mainyard and waited helplessly. She was perhaps stopped, and her very fate hung in the balance, with the black mass of Koh-ring like the gate of the everlasting night towering over her taffrail. What would she do now? Had she way on her yet? I stepped to the side swiftly, and on the shadowy water I could see nothing except a faint phosphorescent flash revealing the glassy smoothness of the sleeping surface. It was impossible to tell—and I had not learned yet the feel of my ship. Was she moving? What I needed was something easily seen, a piece of paper, which I could throw over-board and watch. I had nothing on me. To run down for it I didn't dare. There was no time. All at once my strained, yearning stare distinguished a white object floating within a yard of the ship's side. White on the black water. A phosphorescent flash passed under it. What was that thing? . . . I recognized my own floppy hat. It must have fallen off his head . . . and he didn't bother. Now I had what I wanted—the saving mark for my eyes. But I hardly thought of my other self, now gone from the ship, to be hidden for ever from all friendly faces, to be a fugitive and a vagabond on the earth, with no brand of the curse on his sane forehead to stay a slaying hand . . . too proud to explain.

And I watched the hat—the expression of my sudden pity for his mere flesh. It had been meant to save his homeless head from the dangers of the sun. And now—behold—it was saving the ship, by serving me for a mark to help out the ignorance of my strangeness. Ha! It was drifting forward, warning me just in time that the ship had gathered sternway.

'Shift the helm,' I said in a low voice to the seaman standing still like a statue.

The man's eyes glistened wildly in the binnacle light as he jumped round to the other side and spun round the wheel.

I walked to the break of the poop. On the overshadowed deck all hands stood by the forebraces waiting for my order. The stars ahead seemed to be gliding from right to left. And all was so still in the world that I heard the quiet remark 'She's round,' passed in a tone of intense relief between two seamen.

'Let go and haul.'

The foreyards ran round with a great noise, amidst cheery cries. And now the frightful whiskers made themselves heard giving various orders. Already the ship was drawing ahead. And I was alone with her. Nothing! no one in the world should stand now between us, throwing a shadow on the way of silent knowledge and mute affection, the perfect communion of a seaman with his first command.

Walking to the taffrail, I was in time to make out, on the very edge of a darkness thrown by a towering black mass like the very gateway of Erebus—yes, I was in time to catch an evanescent glimpse of my white hat left behind to mark the spot where the secret sharer of my cabin and of my thoughts, as though he were my second self, had lowered himself into the water to take his punishment: a free man, a proud swimmer striking out for a new destiny.

The Ghost Ship

∽∽∽

RICHARD MIDDLETON

F AIRFIELD is a little village lying near the Portsmouth Road about half-way between London and the sea. Strangers who find it by accident now and then, call it a pretty, old-fashioned place; we, who live in it and call it home, don't find anything very pretty about it, but we should be sorry to live anywhere else. Our minds have taken the shape of the inn and the church and the green, I suppose. At all events we never feel comfortable out of Fairfield.

Of course the Cockneys, with their nasty houses and their noise-ridden streets, can call us rustics if they choose, but for all that Fairfield is a better place to live in than London. Doctor says that when he goes to London his mind is bruised with the weight of the houses, and he was a Cockney born. He had to live there himself when he was a little chap, but he knows better now. You gentlemen may laugh—perhaps some of you come from London way—but it seems to me that a witness like that is worth a gallon of arguments.

Dull? Well, you might find it dull, but I assure you that I've listened to all the London yarns you have spun tonight, and they're absolutely nothing to the things that happen at Fairfield. It's because of our way of thinking and minding our own business. If one of your Londoners were set down on the green of a Saturday night when the ghosts of the lads who died in the war keep tryst with the lasses who lie in the churchyard, he couldn't help being curious and interfering, and then the ghosts would go somewhere where it was quieter. But we just let them come and go and don't make any fuss, and in consequence Fairfield is the ghostiest place in all England. Why, I've seen a headless man sitting on the edge of the well in broad daylight, and the children playing about his feet as if he were their father. Take my word for it, spirits know when they are well off as much as human beings.

Still, I must admit that the thing I'm going to tell you about was queer even for our part of the world, where three packs of ghost-hounds hunt regularly during the season, and blacksmith's great-grandfather is busy all night shoeing the dead gentlemen's horses. Now that's a thing that wouldn't happen in London, because of their interfering ways, but blacksmith he lies up aloft and sleeps as quiet as a lamb. Once when he had a bad head he shouted down to them not to make so much noise, and in the morning he found an old guinea left on the anvil as an apology. He wears it on his watch-chain now. But I must get on with my story; if I start telling you about the queer happenings at Fairfield I'll never stop.

It all came of the great storm in the spring of '97, the year that we had two great storms. This was the first one, and I remember it very well, because I found in the morning that it had lifted the thatch of my pigsty into the widow's garden as clean as a boy's kite. When I looked over the hedge, widow—Tom Lamport's widow that was— was prodding for her nasturtiums with a daisy-grubber. After I had watched her for a little I went down to the Fox and Grapes to tell landlord what she had said to me. Landlord he laughed, being a married man and at ease with the sex. 'Come to that', he said, 'the tempest has blowed something into my field. A kind of a ship I think it would be.'

I was surprised at that until he explained that it was only a ghost-ship and would do no hurt to the turnips. We argued that it had been blown up from the sea at Portsmouth, and then we talked of something else. There were two slates down at the parsonage and a big tree in Lumley's meadow. It was a rare storm.

I reckon the wind had blown our ghosts all over England. They were coming back for days afterwards with foundered horses and as footsore as possible, and they were so glad to get back to Fairfield that some of them walked up the street crying like little children. Squire said that his great-grandfather's great-grandfather hadn't looked so dead-beat since the battle of Naseby, and he's an educated man.

What with one thing and another, I should think it was a week before we got straight again, and then one afternoon I met the landlord on the green and he had a worried face. 'I wish you'd come and have a look at that ship in my field,' he said to me; 'it seems to me it's leaning real hard on the turnips. I can't bear thinking what the missus will say when she sees it.'

I walked down the lane with him, and sure enough there was a ship in the middle of his field, but such a ship as no man had seen on the water for three hundred years, let alone in the middle of a turnip-field. It was all painted black and covered with carvings, and there was a great bay window in the stern for all the world like the Squire's drawing-room. There was a crowd of little black cannon on deck and looking out of her portholes, and she was anchored at each end to the hard ground. I have seen the wonders of the world on picture-post-cards, but I have never seen anything to equal that.

'She seems very solid for a ghost-ship,' I said, seeing the landlord was bothered.

'I should say it's a betwixt and between,' he answered, puzzling it over, 'but it's going to spoil a matter of fifty turnips, and missus she'll want it moved.' We went up to her and touched the side, and it was as hard as a real ship. 'Now there's folks in England would call that very curious,' he said.

Now I don't know much about ships, but I should think that that ghost-ship weighed a solid two hundred tons, and it seemed to me that she had come to stay, so that I felt sorry for landlord, who was a married man. 'All the horses in Fairfield won't move her out of my turnips,' he said, frowning at her.

Just then we heard a noise on her deck, and we looked up and saw that a man had come out of her front cabin and was looking down at us very peaceably. He was dressed in a black uniform set out with rusty gold lace, and he had a great cutlass by his side in a brass sheath. 'I'm Captain Bartolomew Roberts', he said, in a gentleman's voice, 'put in for recruits. I seem to have brought her rather far up the harbour.'

'Harbour!' cried landlord; 'why, you're fifty miles from the sea.'

Captain Roberts didn't turn a hair. 'So much as that, is it?' he said coolly. 'Well, it's of no consequence.'

Landlord was a bit upset at this. 'I don't want to be unneighbourly', he said, 'but I wish you hadn't brought your ship into my field. You see, my wife sets great store on these turnips.'

The captain took a pinch of snuff out of a fine gold box that he pulled out of his pocket, and dusted his fingers with a silk handker-chief in a very genteel fashion. 'I'm only here for a few months,' he said; 'but if a testimony of my esteem would pacify your good lady I should be content,' and with the words he loosed a great gold brooch from the neck of his coat and tossed it down to landlord.

Landlord blushed as red as a strawberry. 'I'm not denying she's fond of jewellery', he said, 'but it's too much for half a sackful of turnips.' And indeed it was a handsome brooch.

The captain laughed. 'Tut, man,' he said, 'it's a forced sale, and you deserve a good price. Say no more about it'; and nodding goodday to us, he turned on his heel and went into the cabin. Landlord walked back up the lane like a man with a weight off his mind. 'That tempest has blowed me a bit of luck,' he said; 'the missus will be main pleased with that brooch. It's better than the blacksmith's guinea any day.'

Ninety-seven was Jubilee year, the year of the second Jubilee, you remember, and we had great doings at Fairfield, so that we hadn't much time to bother about the ghost-ship, though anyhow it isn't our way to meddle in things that don't concern us. Landlord, he saw his tenant once or twice when he was hoeing his turnips and passed the time of day, and landlord's wife wore her new brooch to church every Sunday. But we didn't mix much with the ghosts at any time, all except an idiot lad there was in the village, and he didn't know the difference between a man and a ghost, poor innocent! On Jubilee Day, however, somebody told Captain Roberts why the church bells were ringing, and he hoisted a flag and fired off his guns like a royal Englishman. 'Tis true the guns were shotted, and one of the round shot knocked a hole in Farmer Johnstone's barn, but nobody thought much of that in such a season of rejoicing.

It wasn't till our celebrations were over that we noticed that anything was wrong in Fairfield. 'Twas shoemaker who told me first about it one morning at the Fox and Grapes. 'You know my great great-uncle?' he said to me.

'You mean Joshua, the quiet lad,' I answered, knowing him well.

'Quiet!' said shoemaker indignantly. 'Quiet you call him, coming home at three o'clock every morning as drunk as a magistrate and waking up the whole house with his noise.'

'Why, it can't be Joshua!' I said, for I knew him for one of the most respectable young ghosts in the village.

'Joshua it is,' said shoemaker; 'and one of these nights he'll find himself out in the street if he isn't careful.'

This kind of talk shocked me, I can tell you, for I don't like to hear a man abusing his own family, and I could hardly believe that a steady youngster like Joshua had taken to drink. But just then in came butcher Aylwin in such a temper that he could hardly drink his beer.

'The young puppy! the young puppy!' he kept on saying; and it was some time before shoemaker and I found out that he was talking about his ancestor that fell at Senlac.

'Drink?' said shoemaker hopefully, for we all like company in our misfortunes, and butcher nodded grimly.

'The young noodle,' he said, emptying his tankard.

Well, after that I kept my ears open, and it was the same story all over the village. There was hardly a young man among all the ghosts of Fairfield who didn't roll home in the small hours of the morning the worse for liquor. I used to wake up in the night and hear them stumble past my house, singing outrageous songs. The worst of it was that we couldn't keep the scandal to ourselves, and the folk at Green-hill began to talk of 'sodden Fairfield' and taught their children to sing a song about us:

Sodden Fairfield, sodden Fairfield, has no use for bread-and-butter,
Rum for breakfast, rum for dinner, rum for tea, and rum for supper!

We are easy-going in our village, but we didn't like that.

Of course we soon found out where the young fellows went to get the drink, and landlord was terribly cut up that his tenant should have turned out so badly, but his wife wouldn't hear of parting with the brooch, so that he couldn't give the Captain notice to quit. But as time went on, things grew from bad to worse, and at all hours of the day you would see those young reprobates sleeping it off on the village green. Nearly every afternoon a ghost-wagon used to jolt down to the ship with a lading of rum, and though the older ghosts seemed inclined to give the Captain's hospitality the go-by, the youngsters were neither to hold nor to bind.

So one afternoon when I was taking my nap I heard a knock at the door, and there was parson looking very serious, like a man with a job before him that he didn't altogether relish. 'I'm going down to talk to the Captain about all this drunkenness in the village, and I want you to come with me,' he said straight out.

I can't say that I fancied the visit much myself, and I tried to hint to parson that as, after all, they were only a lot of ghosts, it didn't very much matter.

'Dead or alive, I'm responsible for their good conduct', he said, 'and I'm going to do my duty and put a stop to this continued disorder. And you are coming with me, John Simmons.' So I went, parson being a persuasive kind of man.

We went down to the ship, and as we approached her I could see the Captain tasting the air on deck. When he saw parson he took off his hat very politely, and I can tell you that I was relieved to find that he had a proper respect for the cloth. Parson acknowledged his salute and spoke out stoutly enough. 'Sir, I should be glad to have a word with you.'

'Come on board, sir, come on board,' said the Captain, and I could tell by his voice that he knew why we were there. Parson and I climbed up an uneasy kind of ladder, and the Captain took us into the great cabin at the back of the ship, where the bay-window was. It was the most wonderful place you ever saw in your life, all full of gold and silver plate, swords with jewelled scabbards, carved oak chairs, and great chests that looked as though they were bursting with guineas. Even parson was surprised, and he did not shake his head very hard when the Captain took down some silver cups and poured us out a drink of rum. I tasted mine, and I don't mind saying that it changed my view of things entirely. There was nothing betwixt and between about that rum, and I felt that it was ridiculous to blame the lads for drinking too much of stuff like that. It seemed to fill my veins with honey and fire.

Parson put the case squarely to the Captain, but I didn't listen much to what he said; I was busy sipping my drink and looking through the window at the fishes swimming to and fro over landlord's turnips. Just then it seemed the most natural thing in the world that they should be there, though afterwards, of course, I could see that that proved it was a ghost-ship.

But even then I thought it was queer when I saw a drowned sailor float by in the thin air with his hair and beard all full of bubbles. It was the first time I had seen anything quite like that at Fairfield.

All the time I was regarding the wonders of the deep, parson was telling Captain Roberts how there was no peace or rest in the village owing to the curse of drunkenness, and what a bad example the youngsters were setting to the older ghosts. The Captain listened very attentively, and only put in a word now and then about boys being boys and young men sowing their wild oats. But when parson had finished his speech he filled up our silver cups and said to parson, with a flourish, 'I should be sorry to cause trouble anywhere where I have been made welcome, and you will be glad to hear that I put to sea tomorrow night. And now you must drink me a prosperous voyage.' So we all stood up and drank the toast with honour, and that noble rum was like hot oil in my veins.

After that Captain showed us some of the curiosities he had brought back from foreign parts, and we were greatly amazed, though afterwards I couldn't clearly remember what they were. And then I found myself walking across the turnips with parson, and I was telling him of the glories of the deep that I had seen through the window of the ship. He turned on me severely. 'If I were you, John Simmons,' he said, 'I should go straight home to bed.' He has a way of putting things that wouldn't occur to an ordinary man, has parson, and I did as he told me.

Well, next day it came on to blow, and it blew harder and harder, till about eight o'clock at night I heard a noise and looked out into the garden. I dare say you won't believe me, it seems a bit tall even to me, but the wind had lifted the thatch of my pigsty into the widow's garden a second time. I thought I wouldn't wait to hear what widow had to say about it, so I went across the green to the Fox and Grapes, and the wind was so strong that I danced along on tip-toe like a girl at the fair. When I got to the inn landlord had to help me shut the door; it seemed as though a dozen goats were pushing against it to come in out of the storm.

'It's a powerful tempest,' he said, drawing the beer. 'I hear there's a chimney down at Dickory End.'

'It's a funny thing how these sailors know about the weather,' I answered. 'When Captain said he was going tonight, I was thinking it would take a capful of wind to carry the ship back to sea, but now here's more than a capful.'

'Ah, yes,' said landlord, 'it's tonight he goes true enough, and, mind you, though he treated me handsome over the rent, I'm not sure it's a loss to the village. I don't hold with gentrice who fetch their drink from London instead of helping local traders to get their living.'

'But you haven't got any rum like his,' I said to draw him out.

His neck grew red above his collar, and I was afraid I'd gone too far; but after a while he got his breath with a grunt.

'John Simmons,' he said, 'if you've come down here this windy night to talk a lot of fool's talk, you've wasted a journey.'

Well, of course, then I had to smooth him down with praising his rum and Heaven forgive me for swearing it was better than Captain's. For the like of that rum no living lips have tasted save mine and parson's. But somehow or other I brought landlord round, and presently we must have a glass of his best to prove its quality.

'Beat that if you can!' he cried, and we both raised our glasses to our mouths, only to stop half-way and look at each other in amaze. For the wind that had been howling outside like an outrageous dog had all of a sudden turned as melodious as the carol-boys of a Christmas Eve.

'Surely that's not my Martha,' whispered landlord; Martha being his great-aunt that lived in the loft overhead.

We went to the door, and the wind burst it open so that the handle was driven clean into the plaster of the wall. But we didn't think about that at the time; for over our heads, sailing very comfortably through the windy stars, was the ship that had passed the summer in landlord's field. Her portholes and her bay-window were blazing with lights, and there was a noise of singing and fiddling on her decks. 'He's gone', shouted landlord above the storm, 'and he's taken half the village with him!' I could only nod in answer, not having lungs like bellows of leather.

In the morning we were able to measure the strength of the storm, and over and above my pigsty there was damage enough wrought in the village to keep us busy. True it is that the children had to break down no branches for the firing that autumn, since the wind had strewn the woods with more than they could carry away. Many of our ghosts were scattered abroad, but this time very few came back, all the young men having sailed with Captain; and not only ghosts, for a poor half-witted lad was missing, and we reckoned that he had stowed himself away or perhaps shipped as cabin-boy, not knowing any better.

What with the lamentations of the ghost-girls and the grumblings of families who had lost an ancestor, the village was upset for a while, and the funny thing was that it was the folk who had complained most of the carryings-on of the younsters, who made most noise now that they were gone. I hadn't any sympathy with shoemaker or butcher, who ran about saying how much they missed their lads, but it made me grieve to hear the poor bereaved girls calling their lovers by name on the village green at nightfall. It didn't seem fair to me that they should have lost their men a second time, after giving up life in order to join them, as like as not. Still, not even a spirit can be sorry for ever, and after a few months we made up our mind that the folk who had sailed in the ship were never coming back, and we didn't talk about it any more.

And then one day, I dare say it would be a couple of years after, when the whole business was quite forgotten, who should come trapes-ing along the road from Portsmouth but the daft lad who had gone away with the ship, without waiting till he was dead to become a ghost. You never saw such a boy as that in all your life. He had a great rusty cutlass hanging to a string at his waist, and he was tattooed all over in fine colours, so that even his face looked like a girl's sampler. He had a handkerchief in his hand full of foreign shells and old-fashioned pieces of small money, very curious, and he walked up to the well outside his mother's house and drew himself a drink as if he had been nowhere in particular.

The worst of it was that he had come back as soft-headed as he went, and try as we might we couldn't get anything reasonable out of him. He talked a lot of gibberish about keel-hauling and walking the plank and crimson murders—things which a decent sailor should know nothing about, so that it seemed to me that for all his manners Captain had been more of a pirate than a gentleman mariner. But to draw sense out of that boy was as hard as picking cherries off a crab-tree. One silly tale he had that he kept on drifting back to, and to hear him you would have thought that it was the only thing that happened to him in his life. 'We was at anchor', he would say, 'off an island called the Basket of Flowers, and the sailors had caught a lot of parrots and we were teaching them to swear. Up and down the decks, up and down the decks, and the language they used was dreadful. Then we looked up and saw the masts of the Spanish ship outside the harbour. Outside the harbour they were, so we threw the parrots into the sea and sailed out to fight. And all the parrots were drownded in the sea and the language they used was dreadful.' That's the sort of boy he was, nothing but silly talk of parrots when we asked him about the fighting. And we never had a chance of teaching him better, for two days after he ran away again, and hasn't been seen since.

That's my story, and I assure you that things like that are happening at Fairfield all the time. The ship has never come back, but somehow as people grow older they seem to think that one of these windy nights, she'll come sailing in over the hedges with all the lost ghosts on board. Well, when she comes, she'll be welcome. There's one ghost-lass that has never grown tired of waiting for her lad to return. Every night you'll see her out on the green, straining her poor eyes with looking for the mast-lights among the stars. A faithful lass you'd call her, and I'm thinking you'd be right.

Landlord's field wasn't a penny the worse for the visit, but they do say that since then the turnips that have been grown in it have tasted of rum.

Ambitious Jimmy Hicks

〜〜〜

JOHN MASEFIELD

'WELL,' said the captain of the foretop to me, 'it's our cutter today, and you're the youngest hand, and you'll be bowman. Can you pull an oar?' 'No,' I answered. 'Well, you'd better pull one today, my son, or mind your eye. You'll climb Zion's Hill tonight if you go catching any crabs.' With that he went swaggering along the deck, chewing his quid of sweet-cake. I thought lugubriously of Zion's Hill, a very different place from the one in the Bible, and the longer I thought, the chillier came the sweat on my palms. 'Away cutters,' went the pipe a moment later. 'Down to your boat, foretopmen.' I skidded down the gangway into the bows of the cutter, and cast the turns from the painter, keeping the boat secured by a single turn. A strong tide was running, and the broken water was flying up in spray. Dirty water ran in trickles down my sleeves. The thwarts were wet. A lot of dirty water was slopping about in the well. 'Bowman,' said the captain of the foretop, 'why haven't you cleaned your boat out?' 'I didn't know I had to.' 'Well, next time you don't know we'll jolly well duck you in it. Let go forrard. Back a stroke, starboard. Down port, and shove her off.' 'Where are we going?' asked the stroke. 'We're goin' to the etceteraed slip to get the etceteraed love letters. Now look alive in the bows there. Get your oars out and give way. If I come forward with the tiller your heads'll ache for a week.' I got out my oar, or rather I got out the oar which had been left to me. It was one of the midship oars, the longest and heaviest in the boat. With this I made a shift to pull till we neared the slip, when I had to lay my oar in, gather up the painter, and stand by to leap on to the jetty to make the boat fast as we came alongside. I have known some misery in my time, but the agony of that moment, wondering if I should fall headlong on the slippery green weed, in the sight of the old sailors smoking there, was as bitter as any I have suffered. The cutter's nose rubbed the dangling seaweed. I made a spring, slipped, steadied myself, cast the painter around the mooring-hook, and made the boat fast. 'A round turn and two half hitches', I murmured, as I passed the turns, 'and a third half hitch for luck.' 'Come off with your third half hitch,' said one of the sailors. 'You and your three half hitches. You're like Jimmy Hicks, the come-day go-day. You want to do too much, you do. You'd go dry the keel with a towel, wouldn't you, rather than take a caulk? Come off with your third hitch.'

Late that night I saw the old sailor in the lamp-room, cleaning the heavy copper lamps. I asked if I might help him, for I wished to hear the story of Jimmy Hicks. He gave me half a dozen lamps to clean, with a mass of cotton waste and a few rags, most of them the relics of our soft cloth working caps. 'Heave round, my son', he said, 'and get an appetite for your supper.' When I had cleaned two or three of my lamps I asked him to tell me about Jimmy Hicks.

'Ah,' he said, 'you want to be warned by him. You're too ambitious [i.e. fond of work] altogether. Look at you coming here to clean my lamps. And you after pulling in the cutter. I wouldn't care to be like Jimmy Hicks. No. I wouldn't that. It's only young fellies like you wants to be like Jimmy Hicks.' 'Who was Jimmy Hicks?' I asked; 'and what was it he did?'

'Ah,' said the old man, 'did you ever hear tell of the Black Ball line? Well, there's no ships like them ships now. You think them Cunarders at the buoy there, you think them fine. You should a seen the *Red Jacket*, or the *John James Green*, or the *Thermopylæ*. By dad, that *was* a sight. Spars—talk of spars. And skysail yards on all three masts, and a flying jib-boom the angels could have picked their teeth with. Sixty-six days they took, the Thames to Sydney Heads. It's never been done before nor since. Well, Jimmy Hicks he was a young, ambitious felly, the same as you. And he was in one of them ships. I was shipmates with him myself.

'Well, of all the red-headed ambitious fellies I think Jimmy Hicks was the worst. Yes, sir, I think he was the worst. The day they got to sea the bosun sent him to scrub the fo'c'sle. So he gets some sand and holystone and a three-cornered scraper, and he scrubs that fo'c'sle fit for an admiral. He begun that job at three bells in the morning watch, and he was doing it at eight bells, and half his watch below he was doing it, and when they called him for dinner he was still doing it. Talk about white. White was black alongside them planks. So in the afternoon it came on to blow. Yes, sir, it breezed up. So they had to snug her down. So Jimmy Hicks he went up and made the skysails fast, and then he made the royals fast. And then he come down to see had he got a good furl on them. And then up he went again and put a new stow on the skysail. And then he went up again to tinker the main royal bunt. Them furls of his, by dad, they reminded me of Sefton Park. Yes, sir; they was that like Sunday clothes.

'He was always like that. He wasn't never happy unless he was putting whippings on ropes' ends, or pointing the topgallant and royal braces, or polishing the brass on the ladders till it was as bright as gold. Always doing something. Always doing more than his piece. The last to leave the deck and the first to come up when hands were called. If he was told to whip a rope, he pointed it and gave it a rub of slush and Flemish-coiled it. If he was told to broom down the top of a deckhouse he got it white with holystone. He was like the poet—"Double, double, toil and trouble"—Shakespeare. That was Jimmy Hicks. Yes, sir, that was him. You want to be warned by him. You hear the terrible end he come to.

'Now they was coming home in that ship. And what do you suppose they had on board? Well, they had silks. My word, they was silks. Light as muslin. Worth a pound a fathom. All yellow and blue and red. All the colours. And a gloss. It was like so much moonlight. Well. They had a lot of that. Then they had China tea, and it wasn't none of your skilly. No, sir. It was tea the King of Spain could have drunk in the golden palaces of Rome. There was flaviour. Worth eighteen shillings a pound that tea was. The same as the Queen drank. It was like meat that tea was. You didn't want no meat if you had a cup of that. Worth two hundred thousand pound that ship's freight was. And a general in the army was a passenger. Besides a bishop.

'So as they were coming home they got caught in a cyclone, off of the Mauritius. Whoo! You should have heard the wind. O mommer, it just blew. And the cold green seas they kept coming aboard. Kerwoosh, they kept coming. And the ship she groaned and she strained, and she worked her planking open. So it was all hands to the pumps, general and bishop and all; and they kept pumping out tea, all ready made with salt water. That was all they had to live on for three days. Salt-water tea. Very wholesome it is, too, for them that like it. *And* for them that's inclined to consumption.

'By-and-by the pumps choked. "The silks is in the well," said the mate. "To your prayers, boys. We're gone up." "Hold on with prayers," said the old man. "Get a tackle rigged and hoist the boat out. You can pray afterwards. Work is prayer," he says, "so long as I command." "Lively there," says the mate. "Up there one of you with a block. Out to the mainyard-arm and rig a tackle! Lively now. Stamp and go. She's settling under us." So Jimmy Hicks seizes a tackle and they hook it on to the longboat, and Jimmy nips into the rigging with one of the blocks in his hand. And they clear it away to him as he goes. And she was settling like a stone all the time. "Look slippy there, you!" cries the mate, as Jimmy lays out on the yard. For the sea was crawling across the deck. It was time to be gone out of that.

'And Jimmy gets to the yard-arm, and he takes a round turn with his lashing, and he makes a half hitch, and he makes a second half hitch. "Yard-arm, there!" hails the mate. "May we hoist away?" "Hold on," says Jimmy, "till I make her fast," he says. And just as he makes his third half hitch and yells to them to sway away——Ker-woosh! there comes a great green sea. And down they all go—ship, and tea, and mate, and bishop, and general, and Jimmy, and the whole lash-up. All the whole lot of them. And all because he would wait to take the third half hitch. So you be warned by Jimmy Hicks, my son. And don't you be neither red-headed nor ambitious.'

Poor Old Man!

ᔰᔰᔰ

A. E. DINGLE

T WELVE HUNDRED sheep bleated on the after deck; a thousand cattle swayed and moaned in the pens beneath the hurricane deck. At the forward derrick the bosun and some cattlemen hove out and dumped into the sea a horse that had been cast and trampled to death by his terrified companions.

> Poor Old Man, your horse is dead!
> Oh, poor Old Man.

Chief mate Legge sang the ancient chanty unconsciously as the corpse was let go. He resumed his tramp of the bridge, peering with salt-stung eyes into the furious seas ahead. The long, lean steamer crashed her way homeward, seemingly invulnerable to the sustained hammering of tons of roaring ocean over her forecastle head. In the lee corner of the bridge Tyler, the junior who shared his watch, dodged the sprays rattling against the screen. The mate's old chanty only lived through one stanza, being a habit, as another's habit might be tunelessly whistling under stress; but the mirthless grin on Tyler's gaunt face had an upward twist at the corners at sound of it.

'It's getting dirty,' Legge shouted.

It was necessary to shout to be heard. The seas and the onrush of the big steamer filled the ears with thunder. Other officers could keep their watch in the snug wheelhouse. Legge, with his old-time notions, must stay outside in the bitter weather. That was all right. These square-rigger men were scarcely human anyhow. But there was no good reason why other men, doomed to stand watch with the madman, should shiver and freeze outside. The other watches didn't. It wouldn't hurt to let the Old Man know. Captain Gunter was not in love with his new chief officer. The whole ship knew that. Old

Gunter was suspicious of a new man taken into the employ and pushed at once into a chief officer's birth. Skippers as old as Captain Gunter needed to keep an eye on interlopers. Aged skippers had been supplanted by young mates before.

'Getting dirty,' Legge shouted again, staggering to a rail hold beside Tyler. 'That's the second horse dead. If the Old Man 'ud let me set a bit o' sail aft it 'ud hold her head up to the sea.'

'He doesn't believe in bits o' sail,' grinned Tyler. 'Bits o' sail are out o' date in steam.'

A two-mile-long sea reared up out of the eastward chaos. The steamer reeled, staggered to rise, shook to the keel bolts to the violent racing of her screw, and slid down into the hollow, while a hundred tons of hissing fury shot into the air on her weather bow and fell back roaring. The mate gasped. Only the top of that sea was hurled aboard by sheer impetus; but that top filled the foredeck, killing two more horses, and swept knee-deep over the after deck, washing twelve hundred sheep from their feet and half drowning them. The bosun and his men appeared, dumping the dead horses.

> When he's dead we'll bury him deep,
> Oh, poor Old Man!

Legge sang between his tight teeth, watching for the next sea to roll along. There was no mate to that big one, at least not in sight; but there was that in sky and sea which threatened even worse to come. The quartermaster in the dry wheelhouse snatched glances through brine-incrusted windows and gave queer names to officers who stood out there when they might as well be snug and warm inside.

'Go down and tell the Old Man it's growing dirty,' shouted Legge in Tyler's ear. 'He needn't come up. Ask him if I may set a bit o' canvas aft to ease her. A trys'l on mizzen and jigger 'ud hold her up fine.'

Tyler obeyed gleefully. He knew what the Old Man would say. There would be a fine chance to add a bit of his own when Captain Gunter roared out his ideas about meddlesome mates and sail-setting. So while Legge went on with his singing, and the lean, rusty steamer flogged her four naked masts and gaunt, salt-caked funnel athwart the sullen gale, Tyler listened with seeming meekness to the Old Man's response to his message.

'Says it's his opinion sail 'ud help her, sir,' Tyler interjected slyly. Captain Gunter stood glaring through his misty porthole; saw the derrick swing out. 'Two more horses killed that time, sir. One just before,' Tyler went on. 'If he was in the wheelhouse, where he could watch the course, sir, instead o' staying outside, he might—'

'Tell Mr Legge I'll be up,' snapped the skipper and snatched down his oilskin coat and boots, while Tyler hurried to the bridge full of warmth inside.

Legge stood on sturdy feet, swaying to the dizzy lurching of the ship, watching her every plunge and recovery. Life had not been overkind to him. He had been many dreary years waiting for command. He had been second mate, mate, and back to second mate again, chiefly through lack of influence, interest higher up, or, as he sometimes thought, personal impressiveness. Yet he had not soured. He was first of all a sailorman, and a good one. This cattle steamer, *Arranmore*, was old, slow, and he had been told that promotions went strictly by seniority. He was chief mate. It might be a long, weary climb, by way of chief officerships of all those other steamers, to ultimate command, which after all might be no better than back to the old *Arranmore* again; but if only merit was rewarded fairly and without favour, what could a man want more?

> For three long months I rode him hard,
> Oh, po-oo-oor O-o-old Man!

The gale shrieked, tore the winds into a ragged syllabus. And Captain Gunter clawed his way onto the bridge and appeared beside Mr Legge as the last 'Po-oo-oor O-o-old Man!' was flung broadcast. Tyler grinned, this promised to be good. The Old Man's face was purple. It was several moments before breath relieved him. Then—

'If ye'll stop that silly singing and mind yer ship, Mr Legge, maybe we'll not lose all our livestock this watch!'

'She's hard-mouthed, sir. A bit o' sail might help her,' the mate replied. He chose to let pass the Old Man's harsh comment.

'And I'll tell ye when I want my steamer turned into a trys'l tramp, mister. Mind yer steering and ye'll well enough.'

The Old Man turned to the bridge ladder again.

> When he's dead we'll bury him deep,
> Po-oo-or o-old Man!

Legge possibly was unaware that he was singing. It was a habit, nothing more. But it brought the Old Man back, ramping and raging, shaking his fist, to the great glee of the quartermaster inside and the more secret elation of Mr Tyler behind the dodger.

'Are ye singing that silly song at me?' the Old Man yelled.

'At you, sir?' Legge was staggered.

'Aye, at me! Watch yerself, I warn ye! Yer duty's none too well done in this ship, mister. I'll stand no impudence, mark that!'

Legge could find no words. He watched the skipper go off the bridge as he might have watched a cow climb a mast. Above all the uproar of storm and terrified beasts the slamming of Captain Gunter's stateroom door was audible even in the wheelhouse.

Two minutes after the door slammed the *Arranmore* rolled down heavily. Less than a minute after that a great sea roared up, and she met it with a terrific shock. Like that other earlier sea, but with tenfold the murderous force, just the crest toppled aboard the after deck, and the rails went like straws. Twelve hundred sheep were swept along with the wreckage of their pens to litter the sea and the after deck. Bleeting, kicking, the unfortunate muttons that survived scrambled madly in the welter of sea and pen lumber. Mr Legge was anxiously peering aft from the bridge, hoping for the best. Another great sea rolled up, silent as death, and took the steamer amidships, filling her foredeck, crashing against the bridge, forcing both officers to hang on for their lives.

When that sea passed the sheep deck was clean. The ocean astern was foamy with wool. Mr Legge cleared his eyes of brine and tried to get his bearings. Tyler clung up to windward, white and grinning; he clung to a ragged end of pipe rail, at half distance between the wheelhouse and the bridge end. All beyond that was gone. Steel and teak: all gone like paper and straw. The wheelhouse was smashed in. There was nobody at the wheel. Three spokes only remained of the wheel itself. There was a huddle of blue in the splintered debris of the steering grating; a hand stuck out of the huddle; a spoke was fast in the grip of the hand.

Legge groped along to the wrecked end. He meant to send Tyler for the Old Man. At the same time he wondered why any Old Man could remain below after that sea struck his ship. It had been like striking a cliff. At the beginning of the splintered planking where the bridge was sheared off he stopped in swift dismay. The boat deck, all along the starboard side, was down. Steel beams, stanchions, boats and davits lay flattened in a mad chaos, crushing down upon the bellowing cattle below; and wreckage lay jammed tight across the Old Man's door. The door to the officers' mess-room and cabins was buried in a mass of twisted steel and wreck of boats.

'Get all hands along!' Legge bawled in Tyler's ear. 'Tell Chips to try to get the Old Man out. Send Bosun to me.'

Tyler took the lee side, not sorry to leave that shattered bridge. Legge seized the broken wheel and fought to bring the steamer back to her course. She rolled across the dizzy seas, strewing the leeward side with carcasses and debris. One of the old type of steamers, the *Arranmore* steamed with a hand wheel fitted with a clamp to connect it with steam gear when entering or leaving port. Sea passages were made with the hand gear, and there was plenty of work in the steering of her.

While the mate fought with the broken wheel the bosun came up, his oilskins in ribbons and his face and hands bruised and bleeding. He chewed tobacco busily, and grinned at the bridge wreckage. He was of the same school as Mr Legge.

'Bosun,' said Legge, 'I told Chips to get the Old Man clear. He won't do that for hours. You get tackles rigged and heave the wreckage of the boat deck clear of the cattle. And while you're aft for the tackles set the mizzen and jigger trysails to ease her. Take all hands. Make the cattlemen help. I'll look out for the helm.'

While the seas swept over broken rails and battered decks men clung with one hand and worked tremendously with the other, setting two fluttering straps of dingy canvas aft to steady the ship; crawling among jagged steel ends to fasten tackles; taking falls along to snatch blocks and winches to heave the pressure off the cattle decks.

Chips pried and pounded for ten minutes to free the captain, comments from the open port burning his ears while he worked.

'What's being done to my ship, carpenter?' the Old Man demanded.

'They've set trys'ls aft, to ease her,' Chips grunted, heaving on a pinch bar.

'And this is the result! Get me out of here before they lose me my steamer!'

Another sea lifted aboard on the damaged side. The wreckage surged dangerously, only to be left in a tighter jam. Cattle moaned, cattlemen cursed, a winch clattered, and a tackle set taut. The tackle snapped, the sling of broken iron and shattered wood crashed back upon the cattle pens.

'What's the matter, carpenter? Can't you move that wreck?' the skipper cried, shoving desperately at the door. 'My steamer is being cast away! I can feel her labouring!'

'The ship's all right,' growled Chips. 'Ought to have set them trys'ls before. I'll have to quit this, cap'n. They need all hands down there freein' them steers.'

'Don't you leave me shut up here! That madman on the bridge will drown me!'

'You're safer'n wot he is. The mate knows what he's doing.'

Chips was gone before the Old Man could say more. And for two hours, while the seas roared and leaped, the steamer kept her head towards them with the help of the tiny trysails and the stout-hearted seamanship of the lonely figure on the broken bridge, handling the spokes of the shattered wheel.

Down on the cattle deck men toiled like heroes in the hideous maze of maddened beasts and tangled wreckage, slinging out carcasses, raising imprisoning obstacles, shoring up and securing parts of the ship's structure against further damage.

Every time a man passed the Old Man's porthole in taking along a fall or shifting a tackle the Old Man vociferously commanded him to send somebody to move the wreck from around his door. When the bosun happened along, still grinning as he chewed his quid, Captain Gunter reached out a powerful hand and stopped him.

'Where's Mr Tyler? Send him up here!'

'He's lending a hand along with the rest, sir. Mr Legge is the only man not at the wreckage, and he's holding the hellum.'

'Will you send somebody to me instantly! Send Mr Tyler. Send somebody with sense. I'll log the lot o' ye for insubordination if I'm not out o' here in ten minutes! Ye're all in a conspiracy wi' the mate to ruin me wi' the owners. Away wi' ye, and do as I bid ye!'

Mr Legge stood at the broken wheel. His sou'wester had blown off through the shattered wheelhouse windows. He sweated. His face was salt-crusted and haggard. When he first took the helm he had ordered steam on the steering-gear by means of the speaking-tube, and it had been turned on, with much engine-room profanity at the prodigal waste of steam. Winches and steering-gear using steam at sea! Damned poor seamanship, that! But the steam had come; and in the first whirling of the mutilated wheel Legge had been hurled to the deck by a blow of the broken teak-rim. He had fallen upon the blue-clothed figure whose hand still held a spoke, and the experience had shaken him. He yelled down the tube for the steam to be turned off, disengaged the clamping lever, and thereafter steered by strength and cunning. He kept the labouring steamer to the seas, none but he

knew how, but he was afraid for the remaining cattle. The ship put
her bows deeply into the rolling surges, shaken to the keelson.

The bosun appeared, wiping a brown streak across his humorous
face.

'Old Man says we're all tryin' to bust him, and you're at the head
of it,' he shouted. Mr Legge's teeth clicked. He hove at the wheel.
The bosun grinned, adding, 'Orders me to send somebody to get him
out, sir. What about it? Them steers are in bad shape if we don't get
the boat deck raised.'

'Hold this wheel. I'll take a look myself,' said the mate. 'Watch
yourself, bosun! That ragged wheel kicks like a crippled horse.'

Legge stood outside the captain's room and inspected the damage.
He paid no heed to the scurrilous remarks from the porthole. Satisfied
that the job of extricating the skipper was far bigger than that of
releasing the animals, that the other officers were as completely imprisoned, and that none of the human prisoners was in danger so long
as the steamer floated, he at last gave a scant word of respectful
decision to his captain.

'It'll take an hour to release you, sir. I want every man down on the
cattle deck. The ship's all right. Have you out in a twink when the
cattle are secure. Any orders, sir?'

'My orders are that you get me out! Damme, I'll log ye and break
ye if ye don't, Mr Legge.'

'As soon as the cattle are secure, sir.'

Relieving the bosun at the wheel again, the mate sent him back to
his labours.

'Tell Chips to take another look at the Old Man's door,' Legge
suggested. 'Needn't waste any time, but we don't want to let Captain
Gunter say he was kept prisoner out o' spite; not to the owners,
anyhow.'

Chips looked. Chips went back to cleaning the wreck. And through
seemingly interminable hours, long after watches ought to have been
changed, Legge fought with the jagged wheel, and kept the steamer
to the seas. A steward brought him hot tea and hard biscuits. That
was his dinner. The stewards were having their troubles. A sea had
smashed in the foredeck parts of the saloon, and cracked a bulkhead
plate. The lads were kept busy with in-creeping water. The bosun,
careful of the proprieties, told Tyler the captain had sent for him, and
Tyler went to see what was wanted. It pleased him when the Old Man
ordered him to take men from the other wreck and release him

instantly. But neither bosun nor carpenter would give him men. They had their orders from the chief officer, and their work was more pressing. Tyler had another pleasant moment when he had to tell Captain Gunter he could get no men.

'Mate's orders, sir,' he was scrupulous to explain, and his lips drew back from his teeth at the torrent of deep-sea vituperation that answered him. He could not get into his own room for a smoke any more than he could get the skipper out; so he went back to his job, buoyed up by the certainty that the end of the passage would see one chief mate at least fired out of a job; and that must, logically, mean promotion for juniors.

Legge peered through the driving spray-mists. He felt the strained motion of the steamer, but knew that, as far as was humanly possible, the *Arranmore* was handled perfectly. There was no change in the barometer; night was coming on; the wind shrieked round and through the broken wheelhouse. The sounds from the cattle deck had subsided to a low organ note of beefy moans, with a rare equine squeal to break the drone of it. There were only a few horses left. So far the livestock had suffered sadly. Whatever happened thereafter, the profits of the voyage were about shot to pieces. There would be no reputation made with the *Arranmore* owners. And worst of all, so far as Legge was concerned, he fully believed, and was doubtless right, that the setting of a bit of canvas aft much earlier in the storm would have prevented those two or three vicious seas boarding and sweeping her. He might have felt sorry for the Old Man but for that.

> They say, Old Man, your horse is dead;
> We say so, and we hope so!
> They say, Old Man, your horse is dead;
> Oh, po-oo-or O-old Man!

Though he sang, Legge was not thinking of the song. Even as he droned out the dirge-like lines his glance went towards the dead hand of the blue-clad figure at his feet, still clutching the spoke that had been the handle of its duty while it lived. Hurriedly he took an ensign from the flag locker and carefully laid it over the heap of gratings and wheel and steersman.

Just when darkness began to set in the mate left the wheel for an instant to snatch down the old-fashioned log slate which he personally used in his own watch to record the happenings of that period.

He set down the course. It was several points to the north of the proper course, the divergence made necessary in order to hold the steamer to the seas, so that she took them on the bluff of the bow instead of in the waist. He set down a terse record of damages, dashing a word here, a word there during attention to the wheel. The Old Man and the other officers keeping watches had long ago discarded the slate. Legge held to it, and now was glad, because it was far handier to use while bucking that jagged wheel.

He entered the accident which swept the ship clean of sheep and imperilled half her cattle. A note was made about the imprisonment of the captain and officers. Legge put the pencil into his mouth and let the slate swing by the string while he brought the steamer to her course; and there arose out of the dark east the father of all seas, meeting the ship squarely on the bow. It filled the foredeck. It swept away all that part of the broken boat deck which had been laboriously raised to free the cattle below. It took with it the beasts, maimed or whole, which had just been freed. It swept men with it. The shattered bridge was shorn of another ten feet. The wheel-house suffered, and Legge was hurled headlong across the flag he had lain down. As he picked himself up, dazed, the pencil still in his teeth and the broken string dangling, the bosun staggered up the intact ladder, still grinning through a smear of red and brown, wiping his sleeve across his mouth.

'That was a hot package,' he grinned. Legge groped for the slate with one hand, holding the wheel with the other. 'Won't be enough beasts left to be worth savin' if we ship another like that, sir.'

'How are the pens on the lee side?' Legge asked, still groping.

'All right so far. There's no shelter, though. All the weather side is stove in. Couple o' men hurt. Dead cows floatin' everywhere.'

'Get some sort of shelter rigged, bosun. Got to save all we can.'

Legge found the slate down alongside the dead hand holding the broken spoke. He picked it up, absently moistened the pencil with his tongue, and entered the last catastrophe. The bosun took the wheel unasked while Legge wrote. Then, while the mate sucked his pencil in thought and the bosun experienced the fiendish perversity of the crippled wheel, both gazed out through the gaping front of the wheel-house upon the grey tumultuous seas that leaped to meet the lowering sky round the whole circle of the horizon.

There was not much in the prospect to cheer the mate. No matter how the passage ended, he stood to gain nothing. If the *Arranmore* got any of her live cargo home, the Old Man would get the credit. If none survived, the Old Man would lay it at the mate's door. A gang of men taken from the vital work of preserving cattle might extricate Captain Gunter and the barricaded mates in an hour. The beasts would suffer. Perhaps not one would come through alive. But at any rate the responsibility thereafter would be upon the Old Man's shoulders. Why should a man consider Captain Gunter? That was what Legge was asking of himself as he scanned the tremendous upheaval of sea. More than one man had heard the skipper assert his conviction that Legge was seeking to ruin him with the owners; seeking to keep him prisoner while he worked his devious will on the ship. The bosun, incapable of inventing such a yarn, had reported just such an assertion. Legge turned to the bosun. He would give orders to put a gang to work at the Old Man's door. The bosun was staring ahead. Legge, with a hand outstretched to take the helm, stared too; and into the narrow field of vision between two wrecked windows danced a sizeable speck, vivid against the sky when high-flung on the seas.

'A ship, dismasted!' the bosun shouted. Legge had taken up his glasses. 'White-painted, white masts and yards,' the bosun added. His eyes needed no glasses.

'Looks like one o' them Sierras, sir.'

Legge was peering at the two-flag hoist flying from the mizzen topmast of a three-masted, full-rigged ship. The fore and main lower masts stood, with the lower yards swinging wildly, all above gone. The mizzen lower and topmasts stood, with their yards, and from every spar streamed ribbons of canvas.

'N.C. "In distress! Need assistance!" ' read the mate. He looked a while longer. The derelict, if such she proved to be, lay head to sea by sheer force of wind upon the naked mizzen-mast, a full topmast loftier than the main or fore. 'Lot o' people on her poop. Looks like one's got a skirt on too.'

Dropping the glasses to the strap about his neck, the mate took the helm.

'Can't do nothin' in this sea. Got troubles of our own,' the bosun muttered a bit wistfully, casting a last glance towards the sailing-ship as he turned to put the crew to work again making shelter for the surviving cattle. 'Shall I ask Mr Tyler to try to dig the Old Man out, sir? Can spare a few hands now.'

Legge had spun the wheel. The *Arranmore* sheered a point aside from her course as if to pass clear of the pitiful ship with its pleading two-flag signal. His salt-stung eyes were puckered; his haggard face looked grey in the fading light.

'Leave the Old Man alone. Any man who'll talk about his chief officer like he's done is likely to see ruin in that ship ahead. We'll do without him. Can't spare hands anyhow. Send Mr Tyler to me, bosun; then you leave Chips with half the hands to secure the cattle deck, and take the rest of the men aft. I'm going to take those people off that ship, whatever we may have to do with the ship itself.'

'Boats, sir?' asked the bosun briskly, already at the ladder head, snatching a huge quid of tobacco from a fresh plug. His brown, smeared, humorous face was suddenly all alight.

'No; not boatmen enough. Get gear ready to rig up a breeches-buoy from our jigger-mast to her fore lower-mast. And rouse out a hawser for towing. We can tow her that long anyhow. Send Mr Tyler up right away. I want him to signal that ship.'

Mr Tyler came. He was ragged, wet, wholly miserable. He had worked. Oh, yes. Down there in the mad frenzy of wreck and drowning beasts every man had worked. But the bridge looked good to Mr Tyler; even the poor shattered remnant of it. His teeth gleamed in a grin as he joined the mate.

'Get out the flags. Signal that ship to stand by for line as we steam past,' Legge jerked out rapidly.

Tyler looked, saw for the first time the distressed ship, and his face turned chalk white.

'You can't handle this steamer that close. You'll smash us both!' he stuttered.

'Make up the signal! Tell him we'll not abandon him! Get a move on!'

Tyler took another look, then dropped the flag he had taken from the locker, and ran from the bridge. Legge gauged the distance between the vessels, doubting visibility in the deepening darkness unless the signals were made immediately. Then he took his whistle and blew the piercing summons that would bring a quartermaster running. Not waiting longer, he stepped to the outside bridge and began waving his arms in the flag-wag code towards the wreck, hoping that somebody would be able to see and understand.

The steamer rolled sickeningly. The ship performed antics that took away the breath even of anyone looking on. Legge had keen vision; but the ships were very close before he detected an answering wave to his signal. But once established, communication was swift and certain.

'I'll take your people off in a breeches-buoy forward, and hold you in tow meanwhile,' he signalled; and a weak and watery yell went up across the sea in gratitude. A sailor appeared on the bridge in reply to the whistle, and to him Legge gave the necessary orders to carry aft, so that when the steamer foamed past to windward of the ship no time might be lost in getting a line across.

'Where's Mr Tyler?' he snapped.

'He took three hands to help the captain out, sir,' the man replied, and ran aft with orders.

There was just enough light left to see that somebody was getting a flare ready on the ship. Legge seized the fire axe in the wheelhouse and smashed away all the remaining woodwork between him and the side the ship was on. He could see a man, perhaps the master, leaning out over the near rail as if in alarm at the dangerous course the steamer was taking. But there was no time for sending assurance. Legge snatched time to wigwag a curt 'Stand by!' Then the *Arranmore* roared past the ship's stern, and her bow wave leaped high over the poop as the ship rolled down to meet it. There was a terrified sensation of the two vessels being dragged together. Legge felt a dry lump rise in his throat. Men on the ship screamed warning. Somewhere abaft the bridge a hollow crash sounded. There arose another yell. Somebody shouted something about the funnel. Stubbornly the mate held on his way. He knew the ships touched once, barely touched, but the shudder of it went through the old *Arranmore*. The frantic profanity of Captain Gunter issued from his porthole and reached the bridge.

But another yell went up. The steamer was clear. Six men ran along the ship's clear decks, carrying the heaving line which the bosun had shot across, taking it to the forecastle, where they could handle the heavier line. Then Legge left the wheel long enough to telegraph to the engine room to slow down the engines. He snatched time to look aft. The mizzen-topsail yard of the ship stuck through the *Arranmore*'s funnel like a javelin through a tree-trunk. Braces and fathoms of running gear trailed from it ludicrously. But men were hauling a line in over the ship's forecastle head; men at the *Arranmore*'s stern toiled like fiends under the quiet efficiency of the bosun, paying out line, bending on, paying out again, until, with the steamer wallowing under reduced steam scarcely two lengths clear of the ship, Legge saw the heavy towline slowly snaking over the savage crests, perhaps after all to rob them of their prey.

It was dark then. Only keen, sea-trained eyes could follow what was going on. Somebody lit a flare on the ship, and her people stood out black and animated in the sudden light. Legge switched on the binnacle light and the wheel-house shaded light at the chart board. And he felt an abrupt loss of strength. He felt as if sinking. The strain had been terrific; must continue to be terrific before that ship's crew was safe. He tottered to look aft again. Lights gleamed there. He saw the first man come across on the breeches-buoy. That was to test it. The woman would come next. It was all going to be well. But for what? There would be no glory for him. He had done what any sailorman would do; but he was not master of the steamer. When the master came up, if he were released in time, there would be a show-down. Legge realized at last what the term 'eternal mate' meant. It meant a man doomed never to rise to command; to spend his days a good chief mate without hope. Oh, well:

> When he's dead we'll bury him deep,
> They say so, and hope so.
> When he's dead we'll bury him deep,
> Oh, poo-oo-oor O-old Man!

Somebody had come onto the bridge. Legge sensed it. He did not look up. He needed all his faculties to handle the steamer—a touch ahead, a momentary stop, a bit ahead again to keep the strain on the towline even. But by the heavy breathing and the staccato snort now and then he knew it was the Old Man. Tyler slid into view too, grinning. They just had to come up when he sang that verse. Had Captain Gunter not appeared, Legge might not have been conscious he was singing. But the Old Man was speaking. He was peering round at the shattered bridge, the broken wheel, the pitiful blue-clad figure under the flag, with the spoke still gripped in the dead hand. What was said at the moment mattered little. The reeling steamer groaned and shivered under the burden she dragged. Captain Gunter shoved Tyler aside to look aft, where bundles swinging along the spidery line grew into human beings; where, under a flare still further away in the night, rolled and pitched a fine ship broken by the sea, doomed along with her people but for the *Arranmore* and the stubborn chief mate.

'I see ye have managed to lose a part o' my steamer, Mr Legge,' the Old Man said grimly.

'It was before I set the trys'ls,' answered Legge, without resentment.

'Mr Tyler tells me ye have lost most o' the cattle. Why did ye not let me out sooner?'

The Old Man stood where he could look aft all the time. He saw the shadowy shape of the topsail yard and its gear sticking into the funnel, and his grim old mouth twisted queerly. Who knows what may have been in his canny mind then? He was a good old sailorman at bottom; he had a reputation for shrewdness and regard for the bawbees. Something was working behind his dour exterior. His cattle were lost. But here was a shipwrecked crew being saved by his steamer; and if the steamer could hang on to a towline to take off the people, why might she not hang on a bit longer until the weather moderated? There was much soothing influence in a bit of salvage when an old ship master had suffered a bad voyage. But Legge answered his last query shortly.

'It was none of my orders you were released. I needed all hands to secure the cattle and pass those lines. Tyler went against my orders, wasting time on your door, sir. There were more important things doing.'

Tyler grinned at that. Now Legge was going to hear something. Captain Gunter peered and poked round. He muttered, and to Tyler's astonished ears the mutterings sounded very like, 'I dunno how any one man could ha' done it! Any man! Any man at all!'

Legge went on handling his ship. It was hard to stand on the staggering bridge. He was beyond caring what the Old Man said.

'Mr Legge'—the mate started at the Old Man's tone—'d'ye think if we set yonder fore and main trys'ls, and mebbe a bit o' forestays'l, it might ease her a bit more?'

Tyler gaped. Mr Legge darted a look at the skipper that was more than half suspicious. But the Old Man reached out, taking the wheel.

'Take men as they can be spared and set that canvas, Mr Legge. And ye can stand by to go aboard the tow as soon as her people are all off and carry her into port, if so be we're that lucky. Mr Tyler, I'll have ye stand by the telegraph here. Ye went against the mate's orders, and mebbe lost many a beast for me. Ye're not to be trusted. Away wi' ye, Mr Legge. Pick yer own men when ye go aboard the tow.'

Legge went down the ladder half dazed. But habit was strong.

> If he's not dead we'll ride him again,
> We say so, and we hope so!
> If he's not dead we'll ride him again—

Tyler, standing with his cold hand on the telegraph handle, scarcely yet realizing that Legge was going to the job where the salvage was fattest, was stupefied to hear behind him, in the skipper's queer, rusty old growl, the last line:

Oh, po-oo-oor O-old Man!

Easting Down

෧෧෧

SHALIMAR

T HE *Knightley*, a tramp steamer of about 5,000 tons gross, was lying in Victoria Basin, Cape Town. Breakfast in the saloon was nearly over. It had been eaten in comparative silence, for the *Knightley* was not a particularly happy ship; the captain and the chief officer thought very little of each other, the captain's thoughts being distinctly the less charitable. Toward the end of the meal there had been some talk about the ship's work, which encouraged the chief officer to get on to his favourite subject—cleanliness, and the slapping on of paint.

'The bulwarks on the fore-deck require chipping very badly, but we'll get that done during the next week or so,' he said. 'Luckily we're in for a fine weather passage.'

'Indeed, Mr Wilkins,' the captain remarked loftily.

He rose and left the saloon. The other officers drained their coffee cups and also prepared to go.

'What's he getting at now?' the chief officer growled.

His officers never knew what Captain Hartnell was getting at, and he certainly never took the trouble to enlighten them. He conveyed the impression that he was much too superior to reach down to their meagre intellects, and the impression was probably a correct one; for if ever a man suffered from the reverse of what, in modern jargon, is termed a complex of inferiority, that man was Captain Hartnell. With one exception not a man in the ship could ignore his own inferiority when in the captain's presence—the exception being the young third officer, a Scotsman from the Buchan district of Aberdeenshire, who, placid and laconic, had never discovered any reason why he should be either intimidated or unduly impressed even by a successful shipmaster like his captain who had never suffered the slightest check to his meteoric career and was, without doubt, an outstanding seaman.

Still only thirty-two Captain Hartnell had got command of a barque when he was twenty-four—a great, modern slab-sided barque that sailormen declared was too slow to get out of her own way. He had been chock-full of confidence from the moment he boarded her. Her code number in the International Book of Signals was M.N.B.S., and on the first night out, bowling down-channel before a fresh breeze, he was heard reciting a rhyme of his own composing:

> M.N.B.S.,
> My name brings success,
> Go, you flat-bottomed scow, go.

The 'flat-bottomed scow' went so well that he drove her out to Australia in twenty days less than his predecessor had taken on the previous voyage, and since that first successful passage he had never tired of proving to his owners that he was the smartest man in their employ. Because of that, and the fact that he was at all times acutely aware of his own merits, he was far from being a favourite with his brother shipmasters. Tall and muscular, he carried himself with a swagger, and when he passed over the gangway that forenoon even the independent stevedores on the quay wall greeted him with respect. He was hardly three yards away from the ship, however, when an able seaman named Kelly brushed against him, then lurched unsteadily up the gangway. Kelly had evidently been ashore all night and looked the worse for wear. It was strange, the captain thought, that the miserable chief officer had not reported one of the men absent without leave. Perhaps the chief officer did not even know. Pah!

Captain Hartnell walked on toward the town. The *Knightley* would be ready for sea in a couple of days and he had some preliminary clearing to do at the agent's office. The work done, he strolled along to the Grand Hotel and entered the lounge, where he found an acquaintance, a much older man who commanded another tramp steamer lying in the same dock. Both vessels were bound for Port Pirie, South Australia, in ballast, to load zinc concentrates, and over their drinks the two masters discussed their coming passages.

'I'm a couple of knots faster than you, so I'll probably pick you up somewhere to the nor'a'd of New Amsterdam, even if I sail a week later,' the older man chaffed.

'You won't pick me up at all, and certainly not on that route,' Captain Hartnell retorted. 'I'm going to run the easting down!'

'You're going to do *what?*'

'Run the easting down!'

'You'll regret it,' the other man said seriously.

'Why? I've done it in a barque not a quarter the size of the *Knightley*.'

'Oh, certainly; but the barque hadn't got a racing propeller under her counter,' the other said drily. 'What's the idea?'

'To make a quicker passage. Lots of steamers used to do it—the Aberdeen White Star, Shaw Savill, the New Zealand Shipping Company.'

'Yes, they *used* to, Hartnell; they don't now.' The other man sat back comfortably in his chair. 'Anyhow, a fine weather passage along the edge of the south-east trades as far as the hundredth meridian, then south-eastward toward the Australian coast will do me,' he declared. 'I'll probably get all the bad weather I want off the Leeuwin at this time of the year, and I'm not hankering after any more. By the way, didn't you have wireless that time I met you in New York?'

'Yes, but they took it out of her after the war. Hard times, couldn't afford it, the owners said.'

'H'm! false economy on these long voyages!'

Two days later, on a bright, calm morning, the *Knightley* hauled out of the docks, rounded Green Point, and stood down the shore of the Cape Peninsula. When the chief officer relieved the second on the bridge at four o'clock in the afternoon the faint outline of Cape Point was astern, and no land was visible along the port beam.

'Sou'-sou'-east is the course by the steering compass; south-twenty-east by the standard,' the second officer said.

'Sou'-sou'-east! Here, what the——?' the puzzled chief demanded.

'We're going off to do some exploring in the ruddy Antarctic,' the second answered bitterly. 'Now you know what he was getting at.'

Running the easting down! Sweeping along the troubled four-thousand-mile track that leads from the south of the Cape of Good Hope to the Leeuwin at the south-west corner of Australia, before the furious west winds and the rushing seas that sweep without let or hindrance half-way round the globe. Away down in that trackless waste of waters that lie beyond the parallel of 45° south latitude a succession of beautiful clippers outward bound for Australia to load the wool clip used to storm along under every stitch of canvas they could safely carry. It was there that day after day, week after week, they did the grand sailing that made their wonderful passages possible; today it is deserted except for the fluttering Cape pigeon and the hovering albatross. It was of that gale-swept tract of the great Southern Ocean that Kipling's immortal engineer said if you failed you had time to mend your shaft, even eat it, ere you were spoken, or 'Make Kerguelen under sail, three trysails burned with smoke.'

The *Knightley* had reached a position too far east even for the latter expedient; for the Crozets and Kerguelen were well to windward, and, in any case, she had not got a single sail on board. But failure was, of course, impossible for Captain Hartnell! M.N.B.S.! He was the only happy man on board, as Mr Birnie, the third officer, was the only one who was indifferent. The weather was bitterly cold; squall after squall of hurricane force, and laden with sleet, shrieked out of the north-west, and icy spray lashed the after-deck, even though the vessel was in ballast trim and standing high out of the water. Lurching and pitching wildly, she swept to the eastward, throwing her bows high toward the dark flying scud of the squalls, dipping them till she tossed three-quarters of her rudder out of the sea. She was never still for a single moment, night or day; for the liquid ridges that rushed at her port quarter were of an almost incredible steepness, and the valleys between them were cavernous.

Captain Hartnell was happy because his vessel's progress was even swifter than he imagined it would be when he abandoned the fine weather route further north and stood down into the wild westerlies. The terrific thrust of the favouring wind and waves more than made up for the power wasted by a propeller, the blades of which beat the air almost as much as they churned the sea. He got considerable pleasure from picturing the astonished face of the captain who was going to overhaul the *Knightley* somewhere about New Amsterdam, when he arrived at Port Pirie and found her already half loaded. Mr Birnie, the third officer, was indifferent because, in spite of his placidity and apparent laziness, he was cast in an iron mould; neither cold, wet, nor discomfort worried him in the least. The other two officers were miserable and disgruntled; they felt that life was being incommensurately aggravated by this entirely unnecessary attempt to save a few days on the passage. The chief officer, a much older man than the captain, felt the cold and discomfort particularly; he could not get warm, even in his bunk, and when keeping his watch on the bridge he was weary of staying on his feet. The deck-hands had reached a state bordering on passive insubordination which might easily have deteriorated into something worse, and the chief officer, possessed of a fellow feeling, did little to check them.

If the discomfort on deck was acute, it was worse in the engine-room and stokehold. The engineer on watch had no difficulty about keeping warm; the propeller, with little more solid than air and spray to bite on half the time, would have sent the engines racing so wildly that they must have shaken themselves to pieces but for his unremitting attention to the throttle. Sweat ran in rivulets through the grime on his face, and every limb ached before the four long hours of his watch had passed. In the stokehold the boilers rocked in their saddles with every plunge; coal aimed at an open fire-door rattled against the boiler plates and rebounded; in the bunkers, trimmers were bruised by coal rumbling down on top of them. The chief engineer, stout-hearted in normal times but now feeling for his men, was sullen and resentful—and sufferance, as a rule, is not the badge of the tribe of sea-going engineers. By that time every man on board knew that there was an easier way to Port Pirie further north.

Just after breakfast one day the hands of the watch below, weary after four hours of buffeting, were about to turn into their bunks for a short spell of uneasy sleep. Mr Birnie was on the bridge, pacing unsteadily to and fro, occasionally grabbing a rail to steady himself against a lurch, but keeping a keen look-out. That was Mr Birnie's way, though there was really nothing to look out for; they had not seen a vessel since they left Cape Point and did not expect to see one; Australia was still two thousand miles away. The *Knightley*, her hull sloping upward toward the bows like the roof of a house, climbed to the crest of a huge, foaming roller, wriggled and dropped heavily into the succeeding trough. There followed a violent thud that shook her fore and aft as if she had thumped on a rock. Men and officers, certain that she had struck wreckage, for there were no rocks about in an ocean that was hundreds of fathoms deep, streamed out on deck. Captain Hartnell staggered up to the bridge.

'Put her slow!' he shouted from the top of the ladder.

Mr Birnie rang the telegraph to slow and waited for a reply from the engine-room.

'Get the carpenter along to sound the bilges and peaks,' the captain ordered.

Bewildered men were staring aft striving to get a view of the wake when the stern dipped, but no extraneous object appeared on the crests of the rollers running up behind. The more pessimistic suggested low-lying ice which would be difficult to see but stout enough to damage the vessel. The captain turned on the third officer.

'Were you keeping a good look-out?' he demanded.

Mr Birnie stared him straight in the face and hesitated as if deliberating whether he was keeping a good look-out or not. He was always deliberate and, if possible, his speech was monosyllabic; the 'sir' he had to use when answering the captain seemed to come out reluctantly, as an extra effort, and always after a distinct pause.

'Yes . . . sir,' he replied at last.

'And you saw no wreckage—nor anything else?'

'No . . . sir.'

Even by Captain Hartnell's exacting standard Mr Birnie was a first-class officer and one to be trusted, so there must be another reason for the thud. The chief engineer came up to the bridge with it. He said the young engineer on watch, who was the only person in authority close to the bottom of the ship, considered that the bump was due to the stern plunging into the trough of the sea more violently than usual. The carpenter reported the bilges and peaks dry.

'Put her on full speed again,' the captain said.

For the next seven hours the *Knightley* kept on her tortured eastward way, climbing wind-swept ridges, swooping down into dark, half-sheltered valleys; then, shortly after four o'clock, she again shook fore and aft. The second engineer, who had just relieved the third at the throttle, was startled and badly shaken by a continuous racing of the engines; they jarred and rattled and created a resounding pandemonium in the comparatively restricted space of the engine-room; the maze of glimmering brass and steel—of pistons, connecting rods, and cross-heads—danced and whirled in a frenzy. The second engineer throttled right down, then signalled to the bridge by ringing the engine-room telegraph to stop. The chief engineer discarded the cup of tea he was sipping and dashed down the iron ladder into the engine-room. The engineers off duty followed. They had an idea that the heavy thump of the morning must have fractured the shaft, and made their way into the tunnel. They worked their way aft as far as the stern-gland, but found the shaft intact. The wooden floor was slippery with oil and grease, and they had great difficulty in keeping their feet; for the ship, robbed of her steerage-way and under the influence of her high forecastle-head, had brought the wind and sea dead aft and was pitching very heavily. As it is almost impossible for a tail-shaft to break inside the stern-tube, they came to the conclusion that it had broken just outside—and that, in fact, was what had happened, as Mr Birnie had just discovered. Getting up from his settee, on which he had been trying to sleep, he had made his way aft, got over the taffrail, leaned out as far as he could over the sea, and looked down. The steamer's stern had just dipped downward and over to port; it flew up again with a dizzy lift that exposed the greater part of the arched propeller aperture. It was empty; the propeller had gone.

Darkness was coming down. The two red lights which indicated the vessel was not under command were hoisted. Well down by the stern, she continued to lie with the wind and towering seas almost right aft—pitching heavily but lazily. Captain Hartnell retired to his cabin, shut himself in and pondered, while all over the ship, in cabins and in sailors' and firemen's forecastles, officers and men discussed the awful thing that had happened. They discussed it with voices that were almost reduced to whispering, for the prospect was indeed appalling. The accident which had robbed the ship of her mobility had in it a specially heart-rending quality because of her extreme isolation. The shock to the imagination was cruel. To present discomfort had been added an immediate future that was black and an ultimate future that would not bear thinking about.

On the bridge the chief officer, tired of keeping a useless look-out and with no helmsman to supervise, went into the chart-room and pored over the outspread general chart of the South Indian Ocean. The captain had pencilled a neat cross at the estimated position where the propeller was lost, but beyond that there was little on the chart but parallels of latitude and meridians of longitude. Yes, there *was* something else; much nearer than any land was an irregular dotted line marking the extreme northern limit of icebergs. The chief officer shivered. In that watery abomination of desolation the ship might drift for months without being sighted, unless the drift was suddenly checked by what the irregular dotted line indicated; and if the wind remained in the north-west, that would probably be the result of their drifting. If they got that far! before then they might starve to death! For reasons connected with the Australian customs laws, the captain, a keen business man, had only taken sufficient provisions in Cape Town to allow for unforeseen delays. Total disablement had certainly not been allowed for, and the chief officer doubted if there was sufficient food on board to give full rations for another fortnight. Again he shivered. The silence—broken only by the wind that howled outside and the splash of the waves—and the total lack of vibration were depressing in the extreme. Tears came into his eyes. The weird feeling that he had a dead thing under him almost overwhelmed him; and a black resentment against the captain, who, in the first place, had been given command of the vessel over his head and who had now got her into this plight surged up in his heart. The supper bell went, and the third officer came on the bridge to relieve him.

Supper was eaten in silence. The captain sat like a sphinx at the head of the table, and as soon as he had finished his meal he retired to his cabin again. The chief officer and the second—a somewhat colourless individual—would have liked to know what thoughts were working in that self-reliant brain, what pangs of remorse were tugging at the usually unresponsive heart-strings. They were left in ignorance as far as the pangs were concerned, but the thoughts were soon disclosed. At eight o'clock the captain summoned all the officers and engineers to the saloon. The boatswain was sent on the bridge to allow the third officer to attend; the donkeyman relieved the fourth engineer in the silent engine-room. Very soon those assembled discovered that they had not been called to a conference, or a council of war, or even a discussion of ways and means of carrying out a plan. They had come to be told what the captain proposed to do, and to receive instructions—and what the captain proposed to do was to fit the spare propeller and tail-shaft! Blank incredulity showed on every face but one, and the chief engineer voiced it.

'But . . . but you can't do that at sea,' he spluttered indignantly.

'*Can't!*' the captain repeated with apparent surprise.

'If you knew the trouble it is to fit a propeller and shaft in a sheltered harbour you wouldn't talk like that. With this sea running it's impossible. To begin with, you would have to tip her down by the head till the stern-tube is out of the water.'

'Quite right,' the captain answered calmly. 'We'll make a start at that tomorrow morning. We'll pump out the after-peak and ballast tanks, and flood No. 1 hold.'

'Flood No. 1 hold!' the chief officer shouted hysterically. 'You'll wreck the ship! You'll lose her! What bulk-heads would stand a few hundred tons of water washing against them? If they do, she'll roll over with us!'

'We'll have to chance that—unless you can suggest anything else, Mr Wilkins.'

The chief officer did not reply. There was a silent, puzzled pause and a shuffling of feet. Nobody *could* suggest anything else, and all knew it. The chief engineer broke the silence.

'Look here, Captain Hartnell, I'm as keen as you are to try something; God knows I don't want to end my life down here,' he said soberly. 'But I tell you it's quite impossible, especially with this sea running. If you would wait for a calm, even.'

'Calms are scarce in this latitude and we might wait a month,' the captain replied. 'By that time we should be so weak with starvation that we could do nothing.'

'And you propose to have men working under the counter with a ship jumping like this! It's murder! They'll either drown or have their brains bashed out! I tell you straight, none of *my* men will go over the stern!'

'They had better wait till they're asked,' the captain said drily. 'I'm going over the stern myself.'

'So am I!'

The assembled men looked round in astonishment. The speaker was the young third officer who up till then had remained unobtrusively in the background, apparently studying, with an absent mind, the rivets in the beams overhead.

'Thank you, Mr Birnie,' said the captain.

The third officer's prompt offer had a definite, and remedying, effect. It dropped the seeds of doubt, in some cases of shame, into men's minds; it stirred something in their souls and made an appeal for a more robust attitude toward the crisis that had overtaken them. In a flash it was realized by most of them that resentment, no matter how much it may have been justified, would get them nowhere; that their present attitude was unworthy of British seamen. They shuffled their feet and looked at each other, trying to read each other's minds, as if they were strikers afraid of being suspected of blacklegging. The captain quickly sensed their hesitation.

'Gentlemen, this is the crux of the matter,' he said. 'Are we to remain inactive while the ship drifts helplessly toward the ice-fields, or are we prepared to make an effort to save ourselves?'

The engineers nodded in apparent agreement, but the chief officer remained obdurate.

'I tell you it's madness to try and flood the for'a'd hold,' he moaned.

'That's enough,' Captain Hartnell said curtly. 'At six o'clock tomorrow morning get all your hands on to lifting the ceiling in No. 1 lower hold, and stowing the boards securely in the 'tween-decks. The loose water in the hold *might* damage the bulkhead; if those heavy boards were washing about in it they certainly *would*. Perhaps you didn't think of *that*. Chief, as soon as Mr Wilkins reports to you that the manhole doors have been taken off No. 1 ballast tank top, open the sea-cocks and get the hold flooded. At the same time start the pumps on the after-peak and tanks.'

'Very good, sir,' said the chief engineer.

A quarter of an hour later Captain Hartnell went on the bridge, and after some difficulty, for there was no light even in the useless binnacle, found the third officer leaning against the rail behind the canvas dodger, against which sleet was pattering.

'Mr Birnie, I won't forget this,' he said, 'but there's something else I want you to do. I've been thinking things out, and have come to the conclusion that we shall require another hand under the counter; we can't manage the job by ourselves. Now, I don't want to *order* any man to do the job because the chief engineer certainly didn't exaggerate when he spoke of the dangers of drowning and bashing. We are in for a hell of a time, and if any man refused to take his share in it, he might well be upheld in a court of law. That would place me, as a ship-master, in a very awkward position—but I don't want to *plead* with any man. Will you look around, pick out a suitable man, and broach him on the quiet?'

'Yes . . . sir,' said Mr Birnie.

A wild, wintry dawn was ushered in by the chattering of the winch for No. 1 hold. It was being used for sending a derrick aloft. Down in the bottom of the hold the carpenter, and some of the seamen, were lifting the heavy three-inch ceiling boards. Those were slung and hove up level with the 'tween-decks, where they were stowed and securely lashed. The ballast tank top was exposed along its full length, and the carpenter took off the manhole doors. All was ready for flooding the hold. The engineers opened the sea-cocks for the tank, which was already full, and the water overflowed into the hold. The pumping out of the after ballast tanks and after-peak began. The *Knightley* was drifting with half a gale dead aft, and still pitching heavily; the seas running up astern were rising right up to the counter as her stern dipped, and falling to the middle pintle of the rudder as it rose. It was about noon before the alteration of the trim took effect and she came on to an even keel. Her bow dipped lower in the water, and, no longer holding the wind, came up to windward, bringing the wind and sea abeam so that the steamer lay in the trough and wallowed.

The afternoon watch will long be remembered by every soul on board. The steamer rolled as she had never rolled before, and created a panic. She put a severe strain on the arm and body muscles of men hanging on grimly to keep their footing on the decks; the only reason why those of the watch below were not pitched out of their bunks was that they did not attempt to get into them. They were, however, lucky to *be* below, for there they could feel but not see; they missed the swift swoop of masts and funnel that made men dizzy and brought a sinking feeling to the stoutest heart. Even Captain Hartnell, maintaining his position by the bridge rail by intense, painful, muscular effort, became seriously alarmed, for the first time in his career at sea. From the second his vessel's masts and funnel passed the perpendicular she would lean over somewhat lazily; then, as the great and increasing mass of loose water in the hold washed across, she would fall over with a terrific jerk that almost tore the captain's arms out of their sockets. It was the heavy mass of water that constituted the danger—and absolutely nothing could be done about it; except, perhaps, pump the hold dry again and acknowledge defeat. Even that could not promise immediate relief; and in the meantime, from being almost on her beam-ends on one side, the ship would roll over till she was almost on her beam-ends on the other—and when she was lying over on her side, to men's fevered imagination her masts and funnel seemed to be horizontal.

As she crashed over till the sea lapped over the lee bulwark rail she squashed the water so that it created a smother of foam half an acre in extent; but it was during those seconds, that passed like hours, when she lay on her side as if she would never recover, that her breathless captain suffered his most acute spasms of anxiety. It was then that he cursed illogically the damned chief officer who had prophesied disaster, and assuredly if Mr Wilkins had gone on the bridge with a hint of 'I told you so' on his lips, the captain would have committed manslaughter. Always, however, just as hope had almost died, the powerful righting lever created by her low centre of gravity came reluctantly into play, overcame the mass of water that was listing her and brought her upright—only to fall heavily the other way. Toward four o'clock, when human endurance had almost failed, the rolling eased. The stern rising imperceptibly out of the water was beginning to feel the force of the wind. It blew right off and the *Knightley* came head to wind and sea, and hove herself to. Instead of rolling she pitched, but was comparatively safe. To help to keep her hove-to, two tarpaulins were spread in the main rigging, one on each side, to reinforce the elevated stern which was acting like an after-sail. She was tipped sufficiently soon after dark, when the pumps were stopped and the cocks closed.

Next morning the captain ordered the construction of a sea-anchor. Now, the sea-anchor that will hold a five-thousand-ton steamer, in her ordinary trim of being down by the stern, head to the seas in such a gale as was then blowing, has yet to be constructed; but this was different. The vessel was lying head to sea naturally, and the sea-anchor would serve to steady her. It consisted of a triangular framework of stout awning spars and lifeboat oars, on which was stretched an awning. To the apex of the triangle—which would be inverted when it was in the sea—a five-fathom length of mooring chain with a kedge anchor was attached. A three-and-a-half-inch wire hawser, to which the vessel would ride, was shackled to a triple span of two-inch wire attached to the framework. The end of the hawser had been taken on the forecastle-head, passed out through a chock, and led aft outside the ship before being shackled to the span. The derrick of No. 1 hatch was already up; the end of its fall was attached to the framework by a rope strop; the sea-anchor was lifted and swung out over the sea. The strop was cut, the anchor dropped in the water, and promptly sank. Gradually, as the steamer drifted astern, the wire hawser tightened, and presently the sea-anchor appeared on the crest of a wave about fifty yards ahead. It made an efficient drogue.

Down in the tunnel the engineers had set to work to uncouple and remove the bobbin-piece—a length of shaft in the tunnel recess that connected the main and tail shafts. The broken tail-shaft was withdrawn and the spare one, which had been secured by chains in the tunnel, was run out through the stern-tube till it was level with the stern-post. It was much colder work than engineers usually have to do; for while they were changing the shafts the sea poured in through the stern-tube every time the stern dipped; but, working heroically, they completed the job by midnight. As far as they were concerned all was ready for shipping the spare propeller which was lying in No. 4 'tween-decks. Had the vessel been loaded it would have been lying under tons of cargo, an indication that it was intended for use only after the vessel had been towed, disabled, into a foreign port.

At daybreak next day the chief officer with his men descended into the 'tween-decks and very carefully cast adrift the chain lashings securing it. It weighed over five tons, and there was great danger that it might take charge, surge forward along the steeply sloping deck, and maim or kill any man who tried to control it. In the meantime one of the derricks for No. 4 hatch had been rigged up. A heavy chain sling was passed through the boss of the propeller, and the lower block of a stout tackle suspended from the derrick head hooked on. A winch revolved, and a sleet squall howled at the propeller and the struggling men as it swayed up through the hatchway, with guys attached to prevent it from flying forward when it cleared the coaming. It was landed on the port side of the slippery iron deck, and three more chain slings attached to it, one also through the boss, the others on opposite blades. It was ready for being transported aft. From the counter another stout tackle, with its fall leading in through the port quarter chock on the poop, was taken along outside the poop rail, passed in over the bulwarks, and made fast to the chain sling that had just been passed through the boss.

Again the winch revolved, the propeller was lifted off the deck by means of the derrick and swung out over the heaving sea. It had to be taken aft fully fifty feet. Very carefully the fall of the derrick tackle was surged away, while the tackle from the counter, led to another winch, slowly dragged the propeller aft. More and more the stern tackle took the weight, till eventually the propeller was hanging down from the stern, level with the rudder. It was then the magnitude of the task they were committed to, and its danger, became fully realized. To get the stern tube clear of the water the vessel had been tipped till the eight feet mark was awash; but as her stern dipped, the water was rising to the *sixteen* feet mark, and the sea boiled and swirled under her counter. When the stern rose again there was a violent scour through the propeller aperture. The heavy four-bladed propeller, practically out of control, was surging about; now banging against the rudder, now crunching against the plating round the stern-tube. Darkness was almost on them, yet it could not be left to surge about all night. It was hove up close under the counter and made more secure; and, as there was still some daylight left, the captain decided to complete the preliminaries, all ready for the morning. The job took a good deal longer than he expected, and showed him clearly how laborious the main business was going to be.

Under the counter were two eye-bolts, one on each side, to which lifting tackles were hooked when taking off the propeller for inspection in dry-dock, and for replacing it. Those were the tackles that would be used for suspending the propeller in its correct plumb position, and Captain Hartnell thought there was just time to get them adjusted. Hooking the upper blocks to the eye-bolts was easy enough; the third officer was lowered over the stern in a boatswain's chair, and though on occasion the water rose as high as his waist, he accomplished the job without much difficulty. The lower block on the port side was then attached to one of the chain slings that had been passed round opposite blades; but before the starboard one could be hooked on, it, and the threefold purchase rove through it, had to be passed through the propeller aperture from starboard to port. Down went Mr Birnie, in his boatswain's chair, clinging to the lower block; he was lowered till he gained a footing on the lower plate of the aperture, but the stern dipped and he was washed off it. He regained it with difficulty, and men gasped with horror when the scour through the aperture swung him yards clear. A fathom of ratline line was lowered, and he lashed himself to the rudder-post. A terrific struggle followed. Every time he got the block close to the stern-post the scour tore it from his grasp. The captain shouted to him to give it up till the morning, but he asked for the end of a heaving line, which he passed through the aperture from port to starboard and made fast to the block. By its means the group of men leaning over the taffrail were able to check the block from surging back, and to haul it upward after it had been shoved through the aperture. It was slow work, though; for the third officer had to light the six swollen ropes of the tackle round the stern-post against which it was binding. It became so dark that he could hardly be seen from the taffrail, and the captain grew impatient.

'Come up out of that, Mr Birnie,' he shouted. 'Leave everything as it is till the morning.'

Mr Birnie either could not, or would not hear, for he struggled on.

'He's as full of obstinacy as a mule,' the chief officer, who wanted his supper, cried irritably.

'I prefer to call it determination, and I wish you had some of it, Mr Wilkins,' Captain Hartnell snapped.

The boatswain, who had not heard the captain's order, came along with an iron bucket in which there was a lighted fire made of oakum steeped in tar. By its light the work was continued, and the block pulled up to the taffrail and secured. Mr Birnie, sitting in his boat-swain's chair—a flat board with a rope span attached to it—was hauled up on deck and stood shaking himself like a spaniel.

About nine o'clock Captain Hartnell went along to the third officer's cabin. Mr Birnie had been relieved of watch-keeping, and the captain found him lying in his bunk, chocked off with cushions, smoking his pipe and reading a book.

'Any luck, Mr Birnie?' the captain asked.

'Yes . . . sir.'

'Who did you get?'

'Kelly . . . sir,' said Mr Birnie.

With the first streak of dawn, while the hands assembled the gear aft, the captain stood at the break of the spray-swept forecastle-head. He looked along aft, and his vessel reminded him of a pig rooting for truffles, its nose in the ground and its stern cocked up toward the sky. He looked ahead; stretching out to the invisible sea-anchor the wire hawser became bar-tight as the *Knightley*'s dripping bows lifted, then slackened and splashed into the water as they dipped. The usual dawn squall howled over the grey waste of foaming ridges, but through the sleet he could see the two red lights dangling mournfully in front of the rust-streaked funnel. Even from where he stood he could hear the heavy wash of water, fore and aft, in No. 1 hold—a dismal, menacing sound. With all his self-confidence he had to admit to considerable anxiety, and he prayed that the bulkheads would stand. He was waiting for the carpenter's report, and when it came it was good. The petty officer had sounded the fore-peak and the bilges of No. 2 hold, and found them dry. The bulkheads were not even leaking. Greatly cheered, the captain went aft prepared for a long day of arduous toil.

Among the others waiting on the poop for him he found Kelly, who prided himself on being the hardest case in the ship—a hard case descended from hard cases, indeed a lineal descendant of the Liverpool Irish packet rats who manned the notorious Western Ocean packet ships in the middle of last century. Unlike his officers, who were going over the stern with him wearing sea-boots, oilskins well lashed at the wrists and below the knees, and with small towels tucked under the collars of their coats, Kelly was barefooted and clad only in singlet and dungaree trousers. Erect and jaunty, as if he were going for a stroll along the beach at Southport on a summer day, he stood among his blue-lipped, shivering mates whose bodies were bent to the blast; for on that high exposed afterdeck there was little shelter except two winches on which wire hawsers were wound, a small hatchway, some ventilators, and the emergency hand-steering wheel with its wheel-box.

'Going for a bath this morning, Kelly?' Captain Hartnell asked pleasantly.

'I am, sir; and I hope the maid hasn't forgotten to turn on the hot-water tap,' Kelly answered with a grin.

From the counter the propeller was lowered by the stern tackle till the port lifting tackle took its weight. The other tackle was then unhooked, and the sling removed by the third officer who had already gone over the stern. The propeller was lowered still further and the starboard lifting tackle hove on to drag it into the aperture. The captain and Kelly got into their boatswain's chairs and were slacked down, with life-lines round them, to the eight feet mark. After Mr Birnie's experience of the previous evening they secured themselves to the rudder-post at once. The water was icy cold, and they gasped and choked as it closed over them; from their lower level the foaming crests of the waves seemed to tower to an enormous height.

'Ah, well, we're nice and sheltered down here, sir,' said Kelly.

Under the lee of the counter they were certainly sheltered from the wind, the swirling spray, and the driving snow which swept past the vessel's sides in two streams and united again in the wake a few yards behind them. If the sea was not so infernally cold, the captain thought ruefully as he fingered the heavy hammer slung round his neck. With it he had arranged to make signals by taps on the steel-plating to the engineers in the tunnel, but he did not have occasion to use it that day. Three whole hours elapsed before any of the men were able to lay a hand on the propeller which was grinding and crunching against the stern-post. At first they found the conditions almost terrifying, and all they could do was to hang on grimly and wipe the salt water from their eyes as their heads emerged from the sea. The terrific scour through the propeller aperture seemed as if it would choke the very life out of them. The counter above them would lift dizzily one moment, and dip the next to create a smother of foam in which their heads would be revealed, at the next lift, to the anxious watchers hanging out over the quarters on both sides. A less determined man than Captain Hartnell would have given up the job there and then; men with less powers of endurance than Mr Birnie and Kelly possessed could not have stood it.

It was late in the day before they could get on with the job of coaxing the propeller to lie fairly in the middle of the aperture.

Shouted orders were cut short on the captain's lips, but by signalling to those who were leaning out over the quarter it was possible to convey instructions to heave, or slack, on the tackles. The adjustment became delicate. As the propeller was gradually worked into position, the heaving and slacking got down to a matter of inches. Thrice they got the hole through the boss of the propeller to coincide with the outer flange of the stern-tube, but were unable to stop the winch at the exact moment. Darkness put a stop to the work, and the men had to be hauled up on to the poop—soaked, deafened, and exhausted— with the boss and the stern-tube still out of alignment.

The following daybreak brought no change in the weather. Again the suffering seamen mustered on the bleak, wind-swept poop, appropriately christened by one of them Mount Misery. The three men were lowered into the water and secured to the rudder-post. By noon the five-ton propeller was hanging at the correct height and rubbing against the stern-post, but six inches over to starboard. Mr Birnie and Kelly got over to that side and endeavoured to lever it into position. Their combined efforts, coinciding with a lucky surge of the starboard tackle, got it into the exact position. The captain yelled frantically to the watchers to hold on everything, and hammered his signal to the waiting engineers. The sea closed over him and his companions. His head was the first to emerge from the foam.

'Is it still in its place, Kelly?' he spluttered.

'Faith, sir, it's mighty contagious!' Kelly yelled.

Again the captain hammered on the plating. For moments big as days, as it seemed—at any rate for twenty minutes which passed like hours—the three men watched, and then the thimble point of the tail-shaft making its way through the round aperture in the boss of the propeller appeared. Never did the most anxious of terriers watch a rat-hole more keenly than those men watched that aperture.

'It's coming, Birnie; by God! it's coming!' Captain Hartnell cried in triumph.

'Yes . . . sir,' said Mr Birnie.

Inch by inch the tail-shaft moved outward till six inches of it projected abaft the propeller boss; then the first thread of the worming on which the nut to secure the propeller would be screwed came into view. It would take the engineers some time to put the bobbin-piece in its place and get it connected to the tail-shaft and main-shaft; so, after the propeller had been bound hard against the stern-post by ropes leading from the quarter pipes, the three men in the water were hauled up for a welcome meal. Less than an hour of daylight remained when the chief engineer reported the job complete, and Captain Hartnell decided to wait till next morning before tackling the job of screwing on the propeller nut. At twilight, with a good horizon, the second officer got stellar observations and discovered that the *Knight-ley* had drifted 200 miles to the south-east since she broke down.

Dawn found the little group of watchers mustered on the poop for what they hoped would be the last time; and shortly after that the dauntless three were secured to the rudder-post, waiting for the nut to be lowered to them. It weighed three hundred pounds and had to be slung carefully; for if it had slipped out of the sling and dropped into the sea all the work they had accomplished would have been nothing but wasted effort. There was not another nut on board, and the slinging of it had given the captain food for much anxious thought, till an ingenious engineer came to the rescue by fitting two tap-bolts to it. Even a three-hundred-pound nut takes a lot of handling under the conditions in which the men over the stern worked, and it required their united strength to get it over the thimble point of the shaft and pushed home. With the sling still attached to it they managed to get a round turn of the thread of the nut on to the worm of the shaft; then they reckoned it was safe, and the sling was dispensed with. They got three more turns by hand before the key for tightening the nut was lowered to them.

The key weighed three cwt., and had slats on the rim which fitted projections on the nut. At the end of the handle was a round hole into which the lower block of a tackle from under the counter could be hooked. At first the key could be turned by those above hauling on the tackle by hand, a quarter of a turn of the nut at a time; but to the men in the water fell the almost Herculean task of shifting the position of the key on the nut, ready for another quarter of a turn. To save it from dropping into the sea it was suspended by a line from the port quarter, with a smaller line attached to the rudder-post to enable the men down there to haul it toward them when it swung outward. The sea was running as high as ever; the steamer's stern rose and fell continuously. Up it flew till it brought the nut clear of the water and gave them a brief spell in which to work; down it plunged, and they held on for their lives in the seething vortex.

At times it took them half an hour to shift the grip of the key on the nut. The hours passed; the short spell of winter daylight was drawing to a close, but the soaked, exhausted, yet indomitable men worked on, determined to get the job finished. For the last few turns of the nut the fall of the tackle had to be taken to one of the steam winches. The strain on the tackle became so great that the nut must have been nearly home, but it had to be got into a position where the hole bored through it for the locking pin coincided with the hole in the tail-shaft through which the pin had to pass.

Darkness again overtook them, and once more light was obtained from flares made by burning oakum and tar in buckets. The nut was now turning an eighth of an inch at a time. With the strain on the tackle the moisture was being squeezed out of the rope fall, which on occasion surged back round the winch and was torn out of the frozen hands of the man who was holding on to it down on the main-deck. Captain Hartnell was sitting astride the propeller boss, probing the hole in the nut with the pin held in fingers numb and blue with cold. To avoid losing the pin it was attached to his wrist by a length of spun-yarn. He was almost in despair, when he felt the pin entering the hole in the shaft. He drove it home with his hammer, and the completely exhausted men were hauled up on to the poop. When they reached it they could not stand on their feet.

Later in the evening the captain told the steward to take Kelly along to the saloon and give him a good stiff dram of rum. He then did a thing he had never done since he first took command of a ship. He invited an officer—Mr Birnie—to his cabin for a whisky-and-soda. The whisky-and-soda was followed by others, and by the time Mr Birnie said good night he was almost discursive.

Only Mr Birnie and Kelly went over the stern next morning, and two hours' work sufficed to cover the job of securing the pin and removing the tackles and slings from the propeller. The two men were hauled up to the taffrail for the last time. The captain gave instructions to fill the after-peak and ballast tanks, and at the same time ordered a full head of steam. Gradually the steamer's stern dropped and she came on to a more even keel, but there was still the flooded hold to be pumped out. Without a doubt as her bow rose she would fall off into the trough of the sea, in spite of the pull of the sea-anchor and the tarpaulins in the main rigging, and they might lose her yet through the mass of loose water in the half-empty hold. The memory of that awful rolling was terrifying; she must be kept head on to the seas till the hold was pumped out; the engine-room telegraph on the bridge was rung to 'stand-by'.

It was not answered at once; the engines had to be turned over both ways. Both on deck and in the engine-room there was now considerable anxiety, heart-burning, and searching of mind. Had there been any flaw in the work carried out in the tunnel, or under the counter? Had anything been left undone? Would the first movement of the engines have as a result a crunching jar that would wreck the efforts of the past week and leave them to drift and starve? Only the actual working of the engines could tell. From the bridge Captain Hartnell listened intently. He heard the first wheezings of steam; were the engines vibrating again? They were, and, unknown to him, spray was being tossed up under the counter. The telegraph from the engine-room clanged its reply; the engines were ready.

The sea-anchor was tripped by a line that had been attached to the crown of the kedge, and hoisted on board. The engines were put ahead at half-speed; the *Knightley* was steering again. The captain did not want to drive her too hard into the pounding seas, nor did he wish to steam far back to the westward and lose valuable time. But she had to be kept head to sea at all costs. For most of the time half-speed sufficed; but frequently a touch of full speed was required to give her more steerage-way and straighten her up when she fell off to a dangerous angle and gave a hint that she was about to resume her rolling. The captain's hand was continually on the telegraph handle. At last the flooded hold was pumped out down to the bilges; the manhole doors on top of the ballast tanks were replaced and secured. The chief officer came on the bridge to report, and the captain banged the telegraph handle down to full speed with an air of finality.

'Course . . . sir?' the third officer enquired.

'Keep her north-east, Mr Birnie; you and I and Kelly are due a spell of warm weather,' the captain answered with a partially concealed grin.

The *Knightley* arrived at Port Pirie only a few days overdue. So well had the work been done that she steamed from that port to the United Kingdom, thence to Singapore and back, before the propeller was attended to in the ordinary way at the regular dry-docking. The rival steamer from Cape Town had got to Port Pirie before him; but, later, Captain Hartnell had his compensations. From his owners he received a gold watch, suitably inscribed; and he and his fellow-workers under the counter were awarded Lloyd's medal for meritorious service. That decoration is bestowed irrespective of rank and nationality,

but it is not lightly awarded. It is recognized by seafarers all over the
world as a great honour, and none has earned it more worthily than
those men who were lashed to the *Knightley*'s rudder-post. Captain
Hartnell had his placed in a frame and displayed in a prominent place
in his cabin; Mr Birnie sent his to his mother and thought little more
about it; what happened to Kelly's is unknown; for—true vagabond
of the sea that he was—he left the steamer at her home port and no
one connected with her ever heard of him again.

The Story of the Siren

~~~

E. M. FORSTER

Few things have been more beautiful than my notebook on the Deist Controversy as it fell downward through the waters of the Mediterranean. It dived, like a piece of black slate, but opened soon, disclosing leaves of pale green, which quivered into blue. Now it had vanished, now it was a piece of magical india-rubber stretching out to infinity, now it was a book again, but bigger than the book of all knowledge. It grew more fantastic as it reached the bottom, where a puff of sand welcomed it and obscured it from view. But it reappeared, quite sane though a little tremulous, lying decently open on its back, while unseen fingers fidgeted among its leaves.

'It is such a pity', said my aunt, 'that you will not finish your work in the hotel. Then you would be free to enjoy yourself and this would never have happened.'

'Nothing of it but will change into something rich and strange,' warbled the chaplain, while his sister said, 'Why, it's gone in the water!' As for the boatmen, one of them laughed, while the other, without a word of warning, stood up and began to take his clothes off.

'Holy Moses,' cried the Colonel. 'Is the fellow mad?'

'Yes, thank him, dear,' said my aunt: 'that is to say, tell him he is very kind, but perhaps another time.'

'All the same I do want my book back,' I complained. 'It's for my Fellowship Dissertation. There won't be much left of it by another time.'

'I have an idea,' said some woman or other through her parasol. 'Let us leave this child of nature to dive for the book while we go on to the other grotto. We can land him either on this rock or on the ledge inside, and he will be ready when we return.'

The idea seemed good; and I improved it by saying I would be left behind too, to lighten the boat. So the two of us were deposited outside the little grotto on a great sunlit rock that guarded the harmonies within. Let us call them blue, though they suggest rather the spirit of what is clean—cleanliness passed from the domestic to the sublime, the cleanliness of all the sea gathered together and radiating light. The Blue Grotto at Capri contains only more blue water, not bluer water. That colour and that spirit is the heritage of every cave in the Mediterranean into which the sun can shine and the sea flow.

As soon as the boat left I realized how imprudent I had been to trust myself on a sloping rock with an unknown Sicilian. With a jerk he became alive, seizing my arm and saying, 'Go to the end of the grotto, and I will show you something beautiful.'

He made me jump off the rock on to the ledge over a dazzling crack of sea; he drew me away from the light till I was standing on the tiny beach of sand which emerged like powdered turquoise at the further end. There he left me with his clothes, and returned swiftly to the summit of the entrance rock. For a moment he stood naked in the brilliant sun, looking down at the spot where the book lay. Then he crossed himself, raised his hands above his head, and dived.

If the book was wonderful, the man is past all description. His effect was that of a silver statue, alive beneath the sea, through whom life throbbed in blue and green. Something infinitely happy, infinitely wise—but it was impossible that it should emerge from the depths sunburned and dripping, holding the notebook on the Deist Controversy between its teeth.

A gratuity is generally expected by those who bathe. Whatever I offered, he was sure to want more, and I was disinclined for an argument in a place so beautiful and also so solitary. It was a relief that he should say in conversational tones, 'In a place like this one might see the Siren.'

I was delighted with him for thus falling into the key of his surroundings. We had been left together in a magic world, apart from all the commonplaces that are called reality, a world of blue whose floor was the sea and whose walls and roof of rock trembled with the sea's reflections. Here only the fantastic would be tolerable, and it was in that spirit I echoed his words, 'One might easily see the Siren.'

He watched me curiously while he dressed. I was parting the sticky leaves of the notebook as I sat on the sand.

'Ah,' he said at last. 'You may have read the little book that was printed last year. Who would have thought that our Siren would have given the foreigners pleasure!'

(I read it afterward. Its account is, not unnaturally, incomplete, in spite of there being a woodcut of the young person, and the words of her song.)

'She comes out of this blue water, doesn't she,' I suggested, 'and sits on the rock at the entrance, combing her hair.'

I wanted to draw him out, for I was interested in his sudden gravity, and there was a suggestion of irony in his last remark that puzzled me.

'Have you ever seen her?' he asked.

'Often and often.'

'I, never.'

'But you have heard her sing?'

He put on his coat and said impatiently, 'How can she sing under the water? Who could? She sometimes tries, but nothing comes from her but great bubbles.'

'She should climb on to the rock.'

'How can she?' he cried again, quite angry. 'The priests have blessed the air, so she cannot breathe it, and blessed the rocks, so that she cannot sit on them. But the sea no man can bless, because it is too big, and always changing. So she lives in the sea.'

I was silent.

At that his face took a gentler expression. He looked at me as though something was on his mind, and going out to the entrance rock gazed at the external blue; then returning into our twilight he said, 'As a rule only good people see the Siren.'

I made no comment. There was a pause, and he continued. 'That is a very strange thing, and the priests do not know how to account for it; for she of course is wicked. Not only those who fast and go to Mass are in danger, but even those who are merely good in daily life. No one in the village had seen her for two generations. I am not surprised. We all cross ourselves before we enter the water, but it is unnecessary. Giuseppe, we thought, was safer than most. We loved him, and many of us he loved: but that is a different thing from being good.'

I asked who Giuseppe was.

'That day—I was seventeen and my brother was twenty and a great deal stronger than I was, and it was the year when the visitors, who have brought such prosperity and so many alterations into the village, first began to come. One English lady in particular, of very high birth, came, and has written a book about the place, and it was through her that the Improvement Syndicate was formed, which is about to connect the hotels with the station by a funicular railway.'

'Don't tell me about that lady in here,' I observed.

'That day we took her and her friends to see the grottoes. As we rowed close under the cliffs I put out my hand, as one does, and caught a little crab, and having pulled off its claws offered it as a curiosity. The ladies groaned, but a gentleman was pleased, and held out money. Being inexperienced, I refused it, saying that his pleasure was sufficient reward! Giuseppe, who was rowing behind, was very angry with me and reached out with his hand and hit me on the side of the mouth, so that a tooth cut my lip, and I bled. I tried to hit him back, but he always was too quick for me, and as I stretched round he kicked me under the armpit, so that for a moment I could not even row. There was a great noise among the ladies, and I heard afterward that they were planning to take me away from my brother and train me as a waiter. That, at all events, never came to pass.

'When we reached the grotto—not here, but a larger one—the gentleman was very anxious that one of us should dive for money, and the ladies consented, as they sometimes do. Giuseppe, who had discovered how much pleasure it gives foreigners to see us in the water, refused to dive for anything but silver, and the gentleman threw in a two-lira piece.

'Just before my brother sprang off he caught sight of me holding my bruise, and crying, for I could not help it. He laughed and said, "This time, at all events, I shall not see the Siren!" and went into the water without crossing himself. But he saw her.'

He broke off and accepted a cigarette. I watched the golden entrance rock and the quivering walls and the magic water through which great bubbles constantly rose.

At last he dropped his hot ash into the ripples and turned his head away, and said, 'He came up without the coin. We pulled him into the boat, and he was so large that he seemed to fill it, and so wet that we could not dress him. I have never seen a man so wet. I and the gentleman rowed back, and we covered Giuseppe with sacking and propped him up in the stern.'

'He was drowned, then?' I murmured, supposing that to be the point.

'He was not,' he cried angrily. 'He saw the Siren. I told you.'

I was silenced again.

'We put him to bed, though he was not ill. The doctor came, and took money, and the priest came and spattered him with holy water. But it was no good. He was too big—like a piece of the sea. He kissed the thumb-bones of San Biagio and they never dried till evening.'

'What did he look like?' I ventured.

'Like anyone who has seen the Siren. If you have seen her "often and often" how is it you do not know? Unhappy, unhappy because he knew everything. Every living thing made him unhappy because he knew it would die. And all he cared to do was sleep.'

I bent over my notebook.

'He did no work, he forgot to eat, he forgot whether he had his clothes on. All the work fell on me, and my sister had to go out to service. We tried to make him into a beggar, but he was too robust to inspire pity, and as for an idiot, he had not the right look in his eyes. He would stand in the street looking at people, and the more he looked at them the more unhappy he became. When a child was born he would cover his face with his hands. If anyone was married— he was terrible then, and would frighten them as they came out of church. Who would have believed he would marry himself! I caused that, I. I was reading out of the paper how a girl at Ragusa had "gone mad through bathing in the sea". Giuseppe got up, and in a week he and that girl came in.

'He never told me anything, but it seems that he went straight to her house, broke into her room, and carried her off. She was the daughter of a rich mine-owner, so you may imagine our peril. Her father came down, with a clever lawyer, but they could do no more than I. They argued and they threatened, but at last they had to go back and we lost nothing—that is to say, no money. We took Giuseppe and Maria to the church and had them married. Ugh! that wedding! The priest made no jokes afterward, and coming out the children threw stones . . . I think I would have died to make her happy; but as always happens, one could do nothing.'

'Were they unhappy together then?'

'They loved each other, but love is not happiness. We can all get love. Love is nothing. I had two people to work for now, for she was like him in everything—one never knew which of them was speaking. I had to sell our own boat and work under the bad old man you have today. Worst of all, people began to hate us. The children first—everything begins with them—and then the women and last of all the men. For the cause of every misfortune was—you will not betray me?'

I promised good faith, and immediately he burst into the frantic blasphemy of one who has escaped from supervision, cursing the priests, who had ruined his life, he said. 'Thus are we tricked!' was his cry, and he stood up and kicked at the azure ripples with his feet, till he had obscured them with a cloud of sand.

I too was moved. The story of Giuseppe, for all its absurdity and superstition, came nearer to reality than anything I had known before. I don't know why, but it filled me with desire to help others—the greatest of all our desires, I suppose, and the most fruitless. The desire soon passed.

'She was about to have a child. That was the end of everything. People said to me, "When will your charming nephew be born? What a cheerful, attractive child he will be, with such a father and mother!" I kept my face steady and replied, "I think he may be. Out of sadness shall come gladness"—it is one of our proverbs. And my answer frightened them very much, and they told the priests, who were frightened too. Then the whisper started that the child would be Antichrist. You need not be afraid: he was never born.

'An old witch began to prophesy, and no one stopped her. Giuseppe and the girl, she said, had silent devils, who could do little harm. But the child would always be speaking and laughing and perverting, and last of all he would go into the sea and fetch up the Siren into the air and all the world would see her and hear her sing. As soon as she sang, the Seven Vials would be opened and the Pope would die and Mongibello flame, and the veil of Santa Agata would be burned. Then the boy and the Siren would marry, and together they would rule the world, for ever and ever.

'The whole village was in tumult, and the hotel-keepers became alarmed, for the tourist season was just beginning. They met together and decided that Giuseppe and the girl must be sent inland until the child was born, and they subscribed the money. The night before they were to start there was a full moon and wind from the east, and all along the coast the sea shot up over the cliffs in silver clouds. It is a wonderful sight, and Maria said she must see it once more.

' "Do not go," I said. "I saw the priest go by, and someone with him. And the hotel-keepers do not like you to be seen, and if we displease them also we shall starve."

' "I want to go," she replied. "The sea is stormy, and I may never feel it again."

' "No, he is right," said Giuseppe. "Do not go—or let one of us go with you."

' "I want to go alone," she said; and she went alone.

'I tied up their luggage in a piece of cloth, and then I was so unhappy at thinking I should lose them that I went and sat down by my brother and put my arm round his neck, and he put his arm round me, which he had not done for more than a year, and we remained thus I don't remember how long.

'Suddenly the door flew open and moonlight and wind came in together, and a child's voice said laughing, "They have pushed her over the cliffs into the sea."

'I stepped to the drawer where I keep my knives.

' "Sit down again," said Giuseppe—Giuseppe of all people! "If she is dead, why should others die too?"

' "I guess who it is," I cried, "and I will kill him."

'I was almost out of the door, and he tripped me up and, kneeling upon me, took hold of both my hands and sprained my wrists; first my right one then my left. No one but Giuseppe would have thought of such a thing. It hurt more than you would suppose, and I fainted. When I woke up, he was gone, and I never saw him again.'

But Giuseppe disgusted me.

'I told you he was wicked,' he said. 'No one would have expected him to see the Siren.'

'How do you know he did see her?'

'Because he did not see her "often and often", but once.'

'Why do you love him if he is wicked?'

He laughed for the first time. That was his only reply.

'Is that the end?' I asked.

'I never killed her murderer, for by the time my wrists were well he was in America; and one cannot kill a priest. As for Giuseppe, he went all over the world too, looking for someone else who had seen the Siren—either a man, or, better still, a woman, for then the child might still have been born. At last he came to Liverpool—is the district probable?—and there he began to cough, and spat blood until he died.

'I do not suppose there is anyone living now who has seen her. There has seldom been more than one in a generation, and never in my life will there be both a man and a woman from whom that child can be born, who will fetch up the Siren from the sea, and destroy silence, and save the world!'

'Save the world?' I cried. 'Did the prophecy end like that?'

He leaned back against the rock, breathing deep. Through all the blue-green reflections I saw him colour. I heard him say: 'Silence and loneliness cannot last for ever. It may be a hundred or a thousand years, but the sea lasts longer, and she shall come out of it and sing.' I would have asked him more, but at that moment the whole cave darkened, and there rode in through its narrow entrance the returning boat.

The Rough Crossing

∽∽∽

F. SCOTT FITZGERALD

I

ONCE on the long, covered piers, you have come into a ghostly country that is no longer Here and not yet There. Especially at night. There is a hazy yellow vault full of shouting, echoing voices. There is the rumble of trucks and the clump of trunks, the strident chatter of a crane and the first salt smell of the sea. You hurry through, even though there's time. The past, the continent, is behind you; the future is that glowing mouth in the side of the ship; this dim turbulent alley is too confusedly the present.

Up the gangplank, and the vision of the world adjusts itself, narrows. One is a citizen of a commonwealth smaller than Andorra. One is no longer so sure of anything. Curiously unmoved the men at the purser's desk, cell-like the cabin, disdainful the eyes of voyagers and their friends, solemn the officer who stands on the deserted promenade deck thinking something of his own as he stares at the crowd below. A last odd idea that one didn't really have to come, then the loud, mournful whistles, and the thing—certainly not a boat, but rather a human idea, a frame of mind—pushes forth into the big dark night.

Adrian Smith, one of the celebrities on board—not a very great celebrity, but important enough to be bathed in flash light by a photographer who had been given his name, but wasn't sure what his subject 'did'—Adrian Smith and his blond wife, Eva, went up to the promenade deck, passed the melancholy ship's officer, and, finding a quiet aerie, put their elbows on the rail.

'We're going!' he cried presently, and they both laughed in ecstasy. 'We've escaped. They can't get us now.'

'Who?'

He waved his hand vaguely at the civic tiara.

'All those people out there. They'll come with their posses and their warrants and list of crimes we've committed, and ring the bell at our door on Park Avenue and ask for the Adrian Smiths, but what ho! the Adrian Smiths and their children and nurse are off for France.'

'You make me think we really have committed crimes.'

'They can't have you,' he said, frowning. 'That's one thing they're after me about—they know I haven't got any right to a person like you, and they're furious. That's one reason I'm glad to get away.'

'Darling,' said Eva.

She was twenty-six—five years younger than he. She was something precious to everyone who knew her.

'I like this boat better than the *Majestic* or the *Aquitania*,' she remarked, unfaithful to the ships that had served their honeymoon.

'It's much smaller.'

'But it's very slick and it has all those little shops along the corridors. And I think the staterooms are bigger.'

'The people are very formal—did you notice?—as if they thought everyone else was a card sharp. And in about four days half of them will be calling the other half by their first names.'

Four of the people came by now—a quartet of young girls abreast, making a circuit of the deck. Their eight eyes swept momentarily toward Adrian and Eva, and then swept automatically back, save for one pair which lingered for an instant with a little start. They belonged to one of the girls in the middle, who was, indeed, the only passenger of the four. She was not more than eighteen—a dark little beauty with the fine crystal gloss over her that, in brunettes, takes the place of a blonde's bright glow.

'Now, who's that?' wondered Adrian. 'I've seen her before.'

'She's pretty,' said Eva.

'Yes.' He kept wondering, and Eva deferred momentarily to his distraction; then, smiling up at him, she drew him back into their privacy.

'Tell me more,' she said.

'About what?'

'About us—what a good time we'll have, and how we'll be much better and happier, and very close always.'

'How could we be any closer?' His arm pulled her to him.

'But I mean never even quarrel any more about silly things. You know, I made up my mind when you gave me my birthday present

last week'—her fingers caressed the fine seed pearls at her throat—
'that I'd try never to say a mean thing to you again.'

'You never have, my precious.'

Yet even as he strained her against his side she knew that the
moment of utter isolation had passed almost before it had begun. His
antennæ were already out, feeling over this new world.

'Most of the people look rather awful,' he said—'little and swarthy
and ugly. Americans didn't use to look like that.'

'They look dreary,' she agreed. 'Let's not get to know anybody, but
just say together.'

A gong was beating now, and stewards were shouting down the
decks, 'Visitors ashore, please!' and voices rose to a strident chorus.
For a while the gangplanks were thronged; then they were empty, and
the jostling crowd behind the barrier waved and called unintelligible
things, and kept up a grin of goodwill. As the stevedores began to
work at the ropes a flat-faced, somewhat befuddled young man arrived
in a great hurry and was assisted up the gangplank by a porter and a
taxi driver. The ship having swallowed him as impassively as though
he were a missionary for Beirut, a low, portentous vibration began. The
pier with its faces commenced to slide by, and for a moment the boat
was just a piece accidentally split off from it; then the faces became
remote, voiceless, and the pier was one among many yellow blurs
along the water front. Now the harbour flowed swiftly toward the sea.

On a northern parallel of latitude a hurricane was forming and
moving south by south-east preceded by a strong west wind. On its
course it was destined to swamp the *Peter I. Eudim* of Amsterdam,
with a crew of sixty-six, to break a boom on the largest boat in the world,
and to bring grief and want to the wives of several hundred seamen.
This liner, leaving New York Sunday evening, would enter the zone
of the storm Tuesday, and of the hurricane late Wednesday night.

II

Tuesday afternoon Adrian and Eva paid their first visit to the smoking room. This was not in accord with their intentions—they had
'never wanted to see a cocktail again' after leaving America—but they
had forgotten the staccato loneliness of ships, and all activity centred
about the bar. So they went in for just a minute.

It was full. There were those who had been there since luncheon, and those who would be there until dinner, not to mention a faithful few who had been there since nine this morning. It was a prosperous assembly, taking its recreation at bridge, solitaire, detective stories, alcohol, argument and love. Up to this point you could have matched it in the club or casino life of any country, but over it all played a repressed nervous energy, a barely disguised impatience that extended to old and young alike. The cruise had begun, and they had enjoyed the beginning, but the show was not varied enough to last six days, and already they wanted it to be over.

At a table near them Adrian saw the pretty girl who had stared at him on the deck the first night. Again he was fascinated by her loveliness; there was no mist upon the brilliant gloss that gleamed through the smoky confusion of the room. He and Eva had decided from the passenger list that she was probably 'Miss Elizabeth D'Amido and maid,' and he had heard her called Betsy as he walked past a deck-tennis game. Among the young people with her was the flat-nosed youth who had been 'poured on board' the night of their departure; yesterday he had walked the deck morosely, but he was apparently reviving. Miss D'Amido whispered something to him, and he looked over at the Smiths with curious eyes. Adrian was new enough at being a celebrity to turn self-consciously away.

'There's a little roll. Do you feel it?' Eva demanded.

'Perhaps we'd better split a pint of champagne.'

While he gave the order a short colloquy was taking place at the other table; presently a young man rose and came over to them.

'Isn't this Mr Adrian Smith?'

'Yes.'

'We wondered if we couldn't put you down for the deck-tennis tournament. We're going to have a deck-tennis tournament.'

'Why——' Adrian hesitated.

'My name's Stacomb,' burst out the young man. 'We all know your—your plays or whatever it is, and all that—and we wondered if you wouldn't like to come over to our table.'

Somewhat overwhelmed, Adrian laughed: Mr Stacomb, glib, soft, slouching, waited; evidently under the impression that he had de-livered himself of a graceful compliment.

Adrian, understanding that, too, replied: 'Thanks, but perhaps you'd better come over here.'

'We've got a bigger table.'

'But we're older and more—more settled.'

The young man laughed kindly, as if to say, 'That's all right.'

'Put me down,' said Adrian. 'How much do I owe you?'

'One buck. Call me Stac.'

'Why?' asked Adrian, startled.

'It's shorter.'

When he had gone they smiled broadly.

'Heavens,' Eva gasped, 'I believe they are coming over.'

They were. With a great draining of glasses, calling of waiters, shuffling of chairs, three boys and two girls moved to the Smiths' table. If there was any diffidence, it was confined to the hosts; for the new additions gathered around them eagerly, eyeing Adrian with respect—too much respect—as if to say: 'This was probably a mistake and won't be amusing, but maybe we'll get something out of it to help us in our after life, like at school.'

In a moment Miss D'Amido changed seats with one of the men and placed her radiant self at Adrian's side, looking at him with manifest admiration.

'I fell in love with you the minute I saw you,' she said, audibly and without self-consciousness; 'so I'll take all the blame for butting in. I've seen your play four times.'

Adrian called a waiter to take their orders.

'You see,' continued Miss D'Amido, 'we're going into a storm, and you might be prostrated the rest of the trip, so I couldn't take any chances.'

He saw that there was no undertone or innuendo in what she said, nor the need of any. The words themselves were enough, and the deference with which she neglected the young men and bent her politeness on him was somehow very touching. A little glow went over him; he was having rather more than a pleasant time.

Eva was less entertained; but the flat-nosed young man, whose name was Butterworth, knew people that she did, and that seemed to make the affair less careless and casual. She did not like meeting new people unless they had 'something to contribute', and she was often bored by the great streams of them, of all types and conditions and classes, that passed through Adrian's life. She herself 'had everything'—which is to say that she was well endowed with talents and with charm—and the mere novelty of people did not seem a sufficient reason for eternally offering everything up to them.

Half an hour later when she rose to go and see the children, she was content that the episode was over. It was colder on deck, with a damp that was almost rain, and there was a perceptible motion. Opening the door of her stateroom she was surprised to find the cabin steward sitting languidly on her bed, his head slumped upon the upright pillow. He looked at her listlessly as she came in, but made no move to get up.

'When you've finished your nap you can fetch me a new pillowcase,' she said briskly.

Still the man didn't move. She perceived then that his face was green.

'You can't be seasick in here,' she announced firmly. 'You go and lie down in your own quarters.'

'It's me side,' he said faintly. He tried to rise, gave out a little rasping sound of pain and sank back again. Eva rang for the stewardess.

A steady pitch, toss, roll had begun in earnest and she felt no sympathy for the steward, but only wanted to get him out as quick as possible. It was outrageous for a member of the crew to be seasick. When the stewardess came in Eva tried to explain this, but now her own head was whirring, and throwing herself on the bed, she covered her eyes.

'It's his fault,' she groaned when the man was assisted from the room. 'I was all right and it made me sick to look at him. I wish he'd die.'

In a few minutes Adrian came in.

'Oh, but I'm sick!' she cried.

'Why, you poor baby.' He leaned over and took her in his arms. 'Why didn't you tell me?'

'I was all right upstairs, but there was a steward——Oh, I'm too sick to talk.'

'You'd better have dinner in bed.'

'Dinner! Oh, my heavens!'

He waited solicitously, but she wanted to hear his voice, to have it drown out the complaining sound of the beams.

'Where've you been?'

'Helping to sign up people for the tournament.'

'Will they have it if it's like this? Because if they do I'll just lose for you.'

He didn't answer; opening her eyes, she saw that he was frowning.

'I didn't know you were going in the doubles,' he said.

'Why, that's the only fun.'

'I told the D'Amido girl I'd play with her.'

'Oh.'

'I didn't think. You know I'd much rather play with you.'

'Why didn't you, then?' she asked coolly.

'It never occurred to me.'

She remembered that on their honeymoon they had been in the finals and won a prize. Years passed. But Adrian never frowned in this regretful way unless he felt a little guilty. He stumbled about, getting his dinner clothes out of the trunk, and she shut her eyes.

When a particular violent lurch startled her awake again he was dressed and tying his tie. He looked healthy and fresh, and his eyes were bright.

'Well, how about it?' he enquired. 'Can you make it, or no?'

'No.'

'Can I do anything for you before I go?'

'Where are you going?'

'Meeting those kids in the bar. Can I do anything for you?'

'No.'

'Darling, I hate to leave you like this.'

'Don't be silly. I just want to sleep.'

That solicitous frown—when she knew he was crazy to be out and away from the close cabin. She was glad when the door closed. The thing to do was to sleep, sleep.

Up—down—sideways. Hey there, not so far! Pull her round the corner there! Now roll her, right—left——Crea-eak! Wrench! Swoop!

Some hours later Eva was dimly conscious of Adrian bending over her. She wanted him to put his arms around her and draw her up out of this dizzy lethargy, but by the time she was fully awake the cabin was empty. He had looked in and gone. When she awoke next the cabin was dark and he was in bed.

The morning was fresh and cool, and the sea was just enough calmer to make Eva think she could get up. They breakfasted in the cabin and with Adrian's help she accomplished an unsatisfactory makeshift toilet and they went up on the boat deck. The tennis tournament had already begun and was furnishing action for a dozen amateur movie cameras, but the majority of passengers were represented by lifeless bundles in deck chairs beside untasted trays.

Adrian and Miss D'Amido played their first match. She was deft and graceful; blatantly well. There was even more warmth behind her ivory skin than there had been the day before. The strolling first officer stopped and talked to her; half a dozen men whom she couldn't have known three days ago called her Betsy. She was already the pretty girl of the voyage, the cynosure of starved ship's eyes.

But after a while Eva preferred to watch the gulls in the wireless masts and the slow slide of the roll-top sky. Most of the passengers looked silly with their movie cameras that they had all rushed to get and now didn't know what to use for, but the sailors painting the lifeboat stanchions were quiet and beaten and sympathetic, and probably wished, as she did, that the voyage was over.

Butterworth sat down on the deck beside her chair.

'They're operating on one of the stewards this morning. Must be terrible in this sea.'

'Operating? What for?' she asked listlessly.

'Appendicitis. They have to operate now because we're going into worse weather. That's why they're having the ship's party tonight.'

'Oh, the poor man!' she cried, realizing it must be her steward.

Adrian was showing off now by being very courteous and thoughtful in the game.

'Sorry. Did you hurt yourself? . . . No, it was my fault. . . . You better put on your coat right away, pardner, or you'll catch cold.'

The match was over and they had won. Flushed and hearty, he came up to Eva's chair.

'How do you feel?'

'Terrible.'

'Winners are buying a drink in the bar,' he said apologetically.

'I'm coming, too,' Eva said, but an immediate dizziness made her sink back in her chair.

'You'd better stay here. I'll send you up something.'

She felt that his public manner had hardened toward her slightly.

'You'll come back?'

'Oh, right away.'

She was alone on the boat deck, save for a solitary ship's officer who slanted obliquely as he paced the bridge. When the cocktail arrived she forced herself to drink it, and felt better. Trying to distract her mind with pleasant things, she reached back to the sanguine talks that she and Adrian had had before sailing: There was the little villa in Brittany, the children learning French—that was all she could think of now—the little villa in Brittany, the children learning French—so she repeated the words over and over to herself until they became as meaningless as the wide white sky. The why of their being here had suddenly eluded her; she felt unmotivated, accidental, and she wanted Adrian to come back quick, all responsive and tender, to reassure her. It was in the hope that there was some secret of graceful living, some real compensation for the lost, careless confidence of twenty-one, that they were going to spend a year in France.

The day passed darkly, with fewer people around and a wet sky falling. Suddenly it was five o'clock, and they were all in the bar again, and Mr Butterworth was telling her about his past. She took a good deal of champagne, but she was seasick dimly through it, as if the illness was her soul trying to struggle up through some thickening incrustation of abnormal life.

'You're my idea of a Greek goddess, physically,' Butterworth was saying.

It was pleasant to be Mr Butterworth's idea of a Greek goddess physically, but where was Adrian? He and Miss D'Amido had gone out on a forward deck to feel the spray. Eva heard herself promising to get out her colours and paint the Eiffel Tower on Butterworth's shirt front for the party tonight.

When Adrian and Betsy D'Amido, soaked with spray, opened the door with difficulty against the driving wind and came into the now-covered security of the promenade deck, they stopped and turned toward each other.

'Well?' she said. But he only stood with his back to the rail, looking at her, afraid to speak. She was silent, too, because she wanted him to be first; so for a moment nothing happened. Then she made a step toward him, and he took her in his arms and kissed her forehead.

'You're just sorry for me, that's all.' She began to cry a little. 'You're just being kind.'

'I feel terribly about it.' His voice was taut and trembling.

'Then kiss me.'

The deck was empty. He bent over her swiftly.

'No, really kiss me.'

He could not remember when anything had felt so young and fresh as her lips. The rain lay, like tears shed for him, upon the softly shining porcelain cheeks. She was all new and immaculate, and her eyes were wild.

'I love you,' she whispered. 'I can't help loving you, can I? When I first saw you—oh, not on the boat, but over a year ago—Grace Heally took me to a rehearsal and suddenly you jumped up in the second row and began telling them what to do. I wrote you a letter and tore it up.'

'We've got to go.'

She was weeping as they walked along the deck. Once more, imprudently, she held up her face to him at the door of her cabin. His blood was beating through him in wild tumult as he walked on to the bar.

He was thankful that Eva scarcely seemed to notice him or to know that he had been gone. After a moment he pretended an interest in what she was doing.

'What's that?'

'She's painting the Eiffel Tower on my shirt front for tonight,' explained Butterworth.

'There,' Eva laid away her brush and wiped her hands. 'How's that?'

'A *chef-d'œuvre*.'

Her eyes swept around the watching group, lingered casually upon Adrian.

'You're wet. Go and change.'

'You come too.'

'I want another champagne cocktail.'

'You've had enough. It's time to dress for the party.'

Unwilling she closed her paints and preceded him.

'Stacomb's got a table for nine,' he remarked as they walked along the corridor.

'The younger set,' she said with unnecessary bitterness. 'Oh, the younger set. And you just having the time of your life—with a child.'

They had a long discussion in the cabin, unpleasant on her part and evasive on his, which ended when the ship gave a sudden gigantic heave, and Eva, the edge worn off her champagne, felt ill again. There was nothing to do but to have a cocktail in the cabin, and after that they decided to go to the party—she believed him now, or she didn't care.

Adrian was ready first—he never wore fancy dress.

'I'll go on up. Don't be long.'

'Wait for me, please; it's rocking so.'

He sat down on a bed, concealing his impatience.

'You don't mind waiting, do you? I don't want to parade up there all alone.'

She was taking a tuck in an oriental costume rented from the barber.

'Ships make people feel crazy,' she said. 'I think they're awful.'

'Yes,' he muttered absently.

'When it gets very bad I pretend I'm in the top of a tree, rocking to and fro. But finally I get pretending everything, and finally I have to pretend I'm sane when I know I'm not.'

'If you get thinking that way you will go crazy.'

'Look, Adrian.' She held up the string of pearls before clasping them on. 'Aren't they lovely?'

In Adrian's impatience she seemed to move around the cabin like a figure in a slow-motion picture. After a moment he demanded:

'Are you going to be long? It's stifling in here.'

'You go on!' she fired up.

'I don't want——'

'Go on, please! You just make me nervous trying to hurry me.'

With a show of reluctance he left her. After a moment's hesitation he went down a flight to a deck below and knocked at a door.

'Betsy.'

'Just a minute.'

She came out in the corridor attired in a red pea-jacket and trousers borrowed from the elevator boy.

'Do elevator boys have fleas?' she demanded. 'I've got everything in the world on under this as a precaution.'

'I had to see you,' he said quickly.

'Careful,' she whispered. 'Mrs Worden, who's supposed to be chaperoning me, is across the way. She's sick.'

'I'm sick for you.'

They kissed suddenly, clung close together in the narrow corridor, swaying to and fro with the motion of the ship.

'Don't go away,' she murmured.

'I've got to. I've——'

Her youth seemed to flow into him, bearing him up into a delicate romantic ecstasy that transcended passion. He couldn't relinquish it; he had discovered something that he had thought was lost with his own youth forever. As he walked along the passage he knew that he had stopped thinking, no longer dared to think.

He met Eva going into the bar.

'Where've you been?' she asked with a strained smile.

'To see about the table.'

She was lovely; her cool distinction conquered the trite costume and filled him with a resurgence of approval and pride. They sat down at a table.

The gale was rising hour by hour and the mere traversing of a passage had become a rough matter. In every stateroom trunks were lashed to the washstands, and the *Vestris* disaster was being reviewed in detail by nervous ladies, tossing, ill and wretched, upon their beds. In the smoking room a stout gentleman had been hurled backward and suffered a badly cut head; and now the lighter chairs and tables were stacked and roped against the wall.

The crowd who had donned fancy dress and were dining together had swollen to about sixteen. The only remaining qualification for membership was the ability to reach the smoking room. They ranged from a Groton-Harvard lawyer to an ungrammatical broker they had nicknamed Gyp the Blood, but distinctions had disappeared; for the moment they were samurai, chosen from several hundred for their triumphant resistance to the storm.

The gala dinner, overhung sardonically with lanterns and streamers, was interrupted by great communal slides across the room, precipitate retirements and spilled wine, while the ship roared and complained that under the panoply of a palace it was a ship after all. Upstairs afterward a dozen couples tried to dance, shuffling and galloping here and there in a crazy fandango, thrust around fantastically by a will alien to their own. In view of the condition of tortured hundreds below, there grew to be something indecent about it, like a revel in a house of mourning, and presently there was an egress of the ever-dwindling survivors toward the bar.

As the evening passed, Eva's feeling of unreality increased. Adrian had disappeared—presumably with Miss D'Amido—and her mind, distorted by illness and champagne, began to enlarge upon the fact; annoyance changed slowly to dark and brooding anger, grief to desperation. She had never tried to bind Adrian, never needed to—for they were serious people, with all sorts of mutual interests, and satisfied with each other—but this was a breach of the contract, this was cruel. How could he think that she didn't know?

It seemed several hours later that he leaned over her chair in the bar where she was giving some woman an impassioned lecture upon babies, and said:

'Eva, we'd better turn in.'

Her lip curled. 'So that you can leave me there and then come back to your eighteen-year——'

'Be quiet.'

'I won't come to bed.'

'Very well. Good night.'

More time passed and the people at the table changed. The stewards wanted to close up the room, and thinking of Adrian—her Adrian—off somewhere saying tender things to someone fresh and lovely, Eva began to cry.

'But he's gone to bed,' her last attendants assured her. 'We saw him go.'

She shook her head. She knew better. Adrian was lost. The long seven-year dream was broken. Probably she was punished for something she had done; as this thought occurred to her the shrieking timbers overhead began to mutter that she had guessed at last. This was for the selfishness to her mother, who hadn't wanted her to marry Adrian; for all the sins and omissions of her life. She stood up, saying she must go out and get some air.

The deck was dark and drenched with wind and rain. The ship pounded through valleys, fleeing from black mountains of water that roared toward it. Looking out at the night, Eva saw that there was no chance for them unless she could make atonement, propitiate the storm. It was Adrian's love that was demanded of her. Deliberately she unclasped her pearl necklace, lifted it to her lips—for she knew that with it went the freshest, fairest part of her life—and flung it out into the gale.

III

When Adrian awoke it was lunchtime, but he knew that some heavier sound than the bugle had called him up from his deep sleep. Then he realized that the trunk had broken loose from its lashings and was being thrown back and forth between a wardrobe and Eva's bed. With an exclamation he jumped up, but she was unharmed—still in costume and stretched out in deep sleep. When the steward had helped him secure the trunk, Eva opened a single eye.

'How are you?' he demanded, sitting on the side of her bed.

She closed the eye, opened it again.

'We're in a hurricane now,' he told her. 'The steward says it's the worst he's seen in twenty years.'

'My head,' she muttered. 'Hold my head.'

'How?'

'In front. My eyes are going out. I think I'm dying.'

'Nonsense. Do you want the doctor?'

She gave a funny little gasp that frightened him; he rang and sent the steward for the doctor.

The young doctor was pale and tired. There was a stubble of beard upon his face. He bowed curtly as he came in and, turning to Adrian, said with scant ceremony:

'What's the matter?'

'My wife doesn't feel well.'

'Well, what is it you want—a bromide?'

A little annoyed by his shortness, Adrian said: 'You'd better examine her and see what she needs.'

'She needs a bromide,' said the doctor. 'I've given orders that she is not to have any more to drink on this ship.'

'Why not?' demanded Adrian in astonishment.

'Don't you know what happened last night?'

'Why, no, I was asleep.'

'Mrs Smith wandered around the boat for an hour, not knowing what she was doing. A sailor was sent to follow her, and then the medical stewardess tried to get her to bed, and your wife insulted her.'

'Oh, my heavens!' cried Eva faintly.

'The nurse and I had both been up all night with Steward Carton, who died this morning.' He picked up his case. 'I'll send down a bromide for Mrs Smith. Goodbye.'

For a few minutes there was silence in the cabin. Then Adrian put his arm around her quickly.

'Never mind,' he said. 'We'll straighten it out.'

'I remember now.' Her voice was an awed whisper. 'My pearls. I threw them overboard.'

'Threw them overboard!'

'Then I began looking for you.'

'But I was here in bed.'

'I didn't believe it; I thought you were with that girl.'

'She collapsed during dinner. I was taking a nap down here.'

Frowning, he rang the bell and asked the steward for luncheon and a bottle of beer.

'Sorry, but we can't serve any beer to your cabin, sir.'

When he went out Adrian exploded: 'This is an outrage. You were simply crazy from that storm and they can't be so high-handed. I'll see the captain.'

'Isn't that awful?' Eva murmured. 'The poor man died.'

She turned over and began to sob into her pillow. There was a knock at the door.

'Can I come in?'

The assiduous Mr Butterworth, surprisingly healthy and immaculate, came into the crazily tipping cabin.

'Well, how's the mystic?' he demanded of Eva. 'Do you remember praying to the elements in the bar last night?'

'I don't want to remember anything about last night.'

They told him about the stewardess, and with the telling the situation lightened; they all laughed together.

'I'm going to get you some beer to have with your luncheon,' Butterworth said. 'You ought to get up on deck.'

'Don't go,' Eva said. 'You look so cheerful and nice.'

'Just for ten minutes.'

When he had gone, Adrian rang for two baths.

'The thing is to put on our best clothes and walk proudly three times around the deck,' he said.

'Yes.' After a moment she added abstractedly: 'I like that young man. He was awfully nice to me last night when you'd disappeared.'

The bath steward appeared with the information that bathing was too dangerous today. They were in the midst of the wildest hurricane on the North Atlantic in ten years; there were two broken arms this morning from attempts to take baths. An elderly lady had been thrown down a staircase and was not expected to live. Furthermore, they had received the SOS signal from several boats this morning.

'Will we go to help them?'

'They're all behind us, sir, so we have to leave them to the *Mauretania*. If we tried to turn in this sea the portholes would be smashed.'

This array of calamities minimized their own troubles. Having eaten a sort of luncheon and drunk the beer provided by Butterworth, they dressed and went on deck.

Despite the fact that it was only possible to progress step by step, holding on to rope or rail, more people were abroad than on the day before. Fear had driven them from their cabins, where the trunks bumped and the waves pounded the portholes and they awaited momentarily the call to the boats. Indeed, as Adrian and Eva stood on the transverse deck above the second class, there was a bugle call, followed by a gathering of stewards and stewardesses on the deck below. But the boat was sound; it had outlasted one of its cargo—Steward James Carton was being buried at sea.

It was very British and sad. There were the rows of stiff, disciplined men and women standing in the driving rain, and there was a shape covered by the flag of the Empire that lived by the sea. The chief purser read the service, a hymn was sung, the body slid off into the hurricane. With Eva's burst of wild weeping for this humble end, some last string snapped within her. Now she really didn't care. She responded eagerly when Butterworth suggested that he get some champagne to their cabin. Her mood worried Adrian; she wasn't used to so much drinking and he wondered what he ought to do. At his suggestion that they sleep instead, she merely laughed, and the bromide the doctor had sent stood untouched on the washstand. Pretending to listen to the insipidities of several Mr Stacombs, he watched her; to his surprise and discomfort she seemed on intimate and even sentimental terms with Butterworth, and he wondered if this was a form of revenge for his attention to Betsy D'Amido.

The cabin was full of smoke, the voices went on incessantly, the suspension of activity, the waiting for the storm's end, was getting on his nerves. They had been at sea only four days; it was like a year.

The two Mr Stacombs left finally, but Butterworth remained. Eva was urging him to go for another bottle of champagne.

'We've had enough,' objected Adrian. 'We ought to go to bed.'

'I won't go to bed!' she burst out. 'You must be crazy! You play around all you want, and then, when I find somebody I—I like, you want to put me to bed.'

'You're hysterical.'

'On the contrary, I've never been so sane.'

'I think you'd better leave us, Butterworth,' Adrian said. 'Eva doesn't know what she's saying.'

'He won't go. I won't let him go.' She clasped Butterworth's hand passionately. 'He's the only person that's been half decent to me.'

'You'd better go, Butterworth,' repeated Adrian.

The young man looked at him uncertainly.

'It seems to me you're being unjust to your wife,' he ventured.

'My wife isn't herself.'

'That's no reason for bullying her.'

Adrian lost his temper. 'You get out of here!' he cried.

The two men looked at each other for a moment in silence. Then Butterworth turned to Eva, said, 'I'll be back later,' and left the cabin.

'Eva, you've got to pull yourself together,' said Adrian when the door closed.

She didn't answer, looked at him from sullen, half-closed eyes.

'I'll order dinner here for us both and then we'll try to get some sleep.'

'I want to go up and send a wireless.'

'Who to?'

'Some Paris lawyer. I want a divorce.'

In spite of his annoyance, he laughed. 'Don't be silly.'

'Then I want to see the children.'

'Well, go and see them. I'll order dinner.'

He waited for her in the cabin twenty minutes. Then impatiently he opened the door across the corridor; the nurse told him that Mrs Smith had not been there.

With a sudden prescience of disaster he ran upstairs, glanced in the bar, the salons, even knocked at Butterworth's door. Then a quick round of the decks, feeling his way through the black spray and rain. A sailor stopped him at a network of ropes.

'Orders are no one goes by, sir. A wave has gone over the wireless room.'

'Have you seen a lady?'

'There was a young lady here——' He stopped and glanced around. 'Hello, she's gone.'

'She went up the stairs!' Adrian said anxiously. 'Up to the wireless room!'

The sailor ran up to the boat deck; stumbling and slipping, Adrian followed. As he cleared the protected sides of the companionway, a tremendous body struck the boat a staggering blow and, as she keeled over to an angle of forty-five degrees, he was thrown in a helpless roll down the drenched deck, to bring up dizzy and bruised against a stanchion.

'Eva!' he called. His voice was soundless in the black storm. Against the faint light of the wireless-room window he saw the sailor making his way forward.

'Eva!'

The wind blew him like a sail up against a lifeboat. Then there was another shuddering crash, and high over his head, over the very boat, he saw a gigantic, glittering white wave, and in the split second that it balanced there he became conscious of Eva, standing beside a ventilator twenty feet away. Pushing out from the stanchion, he lunged desperately toward her, just as the wave broke with a smashing roar. For a moment the rushing water was five feet deep, sweeping with enormous force toward the side, and then a human body was washed against him, and frantically he clutched it and was swept with it back toward the rail. He felt his body bump against it, but desperately he held on to his burden; then, as the ship rocked slowly back, the two of them, still joined by his fierce grip, were rolled out exhausted on the wet planks. For a moment he knew no more.

IV

Two days later, as the boat train moved tranquilly south toward Paris, Adrian tried to persuade his children to look out the window at the Norman countryside.

'It's beautiful,' he assured them. 'All the little farms like toys. Why, in heaven's name, won't you look?'

'I like the boat better,' said Estelle.

Her parents exchanged an infanticidal glance.

'The boat is still rocking for me,' Eva said with a shiver. 'Is it for you?'

'No. Somehow, it all seems a long way off. Even the passengers looked unfamiliar going through the customs.'

'Most of them hadn't appeared above ground before.'

He hesitated. 'By the way, I cashed Butterworth's check for him.'

'You're a fool. You'll never see the money again.'

'He must have needed it pretty badly or he would not have come to me.'

A pale and wan girl, passing along the corridor, recognized them and put her head through the doorway.

'How do you feel?'

'Awful.'

'Me, too,' agreed Miss D'Amido. 'I'm vainly hoping my fiancé will recognize me at the Gare du Nord. Do you know two waves went over the wireless room?'

'So we heard,' Adrian answered drily.

She passed gracefully along the corridor and out of their life.

'The real truth is that none of it happened,' said Adrian after a moment. 'It was a nightmare—an incredibly awful nightmare.'

'Then, where are my pearls?'

'Darling, there are better pearls in Paris. I'll take the responsibility for those pearls. My real belief is that you saved the boat.'

'Adrian, let's never get to know anyone else, but just stay together always—just we two.'

He tucked her arm under his and they sat close. 'Who do you suppose those Adrian Smiths on the boat were?' he demanded. 'It certainly wasn't me.'

'Nor me.'

'It was two other people,' he said, nodding to himself. 'There are so many Smiths in this world.'

After the Storm

∽∽∽

ERNEST HEMINGWAY

Iт wasn't about anything, something about making punch,
and then we started fighting and I slipped and he had me down
kneeling on my chest and choking me with both hands like he was
trying to kill me and all the time I was trying to get the knife out of
my pocket to cut him loose. Everybody was too drunk to pull him off
me. He was choking me and hammering my head on the floor and I
got the knife out and opened it up; and I cut the muscle right across his
arm and he let go of me. He couldn't have held on if he wanted to.
Then he rolled and hung on to that arm and started to cry and I said:
  'What the hell you want to choke me for?'
  I'd have killed him. I couldn't swallow for a week. He hurt my
throat bad.
  Well, I went out of there and there were plenty of them with him
and some came out after me and I made a turn and was down by the
docks and I met a fellow and he said somebody killed a man up the
street. I said, 'Who killed him?' and he said, 'I don't know who killed
him but he's dead all right', and it was dark and there was water
standing in the street and no lights and windows broke and boats all
up in the town and trees blown down and everything all blown and I
got a skiff and went out and found my boat where I had her inside
of Mango Key and she was all right only she was full of water. So I
bailed her out and pumped her out and there was a moon but plenty
of clouds and still plenty rough and I took it down along; and when
it was daylight I was off Eastern Harbour.
  Brother, that was some storm. I was the first boat out and you never
saw water like that was. It was just as white as a lye barrel and coming
from Eastern Harbour to Sou'west Key you couldn't recognize the
shore. There was a big channel blown right out through the middle

of the beach. Trees and all blown out and a channel cut through and all the water white as chalk and everything on it; branches and whole trees and dead birds, and all floating. Inside the keys were all the pelicans in the world and all kinds of birds flying. They must have gone inside there when they knew it was coming.

I lay at Sou'west Key a day and nobody came after me. I was the first boat out and I seen a spar floating and I knew there must be a wreck and I started out to look for her. I found her. She was a three-masted schooner and I could just see the stumps of her spars out of water. She was in too deep water and I didn't get anything off of her. So I went on looking for something else. I had the start on all of them and I knew I ought to get whatever there was. I went on down over the sand-bar from where I left that three-masted schooner and I didn't find anything and I went on a long way. I was way out toward the quicksands and I didn't find anything so I went on. Then when I was in sight of the Rebecca Light I saw all kinds of birds making over something and I headed over for them to see what it was and there was a cloud of birds all right.

I could see something looked like a spar up out of the water and when I got over close the birds all went up in the air and stayed all around me. The water was clear out there and there was a spar of some kind sticking out just above the water and when I come up close to it I saw it was all dark under water like a long shadow and I came right over it and there under water was a liner; just lying there all under water as big as the whole world. I drifted over her in the boat. She lay on her side and the stern was deep down. The portholes were all shut tight and I could see the glass shine in the water and the whole of her; the biggest boat I ever saw in my life lying there and I went along the whole length of her and then I went over and anchored and I had the skiff on the deck forward and I shoved it down into the water and sculled over with the birds all around me.

I had a water glass like we use sponging and my hand shook so I could hardly hold it. All the portholes were shut that you could see going along over her but way down below near the bottom something must have been open because there were pieces of things floating out all the time. You couldn't tell what they were. Just pieces. That's what the birds were after. You never saw so many birds. They were all around me; crazy yelling.

I could see everything sharp and clear. I could see her rounded over and she looked a mile long under the water. She was lying on a clear white bank of sand and the spar was a sort of foremast or some sort of tackle that slanted out of water the way she was lying on her side. Her bow wasn't very far under. I could stand on the letters of her name on her bow and my head was just out of water. But the nearest porthole was twelve feet down. I could just reach it with the grains pole and I tried to break it with that but I couldn't. The glass was too stout. So I sculled back to the boat and got a wrench and lashed it to the end of the grains pole and I couldn't break it. There I was looking down through the glass at that liner with everything in her and I was the first one to her and I couldn't get into her. She must have had five million dollars' worth in her.

It made me shaky to think how much she must have in her. Inside the porthole that was closed I could see something but I couldn't make it out through the water glass. I couldn't do any good with the grains pole and I took off my clothes and stood and took a couple of deep breaths and dove over off the stern with the wrench in my hand and swam down. I could hold on for a second to the edge of the porthole, and I could see in and there was woman inside with her hair floating all out. I could see her floating plain and I hit the glass twice with the wrench hard and I heard the noise clink in my ears but it wouldn't break and I had to come up.

I hung on to the dinghy and got my breath and then I climbed in and took a couple of breaths and dove again. I swam down and took hold of the edge of the porthole with my fingers and held it and hit the glass as hard as I could with the wrench. I could see the woman floated in the water through the glass. Her hair was tied once close to her head and it floated all out in the water. I could see the rings on one of her hands. She was right up close to the porthole and I hit the glass twice and I didn't even crack it. When I came up I thought I wouldn't make it to the top before I'd have to breathe.

I went down once more and I cracked the glass, only cracked it, and when I came up my nose was bleeding and I stood on the bow of the liner with my bare feet on the letters of her name and my head just out and rested there and then I swam over to the skiff and pulled up into it and sat there waiting for my head to stop aching and looking down into the water glass, but I bled so I had to wash out the water glass. Then I lay back in the skiff and held my hand under my nose to stop it and I lay there with my head back looking up and there was a million birds above and all around.

When I quit bleeding I took another look through the glass and then I sculled over to the boat to try and find something heavier than the wrench but I couldn't find a thing; not even a sponge hook. I went back and the water was clearer all the time and you could see everything that floated out over that white bank of sand. I looked for sharks but there weren't any. You could have seen a shark a long way away. The water was so clear and the sand white. There was a grapple for an anchor on the skiff and I cut it off and went overboard and down with it. It carried me right down and past the porthole and I grabbed and couldn't hold anything and went on down and down, sliding along the curved side of her. I had to let go of the grapple. I heard it bump once and it seemed like a year before I came up through to the top of the water. The skiff was floated away with the tide and I swam over to her with my nose bleeding in the water while I swam and I was plenty glad there weren't sharks; but I was tired.

My head felt cracked open and I lay in the skiff and rested and then I sculled back. It was getting along in the afternoon. I went down once more with the wrench and it didn't do any good. That wrench was too light. It wasn't any good diving unless you had a big hammer or something heavy enough to do good. Then I lashed the wrench to the grains pole again and I watched through the water glass and pounded on the glass and hammered until the wrench came off and I saw it in the glass, clear and sharp, go sliding down along her and then off and down to the quicksand and go in. Then I couldn't do a thing. The wrench was gone and I'd lost the grappie so I sculled back to the boat. I was too tired to get the skiff aboard and the sun was pretty low. The birds were all pulling out and leaving her and I headed for Sou'west Key towing the skiff and the birds going on ahead of me and behind me. I was plenty tired.

That night it came on to blow and it blew for a week. You couldn't get out to her. They come out from town and told me the fellow I'd had to cut was all right except for his arm and I went back to town and they put me under five hundred dollar bond. It came out all right because some of them, friends of mine, swore he was after me with an axe, but by the time we got back out to her the Greeks had blown her open and cleaned her out. They got the safe out with dynamite. Nobody ever knows how much they got. She carried gold and they got it all. They stripped her clean. I found her and I never got a nickel out of her.

It was a hell of a thing all right. They say she was just outside of Havana harbour when the hurricane hit and she couldn't get in or the owners wouldn't let the captain chance coming in; they say he wanted to try; so she had to go with it and in the dark they were running with it trying to go through the gulf between Rebecca and Tortugas when she struck on the quicksands. Maybe her rudder was carried away. Maybe they weren't even steering. But anyway they couldn't have known they were quicksands and when she struck the captain must have ordered them to open up the ballast tanks so she'd lay solid. But it was quicksand she'd hit and when they opened the tank she went in stern first and then over on her beam ends. There were four hundred and fifty passengers and the crew on board of her and they must all have been aboard of her when I found her. They must have opened the tanks as soon as she struck and the minute she settled on it the quicksands took her down. Then her boilers must have burst and that must have been what made those pieces that came out. It was funny there weren't any sharks though. There wasn't a fish. I could have seen them on that clear white sand.

Plenty of fish now though; jewfish, the biggest kind. The biggest part of her's under the sand now but they live inside of her; the biggest kind of jewfish. Some weigh three to four hundred pounds. Sometime we'll go out and get some. You can see the Rebecca light from where she is. They've got a buoy on her now. She's right at the end of the quicksand right at the edge of the gulf. She only missed going through by about a hundred yards. In the dark in the storm they just missed it; raining the way it was they couldn't have seen the Rebecca. Then they're not used to that sort of thing. The captain of a liner isn't used to scudding that way. They have a course and they tell me they set some sort of a compass and it steers itself. They probably didn't know where they were when they ran with that blow but they come close to making it. Maybe they'd lost the rudder though. Anyway there wasn't another thing for them to hit till they'd get to Mexico once they were in that gulf. Must have been something though when they struck in that rain and wind and he told them to open her tanks. Nobody could have been on deck in that blow and rain. Everybody must have been below. They couldn't have lived on deck. There must have been some scenes inside all right because you know she settled fast. I saw that wrench go into the sand. The captain couldn't have known it was quicksand when she struck unless he knew these waters. He just knew it wasn't rock. He must have seen it all up in the bridge. He must have known what it was about when she settled. I wonder how fast she made it. I wonder if the mate was there with him. Do you think they stayed inside the bridge or do you think they took it outside? They never found any bodies. Not a one. Nobody floating. They float a long way with lifebelts too. They must have took it inside. Well, the Greeks got it all. Everything. They must have come fast all right. They picked her clean. First there was the birds, then me, then the Greeks, and even the birds got more out of her than I did.

The Bravest Boat

〜〜〜

MALCOLM LOWRY

I⊤ was a day of spindrift and blowing sea-foam, with black clouds presaging rain driven over the mountains from the sea by a wild March wind.

But a clean silver sea light came from along the horizon where the sky itself was like glowing silver. And far away over in America the snowy volcanic peak of Mount Hood stood on high, disembodied, cut off from earth, yet much too close, which was an even surer presage of rain, as though the mountains had advanced, or were advancing.

In the park of the seaport the giant trees swayed, and taller than any were the tragic Seven Sisters, a constellation of seven noble red cedars that had grown there for hundreds of years, but were now dying, blasted, with bare peeled tops and stricken boughs. (They were dying rather than live longer near civilization. Yet though everyone had forgotten they were called after the Pleiades and thought they were named with civic pride after the seven daughters of a butcher, who seventy years before when the growing city was named Gaspool had all danced together in a shop window, nobody had the heart to cut them down.)

The angelic wings of the seagulls circling over the tree tops shone very white against the black sky. Fresh snow from the night before lay far down the slopes of the Canadian mountains, whose freezing summits, massed peak behind spire, jaggedly traversed the country northward as far as the eye could reach. And highest of all an eagle, with the poise of a skier, shot endlessly down the world.

In the mirror, reflecting this and much besides, of an old weighing machine with the legend *Your weight and your destiny* encircling its forehead and which stood on the embankment between the streetcar terminus and a hamburger stall, in this mirror along the reedy edge of the stretch of water below known as Lost Lagoon two figures in mackintoshes were approaching, a man and a beautiful passionate-looking girl, both bare-headed, and both extremely fair, and hand-in-hand, so that you would have taken them for young lovers, but that they were alike as brother and sister, and the man, although he walked with youthful nervous speed, now seemed older than the girl.

The man, fine-looking, tall, yet thick-set, very bronzed, and on approaching still closer obviously a good deal older than the girl, and wearing one of those blue belted trenchcoats favoured by merchant marine officers of any country, though without any corresponding cap—moreover, the trenchcoat was rather too short in the sleeve so that you could see some tattooing on his wrist, as he approached nearer still it seemed to be an anchor—whereas the girl's raincoat was of some sort of entrancing forest-green corduroy—the man paused every now and then to gaze into the lovely laughing face of his girl, and once or twice they both stopped, gulping in great draughts of salty clean sea and mountain air. A child smiled at them, and they smiled back. But the child belonged elsewhere, and the couple were unaccompanied.

In the lagoon swam wild swans, and many wild ducks: mallards and buffleheads and scaups, golden eyes, and cackling black coots with carved ivory bills. The little buffleheads often took flight from the water and some of them blew about like doves among the smaller trees. Under these trees lining the bank other ducks were sitting meekly on the sloping lawn, their beaks tucked into their plumage rumpled by the wind. The smaller trees were apples and hawthorns, some just opening into bloom even before they had foliage, and weeping willows, from whose branches small showers from the night's rain were scattered on the two figures as they passed.

A red-breasted merganser cruised in the lagoon, and at this swift and angry sea bird, with his proud disordered crest, the two were now gazing with a special sympathy, perhaps because he looked lonely without his mate. Ah, they were wrong. The red-breasted merganser was now joined by his wife and on a sudden duck's impulse and with immense fuss the two wild creatures flew off to settle on another part of the lagoon. And for some reason this simple fact appeared to make these two good people—for nearly all people are good who walk in parks—very happy again.

Now at a distance they saw a small boy, accompanied by his father who was kneeling on the bank, trying to sail a toy boat in the lagoon. But the blustery March wind soon slanted the tiny yacht into trouble and the father hauled it back, reaching out with his curved stick, and set it on an upright keel again for his son.

Your weight and your destiny.

Suddenly the girl's face, at close quarters in the weighing machine's mirror, seemed struggling with tears: she unbuttoned the top button of her coat to readjust her scarf, revealing, attached to a gold chain around her neck, a small gold cross. They were quite alone now, standing on top of the embankment by the machine, save for a few old men feeding the ducks below, and the father and his son with the toy yacht, all of whom had their backs turned, while an empty tram abruptly city-bound trundled around the minute terminus square; and the man, who had been trying to light his pipe, took her in his arms and tenderly kissed her, and then pressing his face against her cheek, held her a moment closely.

The couple, having gone down obliquely to the lagoon once more, had now passed the boy with his boat and his father. They were smiling again. Or as much as they could while eating hamburgers. And they were smiling still as they passed the slender reeds where a northwestern redwing was trying to pretend he had no notion of nesting, the northwestern redwing who like all birds in these parts may feel superior to man in that he is his own customs official, and can cross the wild border without let.

Along the far side of Lost Lagoon the green dragons grew thickly, their sheathed and cowled leaves giving off their peculiar animal-like odour. The two lovers were approaching the forest in which, ahead, several footpaths threaded the ancient trees. The park, seagirt, was very large, and like many parks throughout the Pacific Northwest, wisely left in places to the original wilderness. In fact, though its beauty was probably unique, it was quite like some American parks, you might have thought, save for the Union Jack that galloped evermore by a pavilion, and but for the apparition, at this moment, passing by on the carefully landscaped road slightly above, which led with its tunnels and detours to a suspension bridge, of a posse of Royal Canadian Mounted Policemen mounted royally upon the cushions of an American Chevrolet.

Nearer the forest were gardens with sheltered beds of snowdrops and here and there a few crocuses lifting their sweet chalices. The man and his girl now seemed lost in thought, breasting the buffeting wind that blew the girl's scarf out behind her like a pennant and blew the man's thick fair hair about his head.

A loudspeaker, enthroned on a wagon, barked from the city of Enochvilleport composed of dilapidated half-skyscrapers, at different levels, some with all kinds of scrap iron, even broken airplanes, on their roofs, others being mouldy stock exchange buildings, new beer parlours crawling with verminous light even in mid-afternoon and resembling gigantic emerald-lit public lavatories for both sexes, masonries containing English tea-shoppes where your fortune could be told by a female relative of Maximilian of Mexico, totem pole factories, drapers' shops with the best Scotch tweed and opium dens in the basement (though no bars, as if, like some hideous old roué shuddering with every unmentionable secret vice, this city without gaiety had cackled 'No, I draw the line at that.—What would our wee laddies come to then?'), cerise conflagrations of cinemas, modern apartment buildings, and other soulless behemoths, housing, it might be, noble invisible struggles, of literature, the drama, art or music, the student's lamp and the rejected manuscript; or indescribable poverty and degradation, between which civic attractions were squeezed occasional lovely dark ivy-clad old houses that seemed weeping, cut off from all light, on their knees, and elsewhere bankrupt hospitals, and one or two solid-stoned old banks, held up that afternoon; and among which appeared too, at infrequent intervals, beyond a melancholy never-striking black and white clock that said three, dwarfed spires belonging to frame façades with blackened rose windows, queer grimed onion-shaped domes, and even Chinese pagodas, so that first you thought you were in the Orient, then Turkey or Russia, though finally, but for the fact that some of these were churches, you would be sure you were in hell: despite that anyone who had ever really been in hell must have given Enochvilleport a nod of recognition, further affirmed by the spectacle, at first not unpicturesque, of the numerous sawmills relentlessly smoking and champing away like demons, Molochs fed by whole mountainsides of forests that never grew again, or by trees that made way for grinning regiments of villas in the background of 'our expanding and fair city', mills that shook the very earth with their tumult, filling the windy air with their sound as of a wailing and gnashing of teeth: all these curious achievements of man, together creating as we say 'the jewel of the Pacific', went as though down a great incline to a harbour more spectacular than Rio de Janeiro and San Francisco put together, with deep-sea freighters moored at every angle for miles in the roadstead, but to whose heroic prospect nearly the only human dwellings visible on this side of the water that had any air of belonging, or in which their inhabitants could be said any longer to participate, were, paradoxically, a few lowly little self-built shacks and floathouses, that might have been driven out of the city altogether, down to the water's edge into the sea itself, where they stood on piles, like fishermen's huts (which several of them apparently were), or on rollers, some dark and tumbledown, others freshly and prettily painted, these last quite evidently built or placed with some human need for beauty in mind, even if under the permanent threat of eviction, and all standing, even the most somber, with their fluted tin chimneys smoking here and there like toy tramp steamers, as though in defiance of the town, before eternity. In Enochvilleport itself some ghastly coloured neon signs had long since been going through their unctuous twitchings and gesticulations that nostalgia and love transform into a poetry of longing: more happily one began to flicker: PALOMAR, LOUIS ARMSTRONG AND HIS ORCHESTRA. A huge new grey dead hotel that at sea might be a landmark of romance, belched smoke out of its turreted haunted-looking roof, as if it had caught fire, and beyond that all the lamps were blazing within the grim courtyard of the law courts, equally at sea a trysting place of the heart, outside which one of the stone lions, having recently been blown up, was covered reverently with a white cloth, and inside which for a month a group of stainless citizens had been trying a sixteen-year-old boy for murder.

Nearer the park the apron lights appeared on a sort of pebble-dashed YMCA-Hall-cum-variety-theatre saying TAMMUZ *The Master Hypnotist, To-nite 8:30*, and running past this the tramlines, down which another parkwise streetcar was approaching, could be seen extending almost to the department store in whose show window Tammuz' subject, perhaps a somnolent descendant of the seven sisters whose fame had eclipsed even that of the Pleiades, but whose announced ambition was to become a female psychiatrist, had been sleeping happily and publicly in a double bed for the last three days as an advance publicity stunt for tonight's performance.

Above Lost Lagoon on the road now mounting toward the suspension bridge in the distance much as a piece of jazz music mounts toward a break, a newsboy cried: 'LASH ORDERED FOR SAINT PIERRE! SIXTEEN YEAR OLD BOY, CHILD-SLAYER, TO HANG! Read all about it!'

The weather too was foreboding. Yet, seeing the wandering lovers, the other passers-by on this side of the lagoon, a wounded soldier lying on a bench smoking a cigarette, and one or two of those destitute souls, the very old who haunt parks—since, faced with a choice, the very old will sometimes prefer, rather than to keep a room and starve, at least in such a city as this, somehow to eat and live outdoors—smiled too.

For as the girl walked along beside the man with her arm through his and as they smiled together and their eyes met with love, or they paused, watching the blowing seagulls, or the ever-changing scene of the snow-freaked Canadian mountains with their fleecy indigo chasms, or to listen to the deep-tongued majesty of a merchantman's echoing roar (these things that made Enochvilleport's ferocious aldermen imagine that it was the city itself that was beautiful, and maybe they were half right), the whistle of a ferryboat as it sidled across the inlet northward, what memories might not be evoked in a poor soldier, in the breasts of the bereaved, the old, even, who knows, in the mounted policemen, not merely of young love, but of lovers, as they seemed to be, so much in love that they were afraid to lose a moment of their time together?

Yet only a guardian angel of these two would have known—and surely they must have possessed a guardian angel—the strangest of all strange things of which they were thinking, save that, since they had spoken of it so often before, and especially, when they had opportunity, on this day of the year, each knew of course that the other was thinking about it, to such an extent indeed that it was no surprise, it only resembled the beginning of a ritual when the man said, as they entered the main path of the forest, through whose branches that shielded them from the wind could be made out, from time to time, suggesting a fragment of music manuscript, a bit of the suspension bridge itself:

'It was a day just like this that I set the boat adrift. It was twenty-nine years ago in June.'

'It was twenty-nine years ago in June, darling. And it was June twenty-seventh.'

'It was five years before you were born, Astrid, and I was ten years old and I came down to the bay with my father.'

'It was five years before I was born, you were ten years old, and you came down to the wharf with your father. Your father and grandfather had made you the boat between them and it was a fine one, ten inches long, smoothly varnished and made of wood from your model airplane box, with a new strong white sail.'

'Yes, it was balsa wood from my model airplane box and my father sat beside me, telling me what to write for a note to put in it.'

'Your father sat beside you, telling you what to write,' Astrid laughed, 'and you wrote:

'Hello.

'My name is Sigurd Storlesen. I am ten years old. Right now I am sitting on the wharf at Fearnought Bay, Clallam County, State of Washington, USA, 5 miles south of Cape Flattery on the Pacific side, and my Dad is beside me telling me what to write. Today is June 27, 1922. My Dad is a forest warden in the Olympic National Forest but my Granddad is the lighthouse keeper at Cape Flattery. Beside me is a small shiny canoe which you now hold in your hand. It is a windy day and my Dad said to put the canoe in the water when I have put this in and glued down the lid which is a piece of balsa wood from my model airplane box.

'Well must close this note now, but first I will ask you to tell the Seattle Star that you have found it, because I am going to start reading the paper from today and looking for a piece that says, who when and where it was found.

'Thanks. Sigurd Storlesen.'

'Yes, then my father and I put the note inside, and we glued down the lid and sealed it and put the boat on the water.'

'You put the boat on the water and the tide was going out and away it went. The current caught it right off and carried it out and you watched it till it was out of sight!'

The two had now reached a clearing in the forest where a few grey squirrels were scampering about on the grass. A dark-browed Indian in a windbreaker, utterly absorbed by his friendly task, stood with a sleek black squirrel sitting on his shoulder nibbling popcorn he was giving it from a bag. This reminded them to get some peanuts to feed the bears, whose cages were over the way.

Ursus Horribilis: and now they tossed peanuts to the sad lumbering sleep-heavy creatures—though at least these two grizzlies were together, they even had a home—maybe still too sleepy to know where they were, still wrapped in a dream of their timberfalls and wild blueberries in the Cordilleras Sigurd and Astrid could see again, straight ahead of them, between the trees, beyond a bay.

But how should they stop thinking of the little boat?

Twelve years it had wandered. Through the tempests of winter, over sunny summer seas, what tide rips had caught it, what wild sea birds, shearwaters, storm petrels, jaegers, that follow the thrashing propellers, the dark albatross of these northern waters, swooped upon it, or warm currents edged it lazily toward land—and blue-water currents sailed it after the albacore, with fishing boats like white giraffes—or glacial drifts tossed it about fuming Cape Flattery itself. Perhaps it had rested, floating in a sheltered cove, where the killer whale smote, lashed, the deep clear water; the eagle and the salmon had seen it, a baby seal stared with her wondering eyes, only for the little boat to be thrown aground, catching the rainy afternoon sun, on cruel barnacled rocks by the waves, lying aground knocked from side to side in an inch of water like a live thing, or a poor old tin can, pushed, pounded ashore, and swung around, reversed again, left high and dry, and then swept another yard up the beach, or carried under a lonely salt-grey shack, to drive a seine fisherman crazy all night with its faint plaintive knocking, before it ebbed out in the dark autumn dawn, and found its way afresh, over the deep, coming through thunder, to who will ever know what fierce and desolate uninhabited shore, known only to the dread Wendigo, where not even an Indian could have found it, unfriended there, lost, until it was borne out to sea once more by the great brimming black tides of January, or the huge calm tides of the mid-summer moon, to start its journey all over again——

Astrid and Sigurd came to a large enclosure, set back from a walk, with two vine-leaved maple trees (their scarlet tassels, delicate precursors of their leaves, already visible) growing through the top, a sheltered cavernous part to one side for a lair, and the whole, save for the barred front, covered with stout large-meshed wire—considered sufficient protection for one of the most Satanic beasts left living on earth.

Two animals inhabited the cage, spotted like deceitful pastel leopards, and in appearance like decorated, maniacal-looking cats: their ears were provided with huge tassels and, as if this were in savage parody of the vine-leaved maples, from the brute's chin tassels also depended. Their legs were as long as a man's arm, and their paws, clothed in grey fur out of which shot claws curved like scimitars, were as big as a man's clenched fist.

And the two beautiful demonic creatures prowled and paced endlessly, searching the base of their cage, between whose bars there was just room to slip a murderous paw—always a hop out of reach an almost invisible sparrow went pecking away in the dust—searching with eternal voraciousness, yet seeking in desperation also some way out, passing and repassing each other rhythmically, as though truly damned and under some compelling enchantment.

And yet as they watched the terrifying Canadian lynx, in which seemed to be embodied in animal form all the pure ferocity of nature, as they watched, crunching peanuts themselves now and passing the bag between them, before the lovers' eyes still sailed that tiny boat, battling with the seas, at the mercy of a wilder ferocity yet, all those years before Astrid was born.

Ah, its absolute loneliness amid those wastes, those wildernesses, of rough rainy seas bereft even of sea birds, between contrary winds, or in the great dead windless swell that comes following a gale; and then with the wind springing up and blowing the spray across the sea like rain, like a vision of creation, blowing the little boat as it climbed the highlands into the skies, from which sizzled cobalt lightnings, and then sank down into the abyss, but already was climbing again, while the whole sea crested with foam like lambs' wool went furling off to leeward, the whole vast moon-driven expanse like the pastures and valleys and snow-capped ranges of a Sierra Madre in delirium, in ceaseless motion, rising and falling, and the little boat rising, and falling into a paralysing sea of white drifting fire and smoking spume by which it seemed overwhelmed: and all this time a sound, like a high sound of singing, yet as sustained in harmony as telegraph wires, or like the unbelievably high perpetual sound of the wind where there is nobody to listen, which perhaps does not exist, or the ghost of the wind in the rigging of ships long lost, and perhaps it was the sound of the wind in its toy rigging, as again the boat slanted onward: but even then what further unfathomed deeps had it oversailed, until what birds of ill omen turned heavenly for it at last, what iron birds with saber wings skimming forever through the murk above the grey immeasurable swells, imparted mysteriously their own homing knowledge to it, the lonely buoyant little craft, nudging it with their beaks under golden sunsets in a blue sky, as it sailed close in to mountainous coasts of clouds with stars over them, or burning coasts at sunset once more, as it rounded not only the terrible spume-drenched rocks, like incinerators in sawmills, of Flattery, but other capes unknown, those twelve years, of giant pinnacles, images of barrenness and desolation, upon which the heart is thrown and impaled eternally!—And strangest of all how many ships themselves had threatened it, during that voyage of only some three score miles as the crow flies from its launching to its final port, looming out of the fog and passing by harmlessly all those years—those years too of the last sailing ships, rigged to the moonsail, sweeping by into their own oblivion—but ships cargoed with guns or iron for impending wars, what freighters now at the bottom of the sea he, Sigurd, had voyaged in for that matter, freighted with old marble and wine and cherries-in-brine, or whose engines even now were still somewhere murmuring: *Frère* Jacques! *Frère* Jacques!

What strange poem of God's mercy was this?

Suddenly across their vision a squirrel ran up a tree beside the cage and then, chattering shrilly, leaped from a branch and darted across the top of the wire mesh. Instantly, swift and deadly as lightning, one of the lynx sprang twenty feet into the air, hurtling straight to the top of the cage toward the squirrel, hitting the wire with a twang like a mammoth guitar, and simultaneously flashing through the wire its scimitar claws: Astrid cried out and covered her face.

But the squirrel, unhurt, untouched, was already running lightly along another branch, down to the tree, and away, while the infuriated lynx sprang straight up, sprang again, and again and again and again, as his mate crouched spitting and snarling below.

Sigurd and Astrid began to laugh. Then this seemed obscurely unfair to the lynx, now solemnly washing his mate's face. The innocent squirrel, for whom they felt such relief, might almost have been showing off, almost, unlike the oblivious sparrow, have been taunting the caged animal. The squirrel's hairbreadth escape—the thousand-to-one chance—that on second thought must take place every day, seemed meaningless. But all at once it did not seem meaningless that they had been there to see it.

'You know how I watched the paper and waited,' Sigurd was saying, stooping to relight his pipe, as they walked on.

'The Seattle *Star*,' Astrid said.

'The Seattle *Star* . . . It was the first newspaper I ever read. Father always declared the boat had gone south—maybe to Mexico, and I seem to remember Granddad saying no, if it didn't break up on Tatoosh, the tide would take it right down Juan de Fuca Strait, maybe into Puget Sound itself. Well, I watched and waited for a long time and finally, as kids will, I stopped looking.'

'And the years went on——'

'And I grew up. Granddad was dead by then. And the old man, you know about him. Well, he's dead too now. But I never forgot. Twelve years! Think of it—! Why, it voyaged around longer than we've been married.'

'And we've been married seven years.'

'Seven years today—'

'It seems like a miracle!'

But their words fell like spent arrows before the target of this fact.

They were walking, as they left the forest, between two long rows of Japanese cherry trees, next month to be an airy avenue of celestial bloom. The cherry trees behind, the forest reappeared, to left and right of the wide clearing, and skirting two arms of the bay. As they approached the Pacific, down the gradual incline, on this side remote from the harbour the wind grew more boisterous: gulls, glaucous and raucous, wheeled and sailed overhead, yelling, and were suddenly far out to sea.

And it was the sea that lay before them, at the end of the slope that changed into the steep beach, the naked sea, running deeply below, without embankment or promenade, or any friendly shacks, though some prettily built homes showed to the left, with one light in a window, glowing warmly through the trees on the edge of the forest itself, as of some stalwart Columbian Adam, who had calmly stolen back with his Eve into Paradise, under the flaming sword of the civic cherubim.

The tide was low. Offshore, white horses were running around a point. The headlong onrush of the tide of beaten silver flashing over its crossflowing underset was so fast the very surface of the sea seemed racing away.

Their path gave place to a cinder track in the familiar lee of an old frame pavilion, a deserted tea house boarded up since last summer. Dead leaves were slithering across the porch, past which on the slope to the right picnic benches, tables, a derelict swing, lay overturned, under a tempestuous grove of birches. It seemed cold, sad, inhuman there, and beyond, with the roar of that deep low tide. Yet there was that between the lovers which moved like a warmth, and might have thrown open the shutters, set the benches and tables aright, and filled the whole grove with the voices and children's laughter of summer. Astrid paused for a moment with a hand on Sigurd's arm while they were sheltered by the pavilion, and said, what she too had often said before, so that they always repeated these things almost like an incantation:

'I'll never forget it. That day when I was seven years old, coming to the park here on a picnic with my father and mother and brother.

After lunch my brother and I came down to the beach to play. It was a fine summer day, and the tide was out, but there'd been this very high tide in the night, and you could see the lines of driftwood and seaweed where it had ebbed. . . . I was playing on the beach, and I found your boat!'

'You were playing on the beach and you found my boat. And the mast was broken.'

'The mast was broken and shreds of sail hung dirty and limp. But your boat was still whole and unhurt, though it was scratched and weatherbeaten and the varnish was gone. I ran to my mother, and she saw the sealing wax over the cockpit, and, darling, I found your note!'

'You found our note, my darling.'

Astrid drew from her pocket a scrap of paper and holding it between them they bent over (though it was hardly legible by now and they knew it off by heart) and read:

Hello.

My name is Sigurd Storlesen. I am ten years old. Right now I am sitting on the wharf at Fearnought Bay, Clallam County, State of Washington, USA, 5 miles south of Cape Flattery on the Pacific side, and my Dad is beside me telling me what to write. Today is June 27, 1922. My Dad is a forest warden in the Olympic National Forest but my Granddad is the lighthouse keeper at Cape Flattery. Beside me is a small shiny canoe which you now hold in your hand. It is a windy day and my Dad said to put the canoe in the water when I have put this in and glued down the lid which is a piece of balsa wood from my model airplane box.

Well must close this note now, but first I will ask you to tell the Seattle Star that you have found it, because I am going to start reading the paper from today and looking for a piece that says, who when and where it was found. Thanks.

SIGURD STORLESEN.

They came to the desolate beach strewn with driftwood, sculptured, whorled, silvered, piled everywhere by tides so immense there was a tideline of seaweed and detritus on the grass behind them, and great logs and shingle-bolts and writhing snags, crucificial, or frozen in a fiery rage—or better, a few bits of lumber almost ready to burn, for someone to take home, and automatically they threw them up beyond the sea's reach for some passing soul, remembering their own winters of need—and more snags there at the foot of the grove and visible high on the sea-scythed forest banks on either side, in which riven trees were growing, yearning over the shore. And everywhere they looked was wreckage, the toll of winter's wrath: wrecked hencoops, wrecked floats, the wrecked side of a fisherman's hut, its boards once hammered together, with its wrenched shiplap and extruding nails. The fury had extended even to the beach itself, formed in hummocks and waves and barriers of shingle and shells they had to climb up in places. And everywhere too was the grotesque macabre fruit of the sea, with its exhilarating iodine smell, nightmarish bulbs of kelp like antiquated motor horns, trailing brown satin streamers twenty feet long, sea wrack like demons, or the discarded casements of evil spirits that had been cleansed. Then more wreckage: boots, a clock, torn fishing nets, a demolished wheelhouse, a smashed wheel lying in the sand.

Nor was it possible to grasp for more than a moment that all this with its feeling of death and destruction and barrenness was only an appearance, that beneath the flotsam, under the very shells they crunched, within the trickling overflows of winterbournes they jumped over, down at the tide margin, existed, just as in the forest, a stirring and stretching of life, a seething of spring.

When Astrid and Sigurd were almost sheltered by an uprooted tree on one of these lower billows of beach they noticed that the clouds had lifted over the sea, though the sky was not blue but still that intense silver, so that they could see right across the Gulf and make out, or thought they could, the line of some Gulf Islands. A lone freighter with upraised derricks shipped seas on the horizon. A hint of the summit of Mount Hood remained, or it might have been clouds. They remarked too, in the southeast, on the sloping base of a hill, a triangle of storm-washed green, as if cut out of the overhanging murk there, in which were four pines, five telegraph posts, and a clearing resembling a cemetery. Behind them the icy mountains of Canada hid their savage peaks and snowfalls under still more savage clouds. And they saw that the sea was grey with whitecaps and currents charging offshore and spray blowing backwards from the rocks.

But when the full force of the wind caught them, looking from the shore, it was like gazing into chaos. The wind blew away their thoughts, their voices, almost their very senses, as they walked, crunching the shells, laughing and stumbling. Nor could they tell whether it was spume or rain that smote and stung their faces, whether spindrift from the sea or rain from which the sea was born, as now finally they were forced to a halt, standing there arm in arm.

... And it was to this shore, through that chaos, by those currents, that their little boat with its innocent message had been brought out of the past finally to safety and a home.

But ah, the storms they had come through!

The Boy Stood on the Burning Deck

⤳⤳⤳

C. S. FORESTER

In this story his name is Ed Jones; his real name is completely different from that. He runs a filling station near where I live, and I often buy gas there; his is not a calling that promises high adventure, nor is it likely to demand selfless devotion to duty. Just after the war the crossroads was quite a lonely point, and Ed's filling station was the only building within half a mile. With the population shift into California there are now great tracts of houses within sight, and there are rows of markets and shops. One might expect in consequence that Ed has made a fortune, but now each of the four corners of the crossroads is occupied by a filling station, and there are plenty of others not far away. Ed agrees that he makes twice as much money as he did when he came here, after his discharge from the Navy; but he points out philosophically that he can buy with his doubled income no more than he could before, and he works four times as hard to earn it.

But I want to write about Ed Jones, and not his filling station; he is the more interesting subject. I have known him for all these years, and I have always liked him, and his sturdy wife Mary, and the four post-war children whom I have seen growing up from babyhood. And I have used the name Ed Jones not only because it is unlike his real one but because it does not suggest heroism or self-sacrifice; neither does he himself—I might otherwise have called him Ironside or Strong. I knew he had served in the Navy during the war, but it was only recently that it came out in casual conversation that he had served in the destroyer *Boon*, and my interest was caught at once, because I have been writing stories about the *Boon*. I did my best to induce him to talk, but without any great success; Ed is not a very communicative man.

It did not call for any great degree of cunning to enlist the services of Mary, his wife. She was on my side almost from the start, but even her coaxing achieved little. Ed only laughed when he did not shrug his shoulders. Then one day when I was about to drive away Mary put her head in at my car window—it was one of the moments (and there are plenty of them) when she looks twenty rather than forty.

'I still have all the letters that he wrote me during the war,' she said breathlessly.

'That's just the sort of thing I'm after,' I said. 'Are you going to let me read them?'

'Hey, hold on a minute,' interposed Ed. 'Those letters—you know—they're not—'

'They're twenty years old,' protested Mary. 'And there's nothing to be ashamed of. And—' she turned back to me '—you wouldn't—'

'I wouldn't read anything I wasn't supposed to read,' I said. 'I'm pretty good at that. I expect they were most of them read by a censor at some time or other, anyway.'

That was how it happened that one evening I found myself sitting in the Jones's house with a cup of coffee at my elbow, and the sound of the television turned down to the lowest limit the children would tolerate, while Mary brought me the letters. She blushed quite charmingly, with a gesture to excuse the sentiment of a young bride who had tied the letters up in pink ribbons (faded now) and packed in a heart-shaped box which had presumably once held a Valentine's Day of chocolates.

To a man who deals with history, letters contemporary with the events he is studying are frequently valuable material. Accounts written later are usually coloured by the knowledge of what actually happened, and are distorted by later prejudices and legends. Wartime letters may be distorted too, admittedly, through the necessity of obeying wartime censorship regulations, and also because husbands, and wives, often wished to appear more cheerful than they actually were. But even the letters that are distorted badly convey an atmosphere, a mood that it is hard to recapture otherwise, and which is important when reconstructing a period; and sometimes they at least give clues that lead to the unearthing of forgotten facts.

'Thank you, Mary,' I said, and shot a glance at Ed before beginning to read; he appeared to have all his attention concentrated upon television.

The letters were love letters, naturally; from the first they contained much of what might be expected to be written by a young sailor who had newly joined the Navy and was newly separated from his young bride. My eyes ran rapidly down paragraphs that had little bearing on the war, trying not to read the tender passages while making sure there was no history buried in them. The letters were all dated and in sequence, and as I read I was conscious of a feeling that I could see into the future, that I had a Cassandra-like ability to prophecy. The letter of December 6th, 1941, written from machinist's mates striker's school, had a light-hearted gaiety that I knew could not endure; and I was ridiculously pleased with myself, as though I had really achieved something, when the next letter of December 8th, written after Pearl Harbor, confirmed my feeling.

It was interesting to read how Fireman Third Class Jones reacted to that news. The earlier letters had breathed a certain patriotism, whose sincerity could be guessed at despite the writer's difficulty in expressing it; this new letter told of a hardened resolve, of a grimmer determination, and it was easy to read into it the writer's certainty that every recruit around him felt similarly inspired. That was a historical fact.

'You don't have to read them if they're not what you want, you know,' said Mary.

'You couldn't stop me,' I answered, reaching for the next letter.

I still felt like Cassandra as I read on; when Fireman Second Class Jones wrote that he was being transferred out of training centre and wondered what was going to happen next. I knew that he was going to the *Boon*, and so he was—here was his new Fleet Post Office to prove it. And when he speculated regarding where *Boon* would be sent, I already knew. I could not merely follow him and the *Boon* across the Pacific and back again—I could travel ahead of them. There were things I knew about which only showed up vaguely in the letters, thanks to the censorship. I knew about the rescuing of the Navy pilot in his rubber boat ten days after the battle of the Coral Sea, and I knew, although Fireman First Class Jones did not, that the *Boon* was going to be one of the ships transferred back to Pearl Harbor to meet the next Japanese thrust, the one that ended in Japanese disaster at Midway, and how the *Boon* comported herself there. The accounts of the battle are so taken up with the action of the carriers, and with the attacks and counter-attacks launched by the aircraft, that there is nothing to spare for an insignificant destroyer like the *Boon*. I wanted to know how much the lower deck knew about the battle; how conscious the men were of having taken part in one of the decisive battles of the war; there was so much I wanted to know. And here was the next letter, just the top and bottom of it, connected by a thin thread of paper, with all the middle of it cut out by the censor's scissors.

My keen anticipation was replaced by a dull disappointment. There it was—Fireman First Class Jones had been promoted to Machinist's Mate Second Class—and then this gap. Some of it I could fill. Jones's rapid promotion was proof of his reliability and of the good opinions which his officers held of him; it was also an indication of the rapid expansion of the United States Navy. Clearly on arrival at Pearl there had been considerable transfers of personnel—skilled ratings had been taken out to help man the flood of new construction. Fresh recruits had been filled by promotions. I could be quite sure of this; but what had *Boon* achieved in the battle? What had been Jones's experiences? I broke in upon his contemplation of television.

'What happened here?' I asked, calling his attention to the gap in the letter. He had a look at it twice before he could be sure which letter it was.

'Oh, that?' he said. 'I didn't know they'd cut all that out of the letter. I wrote that from the sick bay at Pearl.'

'So I see,' I said. 'How did you get in there? What happened?'

He told me in the end, neither willingly nor fluently. To a reader that long-drawn interchange of question and answer would be tedious, no doubt. This is the tale of what happened; this is the completed picture, put together as though Jones's halting answers to any questions were the pieces of a jigsaw puzzle, but with nothing else added.

Machinist's Mate Second Class Ed Jones had the duty, at General Quarters, of attending to the throttle of the port engine in the *Boon*. He stood on a restricted area of iron deck down in the engine-room with the wheel of the valve in his hands and an instrument board on the bulkhead before him. Turning the wheel to the right reduced the amount of steam admitted to the turbine from the boilers; turning to the left increased it, and of course the speed of the turbine—and hence of the port propeller—varied in proportion. On the board in front of him appeared repetitions of the signals from the bridge regarding speed, the five speeds ahead and the three speeds astern and stop. By reference to the tachometer there he could adjust the speed of revolution of the turbine in accordance with the demands made upon him; the control was sensitive enough for him to be able to produce almost exactly any number of revolutions required. And on the board was the dial of another tachometer as well, which registered the revolutions of the starboard-engine, and it was Jones's business to see that the two readings agreed.

So during battle that was where he stood, hands on the wheel, adjusting carefully to left or right, or spinning hurriedly when a large change of speed was ordered; a very solitary and usually monotonous job that demanded unflagging attention. A critic might suggest that all this could be as well done, or better, by a machine; an apparatus that would respond automatically to the signals from the bridge and to the readings of the tachometers. That is perfectly true—such an apparatus would be absurdly simple compared with many employed in ships of war. But that was the real point, this was a ship of war and the regulation of the flow of steam was a vital function, one of overwhelming importance. A shell fragment or a near miss could put such an apparatus out of action easily enough, and that would be a disaster—there would be nothing to replace it. Naturally a shell fragment could put the human operator out of action too, but that would not be such a disaster. Once his dead body had been dragged out of the way another man could take his place.

Boon took her way out from Pearl Harbor along with the two carrier task forces which were going to fight the battle. She was part of the screen, naturally, part of the tight ring thrown about the vital carriers to protect them as much as possible from submarine and air attack. The tighter the ring the more efficient the protection and the greater the demand for good seamanship. The carriers had to make extravagant turns into the wind to fly off their planes and to fly them on again, and then the screen had to dash madly in far wider arcs to maintain their covering positions; there was fuelling at sea to be carried out, pilots to be rescued. A very small miscalculation could mean a collision and disaster. So could a very small mistake by Jones. The best of captains on the bridge could only watch, helpless, as catastrophe loomed ahead, if Jones in an absent-minded moment spun that wheel the wrong way, or did not pay instant attention to the captain's signals.

Boon had hardly secured from General Quarters after sunrise on that historic morning when the warning sounded again and the men had to go back to their battle stations.

'Did you expect a battle?' I asked, when Jones reached this point in his answers to my questions.

'Oh yes,' said Jones. His tone echoed the fatalism of the man under orders, or perhaps the steady determination of the man with a duty to perform.

'So what happened then?'

Jones ran down below down the iron ladders to his station at the throttle valve. He experienced a momentary regret as he did so, for on deck it was a beautiful day of sunshine and occasional cloud, just warm enough and delightful. He wished that fate had made him a gunner at one of the twenty-mm guns, so that he could stay topside and enjoy it. His battle station was too brightly lit to be called gloomy, but it was stark and inhospitable and lonely. He stood there with the steel deck gently swaying under his feet, busy enough after a few moments when the bridge signalled for revolutions for twenty-five knots—about as fast as the old *Boon* could go without straining herself—and then for repeated small variations to keep her in station screening the carrier. The speed itself, with the old destroyer vibrating and trembling, was enough to make Jones quite certain that action was impending, but he knew nothing more than that. He could only stand there, watching his instrument board and moving his wheel, while the fate of the civilized world—and of the uncivilized world—was being decided over his head. He knew nothing of the Japanese carrier force far away over the horizon, of the fleets of planes soaring into the air and returning, of the fighters wheeling overhead maintaining combat air patrol high up in the blue. He could not guess at the decisions that were being reached by the admirals—decisions that might determine his immediate death or survival, but which might affect his whole future life, even if he lived on to old age as a civilian.

He knew nothing of how the Japanese admirals had been tempted into delivering a blow at Midway, so that they were caught off guard by the sudden unexpected appearance of the American carriers within striking distance of their own. The hours went by for Jones in solitary monotony while death rained down on Midway and the bombers from Midway went into heroic death round the Japanese carriers, and while the planes from his own task force avenged them a hundredfold in a new surprise attack. Jones knew almost nothing of what was going on; the breaks in the monotony—the only indications that the task force was engaged in operations—were the occasional sudden turns, when the *Boon* lay over, without warning, under full rudder.

When she did that Jones might well have lost his footing, but he was an experienced seaman by now, and he could steady himself by his hands that gripped the wheel, although even then he sorely wrenched himself in his efforts to combat the sudden inclinations. These told him, however, that the task force was flying planes on and off and that the screen was having to wheel about to shelter the carrier.

Then the monotony was abruptly broken in a new way. Every gun the *Boon* carried suddenly began to fire. The loud bangs of the five-inch and the ear-splitting cracks of the small calibres were carried by the fabric of the ship direct, it almost appeared, to a focus in the steel cell where Jones stood. The concentrated fire was frightful, and the deck on which he stood and the wheel which he held in his hands, leaped and vibrated with the concussions, and it seemed as if every five seconds the ship was making a radical alteration of course, lying over madly first one way and then the other, so rapidly and unexpectedly that this, combined with the vibrations, came nearer than ever to sweeping him off his feet.

He could guess perfectly well what was going on. The suddenness with which it all began, the fact that the small guns were firing as well as the large ones, and the constant alterations of course told him that they were under air attack. Any other kind of daylight battle would have developed more slowly and the small-calibre guns would not have opened fire, as yet, while if they were under submarine attack the guns would not be firing at all, most likely, and certainly not for so long continuously. Being under air attack made no difference to his circumstances; all he had to do still was to attend to his signals and tachometers, and regulate his valve.

The destiny of the world was being decided over his head; the Japanese carrier planes had at last discovered the presence of the task force and were hurtling in to the attack. They were coming in with the speed and skill developed in years of training; their pilots were displaying the courage of their race; some of them more reckless even than usual, for they knew of the disasters that had befallen some of their own carriers and were frantic for revenge—frantic with desperation, some of them, for they guessed that their sinking carriers could no longer provide them with a refuge, so that the pilots' lives at longest could be measured by the gasoline in their tanks.

Suddenly it appeared to Jones as if the *Boon* had leaped clear out of the sea, as if his feet were pressing like ton weights upon the deck beneath them, and as if his thighs were being driven into his body, and then the deck fell away beneath him and the *Boon* rolled and pitched and plunged so that once more only his grip on the wheel saved him from being flung down. He knew, of course, what had caused all this. It was the near miss, the bomb bursting close alongside, which to this day is coldly recorded in the statistical accounts of the battle. It left the *Boon* strained and buckled, although she could still steam and still fight and still cover the aircraft carrier she had to guard with her life. Jones knew that she was strained and buckled—he had cautiously to shift position to keep his footing, and, looking down, he could see that the steel plate on which he stood was inclined slightly upwards from one edge, where it had torn free from its weld to its neighbour, leaving a gap, and it threatened to part altogether and drop him down into the bilges below.

But he could not think about that; his tachometer was registering a declining number of revolutions and he had to spin the wheel hurriedly to bring it back to its proper figure, even while his brain told him that the boiler room must have suffered damage so that the steam pressure had fallen. Not even that deduction had full time to mature. Even as he was thinking along these lines a new hellish noise broke round him. Wango-wango-wango—but much faster than human lips could enunciate those sounds. Some low-flying torpedo plane, its torpedo launched, was doing what further damage it could, and had opened fire with its machine-guns. The bullets beat upon the thin steel plates; the heavy-calibre ones came clean through, and the tracers set the *Boon* on fire. To Jones those seconds were like being in an iron pipe while a dozen men pounded the outside with hammers, but it was only a matter of seconds. The tachometer was behaving erratically, echoing what was going on in the boiler-room, and he had to work hard on the valve to keep it steady.

Then the *Boon* lay over again in another desperate turn and he became aware of a fresh complication. There was a rush of flames up through the gaps in the plating on which he stood, flames licking knee-high around him. He had to shrink to one side to avoid them, and then, as the *Boon* steadied herself on her new course, they died down leaving only a red glow below. The *Boon* lay over again, and the flames lifted their heads again. A ruptured fuel tank had leaked some of its contents into the bilges, and the oil had been set on fire. With the motion of the ship the flaming oil was washing back and forth under Jones's feet, rising higher towards him as the *Boon* turned. Amid the continuous din of the guns Jones was being roasted over an intermittent fire.

Yet whether *Boon* was turning or not, there was still some fire below him; the iron deck on which Jones stood was growing hotter and hotter. Amid the varying stinks that filled the engine room Jones noticed a new one—the acrid smell of burning leather, and at the same time he was conscious of agony in the soles of his feet. The worn-out old shoes that he kept for wear in the engine room were charring against the hot iron. He took his hand from the valve long enough to tear off his outer clothing, and he trampled that under his feet to insulate them from the plating, kicking off the smouldering shoes. That gave him a momentary respite, but momentary only. Soon his jumpers and trousers were smouldering too, as he stood on them. He was leaping with the pain.

'What about damage control?' I asked.

'They'd had a lot of casualties,' explained Jones. 'And there were plenty of other fires to put out, too.'

'How long did this go on?'

'Oh, I don't know. Long enough.'

The guns were silent by now, for the Japanese planes had gone, the pilots to their deaths. *Akagi* and *Kaga*, *Hiryu* and *Soryu*, the four proud Japanese carriers, were sunk or sinking, and the battle of Midway was won. The *Boon* lay over once more, as she turned to help pick up survivors from the sinking *Yorktown*, and Jones was momentarily bathed in flames again. And then came the signal from the bridge.

'Stop.'

Jones spun the wheel as his tortured feet charred on the hot plating, and then down the ladder came clattering the damage control party. It was only a matter of moments for the foam to extinguish the flames in the bilges, and even a brief spraying from a hose cooled the twisted plating on which Jones stood. Nor did he have to bear the agony of standing much longer, for he asked for, and obtained, a relief. He was a vigorous and athletic young man, and he was able to go up the ladder hand over hand without torturing himself further by putting his burned feet on the rungs, and then he could crawl on hands and knees along the deck for a little way before he collapsed. And the task force returned victoriously to Pearl Harbor, and *Boon* went into dry dock, and Jones went into hospital.

'Didn't anybody ask you how your feet got burned?' I asked.

'Not specially. A lot of fellows got burns that day. Worse than mine,' said Jones.

'What about this?' I went on, indicating his mutilated letter.

'Oh, of course I wrote to Mary. I wanted to tell her how I'd come to be in hospital, naturally, and I suppose they cut it out.'

I could picture that part quite well; the weary officer with a hundred letters to censor, reading a description of the flaming bilges of the *Boon*. The damage had not been announced, and this was censorable material. He would take his scissors and cut out the offending passage. His brain would be too numb to think much about the heroism written between the lines of that passage; or perhaps he took heroism for granted.

Now I have told the story. One of the best-known pieces of verse in the English language tells us of a boy standing on a burning deck; I can only write a short story. Of course, in addition, I can go on buying gasoline from Ed Jones.

Turnabout

∽∽∽

WILLIAM FAULKNER

I

THE American—the older one—wore no pink Bedfords. His breeches were of plain whipcord, like the tunic. And the tunic had no long London-cut skirts, so that below the Sam Browne the tail of it stuck straight out like the tunic of a military policeman beneath his holster belt. And he wore simple puttees and the easy shoes of a man of middle age, instead of Savile Row boots, and the shoes and the puttees did not match in shade, and the ordnance belt did not match either of them, and the pilot's wings on his breast were just wings. But the ribbon beneath them was a good ribbon, and the insigne on his shoulders were the twin bars of a captain. He was not tall. His face was thin, a little aquiline; the eyes intelligent and a little tired. He was past twenty-five; looking at him, one thought, not Phi Beta Kappa exactly, but Skull and Bones perhaps, or possibly a Rhodes scholarship.

One of the men who faced him probably could not see him at all. He was being held on his feet by an American military policeman. He was quite drunk, and in contrast with the heavy-jawed policeman who held him erect on his long, slim, boneless legs, he looked like a masquerading girl. He was possibly eighteen, tall, with a pink-and-white face and blue eyes, and a mouth like a girl's mouth. He wore a pea-coat, buttoned awry and stained with recent mud, and upon his blond head, at that unmistakable and rakish swagger which no other people can ever approach or imitate, the cap of a Royal Naval Officer.

'What's this, corporal?' the American captain said. 'What's the trouble? He's an Englishman. You'd better let their MPs take care of him.'

'I know he is,' the policeman said. He spoke heavily, breathing heavily, in the voice of a man under physical strain; for all his girlish delicacy of limb, the English boy was heavier—or more helpless— than he looked. 'Stand up!' the policeman said. 'They're officers!'

The English boy made an effort then. He pulled himself together, focusing his eyes. He swayed, throwing his arms about the policeman's neck, and with the other hand he saluted, his hand flicking, fingers curled a little, to his right ear, already swaying again and catching himself again. 'Cheer-o, sir,' he said. 'Name's not Beatty, I hope.'

'No,' the captain said.

'Ah,' the English boy said. 'Hoped not. My mistake. No offence, what?'

'No offence,' the captain said quietly. But he was looking at the policeman. The second American spoke. He was a lieutenant, also a pilot. But he was not twenty-five and he wore the pink breeches, the London boots, and his tunic might have been a British tunic save for the collar.

'It's one of those navy eggs,' he said. 'They pick them out of the gutters here all night long. You don't come to town often enough.'

'Oh,' the captain said. 'I've heard about them. I remember now.' He also remarked now that, though the street was a busy one—it was just outside a popular café—and there were many passers, soldier, civilian, women, yet none of them so much as paused, as though it were a familiar sight. He was looking at the policeman. 'Can't you take him to his ship?'

'I thought of that before the captain did,' the policeman said. 'He says he can't go aboard his ship after dark because he puts the ship away at sundown.'

'Puts it away?'

'Stand up, sailor!' the policeman said savagely, jerking at his lax burden. 'Maybe the captain can make sense out of it. Damned if I can. He says they keep the boat under the wharf. Run it under the wharf at night, and that they can't get it out again until the tide goes out tomorrow.'

'Under the wharf? A boat? What is this?' He was now speaking to the lieutenant. 'Do they operate some kind of aquatic motorcycles?'

'Something like that,' the lieutenant said. 'You've seen them—the boats. Launches, camouflaged and all. Dashing up and down the harbour. You've seen them. They do that all day and sleep in the gutters here all night.'

'Oh,' the captain said. 'I thought those boats were ship commanders' launches. You mean to tell me they use officers just to—'

'I don't know,' the lieutenant said. 'Maybe they use them to fetch hot water from one ship to another. Or buns. Or maybe to go back and forth fast when they forget napkins or something.'

'Nonsense,' the captain said. He looked at the English boy again.

'That's what they do,' the lieutenant said. 'Town's lousy with them all night long. Gutters full, and their MP's carting them away in batches, like nursemaids in a park. Maybe the French give them the launches to get them out of the gutters during the day.'

'Oh,' the captain said, 'I see.' But it was clear that he didn't see, wasn't listening, didn't believe what he did hear. He looked at the English boy. 'Well, you can't leave him here in that shape,' he said.

Again the English boy tried to pull himself together. 'Quite all right, 'sure you,' he said glassily, his voice pleasant, cheerful almost, quite courteous. 'Used to it. Confounded rough *pavé*, though. Should force French do something about it. Visiting lads jolly well deserve decent field to play on, what?'

'And he was jolly well using all of it too,' the policeman said savagely. 'He must think he's a one-man team, maybe.'

At that moment a fifth man came up. He was a British military policeman. 'Nah then,' he said. 'What's this? What's this?' Then he saw the Americans' shoulder bars. He saluted. At the sound of his voice the English boy turned, swaying, peering.

'Oh, hullo, Albert,' he said.

'Nah then, Mr Hope,' the British policeman said. He said to the American policeman, over his shoulder: 'What is it this time?'

'Likely nothing,' the American said. 'The way you guys run a war. But I'm a stranger here. Here. Take him.'

'What is this, corporal?' the captain said. 'What was he doing?'

'He won't call it nothing,' the American policeman said, jerking his head at the British policeman. 'He'll just call it a thrush or a robin or something. I turn into this street about three blocks back a while ago, and I find it blocked with a line of trucks going up from the docks, and the drivers all hollering ahead what the hell the trouble is. So I come on, and I find it is about three blocks of them, blocking the cross streets too; and I come on to the head of it where the trouble is, and I find about a dozen of the drivers out in front, holding a caucus or something in the middle of the street, and I come up and

I say, "What's going on here?" and they leave me through and I find this egg here laying—'

'Yer talking about one of His Majesty's officers, my man,' the British policeman said.

'Watch yourself, corporal,' the captain said. 'And you found this officer—'

'He had done gone to bed in the middle of the street, with an empty basket for a pillow. Laying there with his hands under his head and his knees crossed, arguing with them about whether he ought to get up and move or not. He said that the trucks could turn back and go around by another street, but that he couldn't use any other street, because this street was his.'

'His street?'

The English boy had listened, interested, pleasant. 'Billet, you see,' he said. 'Must have order, even in war emergency. Billet by lot. This street mine; no poaching, eh? Next street Jamie Wutherspoon's. But trucks can go by that street because Jamie not using it yet. Not in bed yet. Insomnia. Knew so. Told them. Trucks go that way. See now?'

'Was that it, corporal?' the captain said.

'He told you. He wouldn't get up. He just laid there, arguing with them. He was telling one of them to go somewhere and bring back a copy of their articles of war—'

'King's Regulations; yes,' the captain said.

'—and see if the book said whether he had the right of way, or the trucks. And then I got him up, and then the captain come along. And that's all. And with the captain's permission I'll now hand him over to His Majesty's wet nur—'

'That'll do, corporal,' the captain said. 'You can go. I'll see to this.' The policeman saluted and went on. The British policeman was now supporting the English boy. 'Can't you take him?' the captain said. 'Where are their quarters?'

'I don't rightly know, sir, if they have quarters or not. We—I usually see them about the pubs until daylight. They don't seem to use quarters.'

'You mean, they really aren't off of ships?'

'Well, sir, they might be ships, in a manner of speaking. But a man would have to be a bit sleepier than him to sleep in one of them.'

'I see,' the captain said. He looked at the policeman. 'What kind of boats are they?'

This time the policeman's voice was immediate, final and completely inflectionless. It was like a closed door. 'I don't rightly know, sir.'

'Oh,' the captain said. 'Quite. Well, he's in no shape to stay about pubs until daylight this time.'

'Perhaps I can find him a bit of a pub with a back table, where he can sleep,' the policeman said. But the captain was not listening. He was looking across the street, where the lights of another café fell across the pavement. The English boy yawned terrifically, like a child does, his mouth pink and frankly gaped as a child's.

The captain turned to the policeman:

'Would you mind stepping across there and asking for Captain Bogard's driver? I'll take care of Mr Hope.'

The policeman departed. The captain now supported the English boy, his hand beneath the other's arm. Again the boy yawned like a weary child. 'Steady,' the captain said. 'The car will be here in a minute.'

'Right,' the English boy said through the yawn.

II

Once in the car, he went to sleep immediately with the peaceful suddenness of babies, sitting between the two Americans. But though the aerodrome was only thirty minutes away, he was awake when they arrived, apparently quite fresh, and asking for whisky. When they entered the mess he appeared quite sober, only blinking a little in the lighted room, in his raked cap and his awry-buttoned pea-jacket and a soiled silk muffler, embroidered with a club insignia which Bogard recognized to have come from a famous preparatory school, twisted about his throat.

'Ah,' he said, his voice fresh, clear now, not blurred, quite cheerful, quite loud, so that the others in the room turned and looked at him. 'Jolly. Whisky, what?' He went straight as a bird dog to the bar in the corner, the lieutenant following. Bogard had turned and gone on to the other end of the room, where five men sat about a card table.

'What's he admiral of?' one said.

'Of the whole Scotch navy, when I found him,' Bogard said.

Another looked up. 'Oh, I thought I'd seen him in town.' He looked at the guest. 'Maybe it's because he was on his feet that I didn't recognize him when he came in. You usually see them lying down in the gutter.'

'Oh,' the first said. He, too, looked around. 'Is he one of those guys?'

'Sure. You've seen them. Sitting on the curb, you know, with a couple of limey MP's hauling at their arms.'

'Yes. I've seen them,' the other said. They all looked at the English boy. He stood at the bar, talking, his voice loud, cheerful. 'They all look like him too,' the speaker said. 'About seventeen or eighteen. They run those little boats that are always dashing in and out.'

'Is that what they do?' a third said. 'You mean, there's a male marine auxiliary to the Waacs? Good Lord, I sure made a mistake when I enlisted. But this war never was advertised right.'

'I don't know,' Bogard said. 'I guess they do more than just ride around.'

But they were not listening to him. They were looking at the guest. 'They run by clock,' the first said. 'You can see the condition of one of them after sunset and almost tell what time it is. But what I don't see is, how a man that's in that shape at one o'clock every morning can even see a battleship the next day.'

'Maybe when they have a message to send out to a ship', another said, 'they just make duplicates and line the launches up and point them toward the ship and give each one a duplicate of the message and let them go. And the ones that miss the ship just cruise around the harbour until they hit a dock somewhere.'

'It must be more than that,' Bogard said.

He was about to say something else, but at that moment the guest turned from the bar and approached, carrying a glass. He walked steadily enough, but his colour was high and his eyes were bright, and he was talking, loud, cheerful, as he came up.

'I say. Won't you chaps join—' He ceased. He seemed to remark something; he was looking at their breasts. 'Oh, I say. You fly. All of you. Oh, good gad! Find it jolly, eh?'

'Yes,' somebody said. 'Jolly.'

'But dangerous, what?'

'A little faster than tennis,' another said. The guest looked at him, bright, affable, intent.

Another said quickly, 'Bogard says you command a vessel.'

'Hardly a vessel. Thanks, though. And not command. Ronnie does that. Ranks me a bit. Age.'

'Ronnie?'

'Yes. Nice. Good egg. Old, though. Stickler.'

'Stickler?'

'Frightful. You'd not believe it. Whenever we sight smoke and I have the glass, he sheers away. Keeps the ship hull down all the while. No beaver then. Had me two down a fortnight yesterday.'

The Americans glanced at one another. 'No beaver?'

'We play it. With basket masts, you see. See a basket mast. Beaver! One up. The Ergenstrasse doesn't count any more, though.'

The men about the table looked at one another. Bogard spoke. 'I see. When you or Ronnie see a ship with basket masts, you get a beaver on the other. I see. What is the Ergenstrasse?'

'She's German. Interned. Tramp steamer. Foremast rigged so it looks something like a basket mast. Booms, cables, I dare say. I didn't think it looked very much like a basket mast, myself. But Ronnie said yes. Called it one day. Then one day they shifted her across the basin and I called her on Ronnie. So we decided to not count her any more. See now, eh?'

'Oh,' the one who had made the tennis remark said, 'I see. You and Ronnie run about in the launch, playing beaver. H'm'm. That's nice. Did you ever pl—'

'Jerry,' Bogard said. The guest had not moved. He looked down at the speaker, still smiling, his eyes quite wide.

The speaker still looked at the guest. 'Has yours and Ronnie's boat got a yellow stern?'

'A yellow stern?' the English boy said. He had quit smiling, but his face was still pleasant.

'I thought that maybe when the boats had two captains, they might paint the sterns yellow or something.'

'Oh,' the guest said. 'Burt and Reeves aren't officers.'

'Burt and Reeves,' the other said, in a musing tone. 'So they go, too. Do they play beaver too?'

'Jerry,' Bogard said. The other looked at him. Bogard jerked his head a little. 'Come over here.' The other rose. They went aside. 'Lay off of him,' Bogard said. 'I mean it, now. He's just a kid. When you were that age, how much sense did you have? Just about enough to get to chapel on time.'

'My country hadn't been at war going on four years, though,' Jerry said. 'Here we are, spending our money and getting shot at by the clock, and it's not even our fight, and these limeys that would have been goose-stepping twelve months now if it hadn't been—'

'Shut it,' Bogard said. 'You sound like a Liberty Loan.'

'—taking it like it was a fair or something. "Jolly." ' His voice was now falsetto, lilting. ' "But dangerous, what?" '

'Sh-h-h-h,' Bogard said.

'I'd like to catch him and his Ronnie out in the harbour, just once. Any harbour. London's. I wouldn't want anything but a Jenny, either. Jenny? Hell, I'd take a bicycle and a pair of water wings! I'll show him some war.'

'Well, you lay off him now. He'll be gone soon.'

'What are you going to do with him?'

'I'm going to take him along this morning. Let him have Harper's place out front. He says he can handle a Lewis. Says they have one on the boat. Something he was telling me—about how he once shot out a channel-marker light at seven hundred yards.'

'Well, that's your business. Maybe he can beat you.'

'Beat me?'

'Playing beaver. And then you can take on Ronnie.'

'I'll show him some war, anyway,' Bogard said. He looked at the guest. 'His people have been in it three years now, and he seems to take it like a sophomore in town for the big game.' He looked at Jerry again. 'But you lay off him now.'

As they approached the table, the guest's voice was loud and cheerful: 'if he got the glasses first, he would go in close and look, but when I got them first, he'd sheer off where I couldn't see anything but the smoke. Frightful stickler. Frightful. But Ergenstrasse not counting any more. And if you make a mistake and call her, you lose two beaver from your score. If Ronnie were only to forget and call her we'd be even.'

III

At two o'clock the English boy was still talking, his voice bright, innocent and cheerful. He was telling them how Switzerland had been spoiled by 1914, and instead of the vacation which his father had promised him for his sixteenth birthday, when that birthday came he and his tutor had had to do with Wales. But that he and the tutor had got pretty high and that he dared to say—with all due respect to any present who might have had the advantage of Switzerland, of course—that one could see probably as far from Wales as from Switzerland. 'Perspire as much and breathe as hard, anyway,' he added. And about him the Americans sat, a little hard-bitten, a little sober, somewhat older, listening to him with a kind of cold astonishment. They had been getting up for some time now and going out and returning in flying clothes, carrying helmets and goggles. An orderly entered with a tray of coffee cups, and the guest realized that for some time now he had been hearing engines in the darkness outside.

At last Bogard rose. 'Come along,' he said. 'We'll get your togs.' When they emerged from the mess, the sound of the engines was quite loud—an idling thunder. In alignment along the invisible tarmac was a vague rank of short banks of flickering blue-green fire suspended apparently in mid-air. They crossed the aerodrome to Bogard's quarters, where the lieutenant, McGinnis, sat on a cot fastening his flying boots. Bogard reached down a Sidcott suit and threw it across the cot. 'Put this on,' he said.

'Will I need all this?' the guest said. 'Shall we be gone that long?'

'Probably,' Bogard said. 'Better use it. Cold upstairs.'

The guest picked up the suit. 'I say,' he said. 'I say, Ronnie and I have a do ourselves, tomor—today. Do you think Ronnie won't mind if I am a bit late? Might not wait for me.'

'We'll be back before teatime,' McGinnis said. He seemed quite busy with his boot. 'Promise you.' The English boy looked at him.

'What time should you be back?' Bogard said.

'Oh, well,' the English boy said, 'I dare say it will be all right. They let Ronnie say when to go, anyway. He'll wait for me if I should be a bit late.'

'He'll wait,' Bogard said. 'Get your suit on.'

'Right,' the other said. They helped him into the suit. 'Never been up before,' he said, chattily, pleasantly. 'Dare say you can see further than from mountains, eh?'

'See more, anyway,' McGinnis said. 'You'll like it.'

'Oh, rather. If Ronnie only waits for me. Lark. But dangerous, isn't it?'

'Go on,' McGinnis said. 'You're kidding me.'

'Shut your trap, Mac,' Bogard said. 'Come along. Want some more coffee?' He looked at the guest, but McGinnis answered:

'No. Got something better than coffee. Coffee makes such a confounded stain on the wings.'

'On the wings?' the English boy said. 'Why coffee on the wings.'

'Stow it, I said, Mac,' Bogard said. 'Come along.'

They recrossed the aerodrome, approaching the muttering banks of flame. When they drew near, the guest began to discern the shape, the outlines, of the Handley-Page. It looked like a Pullman coach run upslanted aground into the skeleton of the first floor of an incomplete skyscraper. The guest looked at it quietly.

'It's larger than a cruiser,' he said in his bright, interested voice. 'I say, you know. This doesn't fly in one lump. You can't pull my leg. Seen them before. It comes in two parts: Captain Bogard and me in one; Mac and 'nother chap in other. What?'

'No,' McGinnis said. Bogard had vanished. 'It all goes up in one lump. Big lark, eh? Buzzard, what?'

'Buzzard?' the guest murmured. 'Oh, I say. A cruiser. Flying. I say, now.'

'And listen,' McGinnis said. His hand came forth; something cold fumbled against the hand of the English boy—a bottle. 'When you feel yourself getting sick, see? Take a pull at it.'

'Oh, shall I get sick?'

'Sure. We all do. Part of flying. This will stop it. But if it doesn't. See?'

'What? Quite. What?'

'Not overside. Don't spew it overside.'

'Not overside?'

'It'll blow back in Bogy's and my face. Can't see. Bingo. Finished. See?'

'Oh, quite. What shall I do with it?' Their voices were quiet, brief, grave as conspirators.

'Just duck your head and let her go.'

'Oh, quite.'

Bogard returned. 'Show him how to get into the front pit, will you?' he said. McGinnis led the way through the trap. Forward, rising to the slant of the fuselage, the passage narrowed; a man would need to crawl.

'Crawl in there and keep going,' McGinnis said.

'It looks like a dog kennel,' the guest said.

'Doesn't it, though?' McGinnis agreed cheerfully. 'Cut along with you.' Stooping, he could hear the other scuttling forward. 'You'll find a Lewis gun up there, like as not,' he said into the tunnel.

The voice of the guest came back: 'Found it.'

'The gunnery sergeant will be along in a minute and show you if it is loaded.'

'It's loaded,' the guest said; almost on the heels of his words the gun fired, a brief staccato burst. There were shouts, the loudest from the ground beneath the nose of the aeroplane. 'It's quite all right,' the English boy's voice said. 'I pointed it west before I let it off. Nothing back there but Marine office and your brigade headquarters. Ronnie and I always do this before we go anywhere. Sorry if I was too soon. Oh, by the way,' he added, 'my name's Claude. Don't think I mentioned it.'

On the ground, Bogard and two other officers stood. They had come up running. 'Fired it west,' one said. 'How in hell does he know which way is west?'

'He's a sailor,' the other said. 'You forgot that.'

'He seems to be a machine gunner too,' Bogard said.

'Let's hope he doesn't forget that,' the first said.

IV

Nevertheless, Bogard kept an eye on the silhouetted head rising from the round gunpit in the nose ten feet ahead of him. 'He did work that gun, though,' he said to McGinnis beside him. 'He even put the drum on himself, didn't he?'

'Yes,' McGinnis said. 'If he just doesn't forget and think that that gun is him and his tutor looking around from a Welsh alp.'

'Maybe I should not have brought him,' Bogard said. McGinnis didn't answer. Bogard jockeyed the wheel a little. Ahead, in the gunner's pit, the guest's head moved this way and that continuously, looking. 'We'll get there and unload and haul air for home,' Bogard said. 'Maybe in the dark—Confound it, it would be a shame for his country to be in this mess for four years and him not even to see a gun pointed in his direction.'

'He'll see one tonight if he don't keep his head in,' McGinnis said.

But the boy did not do that. Not even when they had reached the objective and McGinnis had crawled down to the bomb toggles. And even when the searchlights found them and Bogard signalled to the other machines and dived, the two engines snarling full speed into and through the bursting shells, he could see the boy's face in the searchlight's glare, leaned far overside, coming sharply out as a spotlighted face on a stage, with an expression upon it of child-like interest and delight. 'But he's firing that Lewis,' Bogard thought. 'Straight too'; nosing the machine further down, watching the pinpoint swing into the sights, his right hand lifted, waiting to drop into McGinnis' sight. He dropped his hand; above the noise of the engines he seemed to hear the click and whistle of the released bombs as the machine, freed of the weight, shot zooming in a long upward bounce that carried it for an instant out of the light. Then he was pretty busy for a time, coming into and through the shells again, shooting athwart another beam that caught and held long enough for him to see the English boy leaning far over the side, looking back and down past the right wing, the undercarriage. 'Maybe he's read about it somewhere,' Bogard thought, turning, looking back to pick up the rest of the flight.

Then it was all over, the darkness cool and empty and peaceful and almost quiet, with only the steady sound of the engines. McGinnis climbed back into the office, and standing up in his seat, he fired the coloured pistol this time and stood for a moment longer, looking backward toward where the searchlights still probed and sabered. He sat down again.

'OK,' he said. 'I counted all four of them. Let's haul air.' Then he looked forward. 'What's become of the King's Own? You didn't hang him onto a bomb release, did you?' Bogard looked. The forward pit was empty. It was in dim silhouette again now, against the stars, but there was nothing there now save the gun. 'No,' McGinnis said: 'there he is. See? Leaning overside. Dammit, I told him not to spew it! There he comes back.' The guest's head came into view again. But again it sank out of sight.

'He's coming back,' Bogard said. 'Stop him. Tell him we're going to have every squadron in the Hun Channel group on top of us in thirty minutes.'

McGinnis swung himself down and stooped at the entrance to the passage. 'Get back!' he shouted. The other was almost out; they squatted so, face to face like two dogs, shouting at one another above the noise of the still-unthrottled engines on either side of the fabric walls. The English boy's voice was thin and high.

'Bomb!' he shrieked.

'Yes,' McGinnis shouted, 'they were bombs! We gave them hell! Get back, I tell you! Have every Hun in France on us in ten minutes! Get back to your gun!'

Again the boy's voice came, high, faint above the noise: 'Bomb! All right?'

'Yes! Yes! All right. Back to your gun, damn you!'

McGinnis climbed back into the office. 'He went back. Want me to take her awhile?'

'All right,' Bogard said. He passed McGinnis the wheel. 'Ease her back some. I'd just as soon it was daylight when they come down on us.'

'Right,' McGinnis said. He moved the wheel suddenly. 'What's the matter with that right wing?' he said. 'Watch it. . . . See? I'm flying on the right aileron and a little rudder. Feel it.'

Bogard took the wheel a moment. 'I didn't notice that. Wire somewhere, I guess. I didn't think any of those shells were that close. Watch her, though.'

'Right,' McGinnis said. 'And so you are going with him on his boat tomorrow—today.'

'Yes. I promised him. Confound it, you can't hurt a kid, you know.'

'Why don't you take Collier along, with his mandolin? Then you could sail around and sing.'

'I promised him,' Bogard said. 'Get that wing up a little.'

'Right,' McGinnis said.

Thirty minutes later it was beginning to be dawn; the sky was grey. Presently McGinnis said: 'Well, here they come. Look at them! They look like mosquitoes in September. I hope he don't get worked up now and think he's playing beaver. If he does he'll just be one down to Ronnie, provided the devil has a beard . . . Want the wheel?'

V

At eight o'clock the beach, the Channel, was beneath them. Throttled back, the machine drifted down as Bogard ruddered it gently into the Channel wind. His face was strained, a little tired.

McGinnis looked tired, too, and he needed a shave.

'What do you guess he is looking at now?' he said. For again the English boy was leaning over the right side of the cockpit, looking backward and downward past the right wing.

'I don't know,' Bogard said. 'Maybe bullet holes.' He blasted the port engine. 'Must have the riggers—'

'He could see some closer than that,' McGinnis said. 'I'll swear I saw tracer going into his back at one time. Or maybe it's the ocean he's looking at. But he must have seen that when he came over from England.' Then Bogard levelled off; the nose rose sharply, the sand, the curling tide edge fled alongside. Yet still the English boy hung far overside, looking backward and downward at something beneath the right wing, his face rapt, with utter and childlike interest. Until the machine was completely stopped he continued to do so. Then he ducked down, and in the abrupt silence of the engines they could hear him crawling in the passage. He emerged just as the two pilots climbed stiffly down from the office, his face bright, eager; his voice high, excited.

'Oh, I say! Oh, good gad! What a chap. What a judge of distance! If Ronnie could only have seen! Oh, good gad! Or maybe they aren't like ours—don't load themselves as soon as the air strikes them.'

The Americans looked at him. 'What don't what?' McGinnis said. 'The bomb. It was magnificent; I say, I shan't forget it. Oh, I say, you know! It was splendid!'

After a while McGinnis said, 'The bomb?' in a fainting voice. Then the two pilots glared at each other; they said in unison: 'That right wing!' Then as one they clawed down through the trap and, with the guest at their heels, they ran around the machine and looked beneath the right wing. The bomb, suspended by its tail, hung straight down like a plumb bob beside the right wheel, its tip just touching the sand. And parallel with the wheel track was the long delicate line in the sand where its ultimate tip had dragged. Behind them the English boy's voice was high, clear, childlike:

'Frightened, myself. Tried to tell you. But realized you knew your business better than I. Skill. Marvellous. Oh, I say, I shan't forget it.'

VI

A marine with a bayoneted rifle passed Bogard on to the wharf and directed him to the boat. The wharf was empty, and he didn't even see the boat until he approached the edge of the wharf and looked directly down into it and upon the backs of two stooping men in greasy dungarees, who rose and glanced briefly at him and stooped again.

It was about thirty feet long and about three feet wide. It was painted with grey-green camouflage. It was quarter-decked forward, with two blunt, raked exhaust stacks. 'Good Lord,' Bogard thought, 'if all that deck is engine—' Just aft the deck was the control seat; he saw a big wheel, an instrument panel. Rising to a height of about a foot above the free-board, and running from the stern forward to where the deck began, and continuing on across the after edge of the deck and thence back down the other gunwale to the stern, was a solid screen, also camouflaged, which enclosed the boat save for the width of the stern, which was open. Facing the steersman's seat like an eye was a hole in the screen about eight inches in diameter. And looking down into the long, narrow, still, vicious shape, he saw a machine gun swivelled at the stern, and he looked at the low screen—including which the whole vessel did not sit much more than a yard above water level—with its single empty forward-staring eye, and he thought quietly: 'It's steel. It's made of steel.' And his face was quite sober, quite thoughtful, and he drew his trench coat about him and buttoned it, as though he were getting cold.

He heard steps behind him and turned. But it was only an orderly from the aerodrome, accompanied by the marine with the rifle. The orderly was carrying a largish bundle wrapped in paper.

'From Lieutenant McGinnis to the captain,' the orderly said.

Bogard took the bundle. The orderly and the marine retreated. He opened the bundle. It contained some objects and a scrawled note. The objects were a new yellow silk sofa cushion and a Japanese parasol, obviously borrowed, and a comb and a roll of toilet paper. The note said:

Couldn't find a camera anywhere and Collier wouldn't let me have his mandolin. But maybe Ronnie can play on the comb.

MAC.

Bogard looked at the objects. But his face was still quite thoughtful, quite grave. He rewrapped the things and carried the bundle on up the wharf and dropped it quietly into the water.

As he returned toward the invisible boat he saw two men approaching. He recognized the boy at once—tall, slender, already talking, voluble, his head bent a little toward his shorter companion, who plodded along beside him, hands in pockets, smoking a pipe. The boy still wore the pea-coat beneath a flapping oilskin, but in place of the rakish and casual cap he now wore an infantryman's soiled Balaclava helmet, with, floating behind him as though upon the sound of his voice, a curtainlike piece of cloth almost as long as a burnous.

'Hullo, there!' he cried, still a hundred yards away.

But it was the second man that Bogard was watching, thinking to himself that he had never in his life seen a more curious figure. There was something stolid about the very shape of his hunched shoulders, his slightly down-looking face. He was a head shorter than the other. His face was ruddy, too, but its mould was of a profound gravity that was almost dour. It was the face of a man of twenty who has been for a year trying, even while asleep, to look twenty-one. He wore a high-necked sweater and dungaree slacks; above this a leather jacket; and above this a soiled naval officer's warmer that reached almost to his heels and which had one shoulder strap missing and not one remaining button at all. On his head was a plaid fore-and-aft deer stalker's cap, tied on by a narrow scarf brought across and down, hiding his ears, and then wrapped once about his throat and knotted with a hangman's noose beneath his left ear. It was unbelievably soiled, and with his hands elbow-deep in his pockets and his hunched shoulders and his bent head, he looked like someone's grandmother hung, say, for a witch. Clamped upside down between his teeth was a short brier pipe.

'Here he is!' the boy cried. 'This is Ronnie. Captain Bogard.'

'How are you?' Bogard said. He extended his hand. The other said no word, but his hand came forth, limp. It was quite cold, but it was hard, calloused. But he said no word; he just glanced briefly at Bogard and then away. But in that instant Bogard caught something in the look, something strange—a flicker; a kind of covert and curious respect, something like a boy of fifteen looking at a circus trapezist.

But he said no word. He ducked on; Bogard watched him drop from sight over the wharf edge as though he had jumped feet first into the sea. He remarked now that the engines in the invisible boat were running.

'We might get aboard too,' the boy said. He started toward the boat, then he stopped. He touched Bogard's arm. 'Yonder!' he hissed. 'See?' His voice was thin with excitement.

'What?' Bogard also whispered; automatically he looked backward and upward, after old habit. The other was gripping his arm and pointing across the harbour.

'There! Over there. The Ergenstrasse. They have shifted her again.' Across the harbour lay an ancient, rusting, sway-backed hulk. It was small and nondescript, and, remembering, Bogard saw that the foremast was a strange mess of cables and booms, resembling—allowing for a great deal of licence or looseness of imagery—a basket mast. Beside him the boy was almost chortling. 'Do you think that Ronnie noticed?' he hissed. 'Do you?'

'I don't know,' Bogard said.

'Oh, good gad! If he should glance up and call her before he notices, we'll be even. Oh, good gad! But come along.' He went on; he was still chortling. 'Careful,' he said. 'Frightful ladder.'

He descended first, the two men in the boat rising and saluting. Ronnie had disappeared, save for his backside, which now filled a small hatch leading forward beneath the deck. Bogard descended gingerly.

'Good Lord,' he said. 'Do you have to climb up and down this every day?'

'Frightful, isn't it?' the other said, in his happy voice. 'But you know yourself. Try to run a war with makeshifts, then wonder why it takes so long.' The narrow hull slid and surged, even with Bogard's added weight. 'Sits right on top, you see,' the boy said. 'Would float on a lawn, in a heavy dew. Goes right over them like a bit of paper.'

'It does?' Bogard said.

'Oh, absolutely. That's why, you see.' Bogard didn't see, but he was too busy letting himself gingerly down to a sitting posture. There were no thwarts; no seats save a long, thick, cylindrical ridge which ran along the bottom of the boat from the driver's seat to the stern. Ronnie had backed into sight. He now sat behind the wheel, bent over the instrument panel. But when he glanced back over his shoulder he did not speak. His face was merely interrogatory. Across his face there was now a long smudge of grease. The boy's face was empty, too, now.

'Right,' he said. He looked forward, where one of the seamen had gone. 'Ready forward?' he said.

'Aye, sir,' the seaman said.

The other seaman was at the stern line. 'Ready aft?'

'Aye, sir.'

'Cast off.' The boat sheered away, purring, a boiling of water under the stern. The boy looked down at Bogard. 'Silly business. Do it shipshape, though. Can't tell when silly fourstriper—' His face changed again, immediate, solicitous. 'I say. Will you be warm? I never thought to fetch—'

'I'll be all right,' Bogard said. But the other was already taking off his oilskin. 'No, no,' Bogard said. 'I won't take it.'

'You'll tell me if you get cold?'

'Yes. Sure.' He was looking down at the cylinder on which he sat. It was a half cylinder—that is, like the hot-water tank to some Gargantuan stove, sliced down the middle and bolted, open side down, to the floor plates. It was twenty feet long and more than two feet thick. Its top rose as high as the gunwales and between it and the hull on either side was just room enough for a man to place his feet to walk.

'That's Muriel,' the boy said.

'Muriel?'

'Yes. The one before that was Agatha. After my aunt. The first one Ronnie and I had was Alice in Wonderland. Ronnie and I were the White Rabbit. Jolly, eh?'

'Oh, you and Ronnie have had three, have you?'

'Oh, yes,' the boy said. He leaned down. 'He didn't notice,' he whispered. His face was again bright, gleeful. 'When we come back,' he said. 'You watch.'

'Oh,' Bogard said. 'The Ergenstrasse.' He looked astern, and then he thought: 'Good Lord! We must be going—travelling.' He looked out now, broadside, and saw the harbour line fleeing past, and he thought to himself that the boat was well-nigh moving at the speed at which the Handley-Page flew, left the ground. They were beginning to bound now, even in the sheltered water, from one wave crest to the next with a distinct shock. His hand still rested on the cylinder on which he sat. He looked down at it again, following it from where it seemed to emerge beneath Ronnie's seat, to where it bevelled into the stern. 'It's the air in her, I suppose,' he said.

'The what?' the boy said.

'The air. Stored up in her. That makes the boat ride high.'

'Oh, yes. I dare say. Very likely. I hadn't thought about it.' He came forward, his burnous whipping in the wind, and sat down beside Bogard. Their heads were below the top of the screen.

Astern the harbour fled, diminishing, sinking into the sea. The boat had begun to lift now, swooping forward and down, shocking almost stationary for a moment, then lifting and swooping again; a gout of spray came aboard over the bows like a flung shovelful of shot. 'I wish you'd take this coat,' the boy said.

Bogard didn't answer. He looked around at the bright face. 'We're outside, aren't we?' he said quietly.

'Yes. . . . Do take it, won't you?'

'Thanks, no. I'll be all right. We won't be long, anyway, I guess.'

'No. We'll turn soon. It won't be so bad then.'

'Yes. I'll be all right when we turn.' Then they did turn. The motion became easier. That is, the boat didn't bang head-on, shuddering, into the swells. They came up beneath now, and the boat fled with increased speed, with a long, sickening, yawing motion, first to one side and then the other. But it fled on, and Bogard looked astern with that same soberness with which he had first looked down into the boat. 'We're going east now,' he said.

'With just a spot of north,' the boy said. 'Makes her ride a bit better, what?'

'Yes,' Bogard said. Astern there was nothing now save empty sea and the delicate needlelike cant of the machine gun against the boiling and slewing wake, and the two seamen crouching quietly in the stern. 'Yes. It's easier.' Then he said: 'How far do we go?'

The boy leaned closer. He moved closer. His voice was happy, confidential, proud, though lowered a little: 'It's Ronnie's show. He thought of it. Not that I wouldn't have, in time. Gratitude and all that. But he's the older, you see. Thinks fast. Courtesy, *noblesse oblige*—all that. Thought of it soon as I told him this morning. I said, "Oh, I say. I've been there. I've seen it"; and he said, "Not flying"; and I said, "Strewth"; and he said "How far? No lying now"; and I said, "Oh, far. Tremendous. Gone all night"; and he said, "Flying all night. That must have been to Berlin"; and I said, "I don't know. I dare say"; and he thought. I could see him thinking. Because he is the older, you see. More experience in courtesy, right thing. And he said, "Berlin. No fun to that chap, dashing out and back with us." And he thought and I waited, and I said, "But we can't take him to Berlin. Too far. Don't know the way, either"; and he said—fast, like a shot—said, "But there's Kiel"; and I knew—'

'What?' Bogard said. Without moving, his whole body sprang. 'Kiel? In this?'

'Absolutely. Ronnie thought of it. Smart, even if he is a stickler. Said at once, "Zeebrugge no show at all for that chap. Must do best we can for him. Berlin," Ronnie said. "My Gad! Berlin."'

'Listen,' Bogard said. He had turned now, facing the other, his face quite grave. 'What is this boat for?'

'For?'

'What does it do?' Then, knowing beforehand the answer to his own question, he said, putting his hand on the cylinder: 'What is this in here? A torpedo, isn't it?'

'I thought you knew,' the boy said.

'No,' Bogard said. 'I didn't know.' His voice seemed to reach him from a distance, dry, cricketlike: 'How do you fire it?'

'Fire it?'

'How do you get it out of the boat? When that hatch was open a while ago I could see the engines. They were right in front of the end of this tube.'

'Oh,' the boy said. 'You pull a gadget there and the torpedo drops out astern. As soon as the screw touches the water it begins to turn, and then the torpedo is ready, loaded. Then all you have to do is turn the boat quickly and the torpedo goes on.'

'You mean—' Bogard said. After a moment his voice obeyed him again. 'You mean you aim the torpedo with the boat and release it and it starts moving, and you turn the boat out of the way and the torpedo passes through the same water that the boat just vacated?'

'Knew you'd catch on,' the boy said. 'Told Ronnie so. Airman. Tamer than yours, though. But can't be helped. Best we can do, just on water. But knew you'd catch on.'

'Listen,' Bogard said. His voice sounded to him quite calm. The boat fled on, yawing over the swells. He sat quite motionless. It seemed to him that he could hear himself talking to himself: 'Go on. Ask him. Ask him what? Ask him how close to the ship do you have to be before you fire. . . . Listen,' he said, in that calm voice. 'Now, you tell Ronnie, you see. You just tell him—just say—' He could feel his voice ratting off on him again, so he stopped it. He sat quite motionless, waiting for it to come back; the boy leaning now, looking at his face. Again the boy's voice was solicitous:

'I say. You're not feeling well. These confounded shallow boats.'

'It's not that,' Bogard said. 'I just— Do your orders say Kiel?'

'Oh, no. They let Ronnie say. Just so we bring the boat back. This is for you. Gratitude. Ronnie's idea. Tame, after flying. But if you'd rather, eh?'

'Yes, some place closer. You see, I—'

'Quite. I see. No vacations in wartime. I'll tell Ronnie.' He went forward. Bogard did not move. The boat fled in long, slewing swoops. Bogard looked quietly astern, at the scudding sea, the sky.

'My God!' he thought. 'Can you beat it? Can you beat it?'

The boy came back; Bogard turned to him a face the colour of dirty paper. 'All right now,' the boy said. 'Not Kiel. Nearer place, hunting probably just as good. Ronnie says he knows you will understand.' He was tugging at his pocket. He brought out a bottle. 'Here. Haven't forgot last night. Do the same for you. Good for the stomach, eh?'

Bogard drank, gulping—a big one. He extended the bottle, but the boy refused. 'Never touch it on duty,' he said. 'Not like you chaps. Tame here.'

The boat fled on. The sun was already down the west. But Bogard had lost all count of time, of distance. Ahead he could see white seas through the round eye opposite Ronnie's face, and Ronnie's hand on the wheel and the granitelike jut of his profiled jaw and the dead upside-down pipe. The boat fled on.

Then the boy leaned and touched his shoulder. He half rose. The boy was pointing. The sun was reddish; against it, outside them and about two miles away, a vessel—a trawler, it looked like—at anchor swung a tall mast.

'Lightship!' the boy shouted. 'Theirs.' Ahead Bogard could see a low, flat mole—the entrance to a harbour. 'Channel!' the boy shouted. He swept his arm in both directions. 'Mines!' His voice swept back on the wind. 'Place filthy with them. All sides. Beneath us too. Lark, eh?'

VII

Against the mole a fair surf was beating. Running before the seas now, the boat seemed to leap from one roller to the next; in the intervals while the screw was in the air the engine seemed to be trying to tear itself out by the roots. But it did not slow; when it passed the end of the mole the boat seemed to be standing almost erect on its rudder, like a sailfish. The mole was a mile away. From the end of it little faint lights began to flicker like fireflies. The boy leaned. 'Down,' he said. 'Machine guns. Might stop a stray.'

'What do I do?' Bogard shouted. 'What can I do?'

'Stout fellow! Give them hell, what? Knew you'd like it!'

Crouching, Bogard looked up at the boy, his face wild. 'I can handle the machine gun!'

'No need,' the boy shouted back. 'Give them first innings. Sporting. Visitors, eh?' He was looking forward. 'There she is. See?' They were in the harbour now, the basin opening before them. Anchored in the channel was a big freighter. Painted midships of the hull was a huge Argentine flag. 'Must get back to stations!' the boy shouted down to him. Then at that moment Ronnie spoke for the first time. The boat was hurtling along now in smoother water. Its speed did not slacken and Ronnie did not turn his head when he spoke. He just swung his jutting jaw and the clamped cold pipe a little, and said from the side of his mouth a single word:

'Beaver.'

The boy, stooped over what he had called his gadget, jerked up, his expression astonished and outraged. Bogard also looked forward and saw Ronnie's arm pointing to starboard. It was a light cruiser at anchor a mile away. She had basket masts, and as he looked a gun flashed from her after turret. 'Oh, damn!' the boy cried. 'Oh, you putt! Oh, confound you, Ronnie! Now I'm three down!' But he had already stooped again over his gadget, his face bright and empty and alert again; not sober; just calm, waiting. Again Bogard looked forward and felt the boat pivot on its rudder and head directly for the freighter at terrific speed, Ronnie now with one hand on the wheel and the other lifted and extended at the height of his head.

But it seemed to Bogard that the hand would never drop. He crouched, not sitting, watching with a kind of quiet horror the painted flag increase like a moving picture of a locomotive taken from between the rails. Again the gun crashed from the cruiser behind them, and the freighter fired point-blank at them from its poop. Bogard heard neither shot.

'Man, man!' he shouted. 'For God's sake!'

Ronnie's hand dropped. Again the boat spun on its rudder. Bogard saw the bow rise, pivoting; he expected the hull to slam broadside on into the ship. But it didn't. It shot off on a long tangent. He was waiting for it to make a wide sweep, heading seaward, putting the freighter astern, and he thought of the cruiser again. 'Get a broadside, this time, once we clear the freighter,' he thought. Then he remembered the freighter, the torpedo, and he looked back toward the freighter to watch the torpedo strike, and saw to his horror that the boat was now bearing down on the freighter again, in a skidding turn. Like a man in a dream, he watched himself rush down upon the ship and shoot past under her counter, still skidding, close enough to see the faces on her decks. 'They missed and they are going to run down the torpedo and catch it and shoot it again,' he thought idiotically.

So the boy had to touch his shoulder before he knew he was behind him. The boy's voice was quite calm: 'Under Ronnie's seat there. A bit of a crank handle. If you'll just hand it to me—'

He found the crank. He passed it back; he was thinking dreamily: 'Mac would say they had a telephone on board.' But he didn't look at once to see what the boy was doing with it, for in that still and peaceful horror he was watching Ronnie, the cold pipe rigid in his jaw, hurling the boat at top speed round and round the freighter, so near that he could see the rivets in the plates. Then he looked aft, his face wild, importunate, and he saw what the boy was doing with the crank. He had fitted it into what was obviously a small windlass low on one flank of the tube near the head. He glanced up and saw Bogard's face. 'Didn't go that time!' he shouted cheerfully.

'Go?' Bogard shouted. 'It didn't—The torpedo—'

The boy and one of the seamen were quite busy, stooping over the windlass and the tube. 'No. Clumsy. Always happening. Should think clever chaps like engineers— Happens, though. Draw her in and try her again.'

'But the nose, the cap!' Bogard shouted. 'It's still in the tube, isn't it? It's all right, isn't it?'

'Absolutely. But it's working now. Loaded. Screw's started turning. Get it back and drop it clear. If we should stop or slow up it would overtake us. Drive back into the tube. Bingo! What?'

Bogard was on his feet now, turned, braced to the terrific merry-go-round of the boat. High above them the freighter seemed to be spinning on her heel like a trick picture in the movies. 'Let me have that winch!' he cried.

'Steady!' the boy said. 'Mustn't draw her back too fast. Jam her into the head of the tube ourselves. Same bingo! Best let us. Every cobbler to his last, what?'

'Oh, quite,' Bogard said. 'Oh, absolutely.' It was like someone else was using his mouth. He leaned, braced, his hands on the cold tube, beside the others. He was hot inside, but his outside was cold. He could feel all his flesh jerking with cold as he watched the blunt, grained hand of the seaman turning the windlass in short, easy, inch-long arcs, while at the head of the tube the boy bent, tapping the cylinder with a spanner, lightly, his head turned with listening delicate and deliberate as a watchmaker. The boat rushed on in those furious, slewing turns. Bogard saw a long, drooping thread loop down from somebody's mouth, between his hands, and he found that the thread came from his own mouth.

He didn't hear the boy speak, nor notice when he stood up. He just felt the boat straighten out, flinging him to his knees beside the tube.

The seaman had gone back to the stern and the boy stooped again over his gadget. Bogard knelt now, quite sick. He did not feel the boat when it swung again, nor hear the gun from the cruiser which had not dared to fire and the freighter which had not been able to fire, firing again. He did not feel anything at all when he saw the huge, painted flag directly ahead and increasing with locomotive speed, and Ronnie's lifted hand drop. But this time he knew that the torpedo was gone; in pivoting and spinning this time the whole boat seemed to leave the water; he saw the bow of the boat shoot skyward like the nose of a pursuit ship going into a wingover. Then his outraged stomach denied him. He saw neither the geyser nor heard the detonation as he sprawled over the tube. He felt only a hand grasp him by the slack of his coat, and the voice of one of the seamen: 'Steady all, sir. I've got you.'

VIII

A voice roused him, a hand. He was half sitting in the narrow starboard runway, half lying across the tube. He had been there for quite a while; quite a while ago he had felt someone spread a garment over him. But he had not raised his head. 'I'm all right,' he had said. 'You keep it.'

'Don't need it,' the boy said. 'Going home now.'

'I'm sorry I—' Bogard said.

'Quite. Confounded shallow boats. Turn any stomach until you get used to them. Ronnie and I both, at first. Each time. You wouldn't believe it. Believe human stomach hold so much. Here.' It was the bottle. 'Good drink. Take enormous one. Good for stomach.'

Bogard drank. Soon he did feel better, warmer. When the hand touched him later, he found that he had been asleep.

It was the boy again. The pea-coat was too small for him; shrunken, perhaps. Below the cuffs his long, slender, girl's wrists were blue with cold. Then Bogard realized what the garment was that had been laid over him. But before Bogard could speak, the boy leaned down, whispering; his face was gleeful: 'He didn't notice!'

'What?'

'Ergenstrasse! He didn't notice that they had shifted her. Gad, I'd be just one down, then.' He watched Bogard's face with bright, eager eyes. 'Beaver, you know. I say. Feeling better, eh?'

'Yes,' Bogard said, 'I am.'

'He didn't notice at all. Oh, gad! Oh, Jove!'

Bogard rose and sat on the tube. The entrance to the harbour was just ahead; the boat had slowed a little. It was just dusk. He said quietly: 'Does this often happen?' The boy looked at him. Bogard touched the tube. 'This. Failing to go out.'

'Oh, yes. Why they put the windlass on them. That was later. Made first boat; whole thing blew up one day. So put on windlass.'

'But it happens sometimes, even now? I mean, sometimes they blow up, even with the windlass?'

'Well, can't say, of course. Boats go out. Not come back. Possible. Not ever know, of course. Not heard of one captured yet, though. Possible. Not to us, though. Not yet.'

'Yes,' Bogard said. 'Yes.' They entered the harbour, the boat moving still fast, but throttled now and smooth, across the dusk-filled basin. Again the boy leaned down, his voice gleeful.

'Not a word, now!' he hissed. 'Steady all!' He stood up; he raised his voice: 'I say, Ronnie.' Ronnie did not turn his head, but Bogard could tell that he was listening. 'That Argentine ship was amusing, eh? In there. How do you suppose it got past us here? Might have stopped here as well. French would buy the wheat.' He paused, diabolical—Machiavelli with the face of a strayed angel. 'I say. How long has it been since we had a strange ship in here? Been months, eh?' Again he leaned, hissing. 'Watch, now!' But Bogard could not see Ronnie's head move at all. 'He's looking, though!' the boy whispered, breathed. And Ronnie was looking, though his head had not moved at all. Then there came into view, in silhouette against the dusk-filled sky, the vague, basket-like shape of the interned vessel's foremast. At once Ronnie's arm rose, pointing; again he spoke without turning his head, out of the side of his mouth, past the cold, clamped pipe, a single word:

'Beaver.'

The boy moved like a released spring, like a heeled dog freed. 'Oh, damn you!' he cried. 'Oh, you putt! It's the Ergenstrasse! Oh, confound you! I'm just one down now!' He had stepped in one stride completely over Bogard, and he now leaned down over Ronnie. 'What?' The boat was slowing in toward the wharf, the engine idle. 'Aren't I, Ronnie? Just one down now?'

The boat drifted in; the seaman had again crawled forward onto the deck. Ronnie spoke for the third and last time. 'Right,' he said.

IX

'I want', Bogard said, 'a case of Scotch. The best we've got. And fix it up good. It's to go to town. And I want a responsible man to deliver it.' The responsible man came. 'This is for a child,' Bogard said, indicating the package. 'You'll find him in the Street of the Twelve Hours, somewhere near the Café Twelve Hours. He'll be in the gutter. You'll know him. A child about six feet long. Any English MP will show him to you. If he is asleep, don't wake him. Just sit there and wait until he wakes up. Then give him this. Tell him it is from Captain Bogard.'

X

About a month later a copy of the English Gazette which had strayed onto an American aerodrome carried the following item in the casualty lists:

MISSING: Torpedo Boat XOOI. Midshipmen R. Boyce Smith and L. C. W. Hope, R.N.R., Boatswain's Mate Burt and Able Seaman Reeves. Channel Fleet, Light Torpedo Division. Failed to return from coast patrol duty.

Shortly after that the American Air Service headquarters also issued a bulletin:

For extraordinary valor over and beyond the routine of duty, Captain H. S. Bogard, with his crew, composed of Second Lieutenant Darrel McGinnis and Aviation Gunners Watts and Harper, on a daylight raid and without scout protection, destroyed with bombs an ammunition depot several miles behind the enemy's lines. From here, beset by enemy aircraft in superior numbers, these men proceeded with what bombs remained to the enemy's corps headquarters at Blank and partially demolished this château, and then returned safely without loss of a man.

And regarding which exploit, it might have added, had it failed and had Captain Bogard come out of it alive, he would have been immediately and thoroughly court-martialled.

Carrying his remaining two bombs, he had dived the Handley-Page at the château where the generals sat at lunch, until McGinnis, at the toggles below him, began to shout at him, before he ever signalled.

He didn't signal until he could discern separately the slate tiles of the roof. Then his hand dropped and he zoomed, and he held the aeroplane so, in its wild snarl, his lips parted, his breath hissing, thinking: 'God! God! If they were all there—all the generals, the admirals, the presidents and the kings—theirs, ours—all of them.'

The Frontiers of the Sea

〜〜〜

PETER USTINOV

O LD men sit on walls and watch the sea; young men do it too, but dutifully. Among the nets and green glass baubles they do it, and seem to read the sky like a newspaper. At all points of the compass they sit on walls, as though the sea were a vast arena full of spectacle and pageantry and meaning, which, for them, it is. The smell of tar and rancid water, thick as blood to the nostril, hovers round the edges of the arena, and the old men no longer notice it. They have travelled beyond the trifling bits of observation a landlubber may pick proudly up on holiday; they have travelled beyond prose and poetry into that ultimate simplicity which separates them from life as surely as luck and seamanship have always separated them from death. They spend their days in a wordless limbo of comprehension. They think of nothing and understand.

Planted like trees, or rather like masts, on the best seats, they stare with the lustrous patience of old dogs at the vast hunting ground. They are part of the seascape, and it often appears that, ashes to ashes, dust to dust, they are slowly returning into nature without surprise or fear of death. The shells on the beach look like the abandoned toenails of these old men, and they are more beautiful there than on the foot, among all the other vestiges of decay, the broken wings, the sand-logged crabs, the silver fish with the surprised eyes, the woman's lonely shoe, the rusty toy. Cleansed and sterilized by salt and iodine in the great hospital of the sea, decay and corruption are as evocative on the shore as broken columns and noseless gods are inland, and they are older still. There are no compromises. No need is there to subtract the television aerials which are silhouetted against the peachy sky behind the Colosseum; no need to half close the eyes in order to eliminate the Autostrada which sweeps by the crumbling temples and frozen palisades. The sea is as it was, and if an airliner whistles and sobs above it for a while, it brushes it away as a horse wearily dispenses with a fly. Landed man has not yet found a way to take possession of the sea, to tame it and bend it to the glacial mechanical will, while the seaman knows better than to try.

With great indulgence, without comment, the old men watch the holidaymakers: the varicose columns of white flesh which stand in the shallows like chunks of veined marble under their canopies of gathered skirts; the opulent stomachs rising softly to the crater of the navel; the tiny children (the only sensible ones in the old men's unexpressed opinions) yelling their heads off with rage and fright as their laughing parents (the idiots) try to force them to learn to swim; the brown ladies, aglisten with pungent unguents, praying to the sun with that intense application which their forebears used to reserve for God, and with only an exorbitantly expensive handkerchief between them and scandal.

There go the cabin cruisers, the owners wearing rakish caps with anchors on them and braid on the peaks, and not a knot or a hazy knowledge of celestial navigation among the lot of them. And there the playboys and playgirls screech by on parallel planks of wood, standing first on one leg, then on the other, then sinking between the ridges of high water.

This was all midsummer madness, a malady of heat. The people from beyond the hills have skulls like eggshells, and the first ray of sun and the first whiff of sea air send them off in this lunatic flirtation with shallow water. The old men look through all this with the air of great lovers who have suffered a lifelong passion for a demanding mistress. They have delved to the depths of sorrow and vaulted to the pinnacles of delight, in silence and in solitude and for life. They vaguely notice these flatulent little outbursts, this unworthy bottom-pinching seduction of their element. They hardly hear the saxophones and hissing percussion as the tiny tumult barks away from the bars all night. They wait patiently for the autumn.

In the village of San Jorge de Bayona, one such old man was Vicente Mendendez Balestreros, and for a man with such a sonorous name, it was perhaps surprising that he could neither read nor write. In truth, he didn't have to, since he had only received letters from the government, and such letters were not worthy of an answer. He didn't talk very much, but his thoughts, while rare, were mysterious and abstract. He had no wife, since he had little enough without having to share it. Money never worried him, but he was jealous of the silent cathedral of his mind.

The summer was over, thanks be to God, creator of all seasons. The bars were either locked or else humbled by presence of the locals. The little boutiques, with names like Conchita and Eros, were shut, their minute windows empty. The two modern residential hotels, El Fandango and the Hacienda Goya, had their slatted blinds down, and there were no bikinis to hang on the balconies. The nights were noiseless once again, apart from the deep breathing of the sea.

Vicente, with no wife to scold him, and no great appetite, sat on the wall longer than the other old men. Whenever they left him, they never said good night, so envious were they of his liberty. Somehow in old age he had preserved his youth, while they had voluminous ladies waiting for them in their two-room cottages, venerable hags with hairs on their chins like the hairs that leak out of overripe sofas, and with breasts that hang like a donkey's burden. They also had holy lithographs all over the place. Religion comes into the house with the women. The priest even dresses like a woman to propagate it.

Vicente was Catholic, but he didn't believe in God, unless belief in the sea can be conceived as a form of belief in God. He would mutter and make signs and kneel and kiss like all the others because he had been brought up that way, but when it came to belief, he could only subject himself to the guidance of reason, tempered always by the bitter paradoxes of experience. Priests he regarded not as men with a divine vocation, but rather as men skilfully avoiding work. The organ gave him earache, and the better it was played, the more intense was his pain. At the same time, he had no patience with members of other religions, unless they happened to be sailors, in which case they had better things to do than to bother with dogma.

One evening—it was well after ten, the moon was full, with black clouds scudding in ordered masses across the sky—Vicente was still on his wall, and all alone. Suddenly he shivered, and the toes of his bare feet curled up as though at bay. A cool wind sighed from an unexpected quarter, and a noise like a distant cavalry charge began to grow fitfully from the horizon. A sheet on a clothes line flapped like a sail when a ship changes direction. He rose. His face creased up as his hazel eyes looked into the distance, where the last colours of the day were by now only suggested by a trace of green, a touch of mauve, a tortoiseshell patch of black and orange.

He hobbled to the nearest cottage and banged on the door. One of the monumental women opened up and asked what he wanted. At this hour of the night, at almost any hour, there was a barrier between each man and the outer world. Vicente didn't say what he wanted. He merely pointed at the horizon with his chin. Eventually the man appeared. It was Paco Miranda Ramirez.

'What can you see there?' he asked.

Since Vicente couldn't be bothered to say, Paco walked out in his underwear, brushing away the stridencies of his wife, and had a good look at the horizon himself.

'It's too dark to see,' Paco said.

Vicente shook his head briefly in disagreement.

'What do you see?'

'Come indoors,' cried the wife.

'Silence, woman,' countercried the husband, who was always courageous in front of another man.

'A boat?' he asked Vicente.

Vicente nodded.

'In trouble?'

Vicente made a gesture, a languorous sweeping movement of his arm and a bridling of his head to suggest the enormity of the trouble.

Paco paddled off barefoot and woke some of the other men with talk of shipwreck. The reason the men responded with such alacrity to the call was that almost twenty years back they had towed a Belgian yacht to safety and been compensated by half its purchase price, which is a rule of the sea. This prize money had brought great happiness to the village, and one man, a certain Diego Liñares Montoya, had even been able to fulfil his life's ambition and die of cirrhosis of the liver as a consequence of this heaven-sent bounty. At that time too it had been Vicente, then a newcomer to the wall, who had peered into the inky night and sensed distress. His senses were respected throughout the local countryside, as the others knew, not without bitterness, that he had had the courage to remain a bachelor, and that in reward his eye and ear and especially his telepathy had remained unimpaired and pure.

'I bet he's made a mistake this time,' grumbled José Machado Jaen, as he helped push the heavy rowing boat into the water.

'And when you get your ten thousand pesetas, I'll be there to watch you eat your words,' said Paco.

The women stood in a wailing phalanx at the edge of the beach, their handkerchiefs to their mouths, avid for tragedy, praying. Vicente was like Ulysses at the stern, tiller in hand, guiding the boat as it entered the zone of the sudden wind. The women saw their men disappear, reappear, disappear, reappear, and finally disappear into the darkness. Only the rhythm of the oars could be vaguely heard for a moment, and then it was swallowed by the gathering storm.

The seas became mountainous, but the men hardly noticed. It was only when Vicente held up his hand and they stopped rowing that they became conscious of the folly of it all. It was raining now, and the waves broke over them, covering their feet and even their calves with galloping streams of hysterical water. There was nothing in sight.

'We'll all drown, and it'll be a magnificent funeral,' shouted José Machado Jaen.

'The old man knows what he's doing,' cried Paco.

Vicente's expression never changed as he looked around him, his gnarled face wet with spray. He pointed, and they turned the boat briefly sideways to the waves, almost capsizing. There was no sign of mast or hull, no sound but the joyful anger of the sea. The men looked anxiously at Vicente, and he suddenly grew tense. They followed his gaze, and a dark object appeared momentarily, only to sink again in a deep trench of water. Try as they might, they seemed unable to approach it. High seas destroy all sense of distance. The dark object drifted away at one moment, the next it was upon them, sucked up against the side of the boat. It was a man.

These Spanish fishermen had cultivated feet as adroit as those of monkeys, and now they suspended themselves at dangerous angles over the side of their craft, and although they were often submerged by the elemental panic, they held on to the poor fellow, and eventually succeeded in dragging him aboard. Nobody could blame Vicente for the fact that there was no prize this time. His senses were as keen as ever, and he appeared to be doing the work of God, which, even if less lucrative than the best works of man, salved the conscience in advance against the next mortal sin. The crew felt humble and virtuous as they rowed strongly back to their village. They could be sure that they had taken part in a miracle.

The rescued man was half dead when they carried him ashore. He had a pair of rough white canvas trousers on, but his torso was bare, and dramatically thin. He was dark, but one could tell he was a stranger. His grey eyebrows met over the bridge of his aquiline nose, and his full lips expressed that sensuous disgust which people from the Eastern Mediterranean often share with their camels.

A feeling of biblical wonder had now so gripped them that it spread to the women, and couples vied with each other to offer hospitality to this half-dead man, and they almost came to blows in their struggle for visas to heaven. Eventually it was decided that the honour of giving up his sofa to this dripping shred of humanity should fall to one Antonio Martinez Mariscal, who was the oldest of the rescuers, and who would therefore presumably have need of this good mark in the profit column of the soul's account earlier than the others. The nearest doctor was in the small town of Maera de las Victorias some eighteen kilometres inland. Paco Miranda Ramirez set off on a rusty bicycle without lights to fetch him. Deprived of one good deed, he eagerly volunteered for another, more difficult and more exhausting. Eusebio Sanchez Marin decided to go on foot to the neighbouring village of Santa Maria de la Inmaculada Conception to fetch the priest, so that all this virtue could be registered with the proper Authority. The others saw the volunteers off with the jealousy of Holy Week flagellants who find there aren't enough scourges to go round.

It was almost four in the morning and a feather of light lay gently on the horizon by the time Dr Valdes arrived in his rickety car, Paco Miranda Ramirez standing on the running board and holding his bicycle on the roof. One feeble headlight of the doctor's car kept winking suggestively like an aged roué at a stage door. As dogs and masters grow to resemble each other, so had the doctor and his car.

'Let's have a look at this miracle,' he wheezed as he entered Antonio's house. Vicente pointed at the stranger with his chin. The women made way, and the doctor saw a frightened little man in a shirt several sizes too large for him lying on Antonio's bed, as though he had been the subject of some Rembrandtian lesson in anatomy. Certainly his expression of fear was in large measure due to the circle of impressive women who had sat round him all night, muttering, telling their beads, and searching his face for a sign. All in all, it had been more nerve-racking than any shipwreck.

'He's not Spanish,' suggested Paco's wife darkly, meaning that access to the True Cross might be denied him. The other wives were not willing to go so far, and thought that perhaps he was a Basque, or a Portuguese from some remote province, or perhaps a South American. The doctor asked him how he felt. He grinned in a meaningless way, since he realized from the tone of voice that he was being addressed, but didn't seem interested or capable of answering.

'If you ask me,' said the doctor, 'he's as Spanish as everyone else here, but of a backward mentality—or else he has been the victim of a traumatic shock which has affected his powers of speech.'

'Is he Spanish, Vicente?' asked Paco.

Vicente shook his head negatively.

'What the devil does he know about it?' cried the doctor angrily. 'Can't even read or write, and suddenly he's an authority on whether a man is Spanish or not!'

Vicente shrugged his shoulders like a child who pretends he doesn't mind being punished.

By the time Father Ignacio arrived, Paco's wife was able to tell him that the miracle involved one of God's idiots, who had been saved from a roaring sea by those of solid mind. This, she suggested, had a comfortable ring of Christian charity to it, with a soupçon of celestial embellishment for good measure.

Father Ignacio, a man narrow both in body and in mind, knew intimately that no-man's-land of sceptical expectation in which many country priests put out their thoughts to pasture. He lived in the knowledge that miracles had occurred in other places and at other times, and yet he had the saddening but certain conviction that nothing extraordinary would ever happen to him. If it did, he certainly wouldn't know how to react.

The little man spoke suddenly, saying something like 'Shkipra'.

'Shkipra, Shkipra,' he repeated insistently when they asked him to elucidate.

The ladies made quite a few wild guesses at the meaning of this elusive word, eventually settling for madness as its most probable source. Dr Valdes racked his brain for any malady of that name, but he had taken his exams very long ago, and he required of his patients that they fall ill with a few well-defined complaints. Shkipra was not one of them.

Father Ignacio sharpened abruptly, and said, quite out of the blue, 'Senatus Populusque Romanus.'

The ladies looked at him enquiringly.

'What did the Reverend Father say?' asked Paco's wife.

'SPQR.'

'Shkipra,' agreed the little man excitedly, pointing to his own chest.

'The man is, no doubt, a Roman,' declared Father Ignacio, glaring through his metal-rimmed glasses. 'That is what he has been trying to tell us.'

'A Roman,' spluttered Dr Valdes, 'how d'you make that out?'

'Senatus Populusque Romanus,' replied Father Ignacio, 'the Senate and the People of Rome . . . I remember seeing it on every dustbin in the eternal city.'

'And Rome is the seat of Mother Church,' reminded Paco's wife with a sallow look of sanctity, 'the home of all miracles.'

'What the devil d'you mean!' Dr Valdes protested. He had served in the Legion of Death and survived. 'Spain produces more miracles than any other country in the world, and without foreign assistance. The weeping Virgin of Fuenteleal, the Fountain of San Leandro, which spouts blood, the nodding Christ of the Thorns.'

Father Ignacio held up an indulgent yet peremptory hand.

'It is unseemly to enter into worldly arguments about the extent of our and other people's miracles, especially since our richness in these divine phenomena should make us tolerant towards those less endowed. It remains that this simple Roman peasant owes his life to the fact that some celestial force visited our good friend Vicente and directed his eye to a specific point in the heaving waters. It is enough that we and this poor peasant share the true faith. The story is a perfect one; its moral is as symmetrical and as lovely as a flower. *Deo gratias.*'

'Amen,' murmured the ladies.

The door burst open, and Sergeant Cuenca Loyola of the Guardia Civil stood there, his sinister patent-leather hat reflecting the unsteady light of the candles. Behind him stood Baez, his assistant.

'What's going on here?' growled Sergeant Cuenca Loyola.

'A miracle,' crowed the ladies.

'A miracle? I'm surprised. Surprised with the lot of you, and disgusted. Father Ignacio, Dr Valdes, Paco, Vicente. Don't you people know that I should have been the first to have been informed of a new arrival? I could arrest you all for attempting to smuggle a person into Spain.'

'If I hadn't been summoned, he'd have been a corpse by now,' snarled Dr Valdes.

'I should have been called at the same time!' Sergeant Cuenca Loyola was willing to make that concession. 'Now, to work,' Baez took out a notebook and pencil.

'I'd like to know what the devil you think you're going to write down on that pad,' Dr Valdes cackled.

'What we put down on our pad is official business, and I want no reflection or comment on it whatsoever,' declared the sergeant.

'I served with the Legion of Death', protested Dr Valdes, 'and I know and respect the regulations, but even General Millan-Astray himself, genius that he was, may his soul repose in peace, couldn't have made sense of the silence of this fellow.'

'We'll make him talk,' said the sergeant, who believed that even the ignorant were in the habit of deliberately attempting to conceal their ignorance. 'Now, your name!'

The stranger smiled, and nodded.

Dr Valdes began spluttering with asthmatic laughter.

'He agrees with you, Sergeant!'

'Silence. I asked you for your name!'

'Shkipra.'

'Now we're getting somewhere,' remarked the sergeant with satisfaction.

'How do you spell it?' asked Baez.

'Shkipra.'

'He's illiterate,' declared the sergeant. 'Baez, write it phonetically. Now, in which province were you born?'

'Shkipra.'

'Date?'

'Shkipra.'

The sergeant exploded. 'And I suppose your father's name, your profession, the unit in which you performed your military service, they're all Shkipra!'

'Shkipra.'

Later in the morning, a handsome car pulled up outside the police station of San Jorge de Bayona, and three officers stepped out. They had been summoned urgently by Sergeant Cuenca Loyola. As they entered the chalk-white room, silent but for the baleful buzzing of imprisoned flies, the sergeant sprang to his feet, and indicated to Shkipra to do likewise. Poor Shkipra had begun to find the inability to communicate oppressive, and he just sat in pained silence, staring at the floor as though fascinated by something going on there.

'Never mind, never mind,' said Major Gallego y Gallego good-naturedly, sitting on a wooden form and beckoning to his colleagues to do likewise.

'Now, Sergeant, what is the trouble?'

The sergeant glanced up. His style was going to be inhibited by the sudden appearance of the entire village at the tiny barred window, to say nothing of Vicente, who stood leaning on the frame of the open door, sullenly minding his own business in a place he had no business to mind.

'Clear away there, clear away from the window! Out! Out!' cried the sergeant.

'Easy,' said the major. 'Let us retain our composure, please. Now, Sergeant, let's have your report.'

Acutely aware of his loss of face, and cursing those administrative necessities which at times forced a man to have recourse to higher authorities, the sergeant cleared his throat.

'Well, sir, as I understand it, this person landed on Spanish soil in an unauthorized manner between twenty-three hours and twenty-four hours last night.'

The major smiled.

'Did he report to the police?' asked Captain Zuñiga.

'Did he have anything to declare?' asked Lieutenant Quiroga, local chief of customs.

The major asked for quiet with a gesture of his hand.

'What do you mean by an unauthorized manner?' he enquired.

The sergeant hesitated.

'A manner not in accordance with the usual manner of entering the country,' he said.

'If an angel from heaven suddenly landed on your roof, would you describe that as an unauthorized manner of entering the country?'

'No, sir.'

'Why not?'

The sergeant looked nervously at the priest, who, being outside the closed window with the rest of the village could hear none of this.

'Well, yes, sir, I would regard it as unauthorized, unless I had previous instructions to that effect.'

'From whom?'

'From you, sir.'

The sergeant wiped his brow with a rag.

'Why from me?'

'From you or from Father Ignacio.'

'Very good.' The major chuckled.

'Now, suppose you tell me exactly how this invading army crossed the Spanish frontier.'

'This invading army—?'

'This man.'

'He was brought in by the men of the village—on a boat.'

'On a boat? In other words, he was drowning in Spanish territorial waters?'

The sergeant heartily detested the major's tone without being able to understand it.

'He was drowning, yes—or at least swimming, sir.'

'It's one way of avoiding the expense of the more conventional means of transportation, although it can be wearying if you have a lot of baggage.'

The major turned to Quiroga of the customs. 'I think we can be fairly sure, Quiroga, that he had very little to declare—how was he dressed on arrival, Sergeant?'

'In trousers, sir.'

'Trousers, that's all?'

'Just trousers, sir.'

'If he had anything to declare, Quiroga, it's probably what any member of the male sex might decently be concealing.' He laughed at his own levity, and then, with a touch of mock concern, he asked, 'The priest can't hear through that window, I hope?'

'No, sir,'

'Good, good. Now, Sergeant, what did this man have to say for himself when he had sufficiently recovered to talk?'

'Nothing, sir. He has consistently refused to say anything.'

'Refused? Has it occurred to you that he might be incapable of saying anything?'

'He has said one word, sir.'

'One word, Sergeant? Then he has not said nothing, as you have stated.'

The sergeant wiped his brow again.

'What is that word, Sergeant?'

'Shkipra.'

'And what does that signify?'

'I don't know, sir.'

The major sighed.

'To what question did he reply—what is the word again?'

'Shkipra, sir. To all questions, sir.' The sergeant held up his two-page questionnaire. 'I have no idea in which column to place the reply, sir.'

The major turned to the stranger.

'What is your name?' he asked.

'Shkipra.'

'And how do you enjoy Spain?'

'Shkipra.'

'I see what you mean.'

The major drew a packet of Bisonte from his pocket and offered the man one.

'Cigarette!' said the man delightedly, accepting it.

'He says at least two words, Sergeant,' the major said menacingly, ' "Shkipra" and "cigarette".'

'Cigarette,' agreed the man.

'Now I tell you what we'll do. *Olá*, you!' called the major to Vicente, in the doorway. 'Run over to the schoolhouse, and bring an atlas of the world.'

'He's no use,' growled the sergeant; 'he can't read or write.'

'Did you understand me?'

Vicente didn't deign to reply, but left slowly.

The awe of the villagers was considerable when they saw Vicente return a few minutes later holding a globe high over his head like a beacon.

'That's not exactly what I meant,' said the major, 'but it'll do,' and he turned to the stranger. 'Shkipra?' he said.

The stranger knitted his brow with effort, and had difficulty with the curious shape of the map, but he worked his way around it. Suddenly he stopped gyrating the atlas and pointed at a small area with his finger, shouting excitedly, 'Shkipra, Shkipra!'

The major put his glasses on and examined the indicated area.

'Albania,' he announced.

'Albania,' echoed round the room.

'Impossible,' said Zuñiga.

'Tirana?' asked the major.

'Tirana,' replied the stranger, 'Dürres, Elbasan, Shkoder.'

'It is Albania,' declared the major, folding his glasses.

'But that's a Communist country,' said Quiroga.

'It's also very far away for a lonely swimmer,' reflected the major.

'What do we do now?'

After a pause, the major said, 'This is a matter for Madrid.'

'Meanwhile, I'd best lock him up,' the sergeant volunteered.

The major studied his man.

'Oh, I hardly think that's necessary, Sergeant; find him a few little things to do. I don't think he's much of a threat to our security.'

Before the sergeant could remonstrate, Vicente had made a clicking noise which attracted their attention. Looking at the stranger, Vicente invited him to come along with him. Without consulting anyone, or asking permission, the stranger rose and left with Vicente.

'Who is that?' asked the major in admiration. He had rarely seen such a display of authority as that exercised by Vicente.

'A poor ignoramus,' replied the sergeant, with ill-disguised hatred; 'it was he who rescued El Albanes.'

'Then it is only just that they should become friends,' said the major. 'Come, we must make out our report to Madrid.'

And with a parting piece of charity, which proved he wasn't such a terrible man after all, he told the sergeant that there would be no need to fill up his questionnaire. The sergeant, however, was in no mood to see this gesture as anything but an insult, since in his opinion a sergeant with an unfilled questionnaire is only half a sergeant.

Every day, and sometimes far into the night, there were two figures on the sea-wall. They never spoke, since there was nothing to say and no common language to say anything in. Their eyes were unblinkingly fixed on the huge winter canvas, with its shrieking gulls and its vast uncertainties. Sometimes they would roll a cigarette or two; at other times they would tie and untie knots idly in stray bits of discarded fishing net. Occasionally some abstraction would make them both sit up and take notice or even smile. The nearest they ever came to any conventional communication was when one of them would stare quizzically at a cloud, and the other would nod slowly or shrug half a shoulder. The villagers were loath to intrude into the great silence, which became like a fount of peace, an influence on all who were open to it and who knew its history. Every now and then Paco's wife or one of the other ample señoras would arrive with some goody which had been left over or else specifically cooked for the two old men. And every morning, the fishermen would ask Vicente's advice on the weather conditions, and he would reply with an affirmative or negative gesture. The Albanian, who understood the nature of the questions without any difficulty, would always reply silently in exactly the same way as Vicente, but with movements more suave and less austere than those of his friend, movements which had their roots in a more mellifluous choreography.

Noisy children would become quiet for a while when they passed the sea-wall on their way to and from school, and scrofulous dogs with degrading habits, who encircled people in vast untrustworthy patterns, shot with yellow looks and dishonest trepidation, would go straight up to the two old men, their sparse tails wagging and their eyes alert. The Albanian liked to pretend to throw stones. The dogs would turn in a single leap, waiting for the sound of the falling stone, and when it failed to register, they turned again slowly, with the patient look of one who is being teased. Every now and then a stone would whistle through the air, and smash itself giddily on other stones while the dogs flew howling after it, stopping perplexed in an ocean of pebbles, unable to identify the one that had been thrown, now as motionless as all the rest. Then some innocent villager would pass that way, and once again the dogs would hang their heads, bare their unhealthy fangs, and seem to tiptoe in apparently aimless but hate-filled circles round the intruder.

One day, Major Gallego y Gallego turned up again in his blue car, and they took El Albanes away. Vicente leaned against the car, and the major had to ask him to leave. He refused. They accelerated away, and Vicente stumbled into the road behind them. He walked a kilometre in the direction the car had taken, and then stopped in the naked Spanish landscape. Far from the sea, he was lost. Roads led nowhere. Here all was dust and dryness. Trees seemed to be dying, the bushes were grey with lack of moisture. Even the hot cackle of the cicadas sounded to Vicente like the noise of fleshless beings, the grinding of bones, the semaphore of death. Defeated, he turned back towards the sea and the confines of his understanding. His face was suffused with sorrow, an emotion more terrible than pain because of its longevity.

When he returned, he lay down on the beach and slept. At dawn, he woke, but did not rise. The villagers were upset to find him there; even more upset not to see him on the wall. Paco's wife cooked him some food, which he refused to eat. Their concern turned to anger on his behalf.

'Why couldn't they leave El Albanes here?' cried the wife of José Machado Jaen; 'what harm had he done?'

The men were less emotional, since they were tolerant about questionnaires and forms and applications and military service and war, and even if they didn't fully understand them, they recognized them as the barriers beyond which a woman's influence cannot penetrate. The zones of masculine folly are well guarded.

It was only when it became clear that Vicente had decided to die that the men joined in the chorus of complaint.

Father Ignacio came down to the beach, tripping on his cassock, to try and convince Vicente that suicide was a sin, but all his suggestions of hellfire and brimstone seemed like a relief from the useless wounds of this world.

Dr Valdes paid the beach a visit, at the behest of Paco Miranda Ramirez, who bicycled all the way to Maera de las Victorias again.

'If you don't eat,' Dr Valdes wheezed, 'I'll take you away, and in the hospital at Maera, a wizened and terrifying nun will practise what's known as intravenous feeding on you. D'you know what that is, Don Vicente? They make a hole in your arm the size of a finger, and pump beef tea in there until it begins to come out of your eyes. I knew a woman who, every time she cried, had beef tea rolling down her cheeks, so that everyone knew her shame, and instead of sympathizing with her tragedies, people used to say, "Aha, she has been to the good sisters in the hospital in Maera and had intravenous feeding, the wicked soul"!'

It was of no avail. Every time Dr Valdes wanted to feel Vicente's pulse, or look at his face, he rolled over on his stomach. Eventually the doctor left, discouraged, asking who was going to pay his fee.

As a last resort, the villagers urged Sergeant Cuenca Loyola to come down to the beach. It was hardly to be expected that he would succeed where the others had failed, but he made a brave attempt at gentleness all the same.

'Look here, hombre,' he said, trying to kneel in such a way that his uniform would not be soiled, 'there's nothing to get so upset about. El Albanes wants to go home to Albanera, or wherever it is those people live—Shkipra—how would you like to be in a country in which you don't know the language?'

Vicente looked at him feebly, nevertheless suggesting by his expression that it would be marvellous.

'Get some food inside you. Don't be a fool. I can't order you to eat, I can only ask you, which I do, you see. You've done a fine job with El Albanes, don't ruin it all. I've written a report, and you're in it. It's gone to Madrid. Your name is at this very moment on a desk in Madrid.'

The business of saving Vicente had become so perplexing that there was never a moment during the day when there wasn't someone hovering about, even local journalists, and the first person up in the morning would report to all the others that he was still alive.

'What can we do?' asked Major Gallego y Gallego, who had been consulted, and who was attracted, as ever, by the quirks of human nature and the inexhaustible stupidity of men. 'It has never been possible to prevent a man from dying when he wants to die. I'm not even sure that such an attempt isn't an invasion of personal liberty, whatever Mother Church may say. But what a reason for dying! It seems ridiculous to any lucid and educated man—and yet, if we think for a moment, isn't there something ennobling in the purity and simplicity of such a desire in this case? It is like the adolescent love of two schoolchildren, or even more, like the unquestioning and silent devotion of a dog. A dog? It sounds like a pejorative comparison, and yet, much as I love my wife and children, the only being in the world I can always trust is my dog, precisely because he is silent. Words complicate and betray. I wouldn't know how to live without them, but blessed are those that can.'

'Yes, but isn't there something we can tell him?' asked Father Ignacio, who blushed at any vaguely Voltairean sentiment, and who was eager not to be involved in an argument with the major, more from a fear of contamination than anything else.

'I feel sure that if we could say that El Albanes has safely reached his homeland, or something of the sort, that he has been happily reunited with his family, it would influence our poor friend.'

'We can say that, but it would be a lie,' said the major.

'We mustn't lie, of course we mustn't,' retorted Father Ignacio. 'But isn't there any happy aspect of the truth we can render even happier?'

'Not yet. He was taken to Madrid. There are only two Albanian refugees in Spain, and they are both classified as unreliable by the police, so we still are totally in the dark as to how he swam into our territorial waters. There is, of course, no diplomatic representation, and we have had to rely upon the Swiss, as usual, to find a way of returning him to where he came from. Unfortunately, there was a delay, since the Americans got to hear of this, and wished to interview him, believing that Albanian or even Chinese submarines might be operating in waters frequented by an American fleet. An admiral, a vice-admiral, and three rear-admirals grilled him for three hours.'

'What did he reveal?' asked Zuñiga.

The major smiled.

'Shkipra,' he said.

One morning, when all hope had been given up, and the case of the man on the beach was beginning to excite the entire Spanish Press, and even the international news agencies, and when the police had taken the decision to drag Vicente forcibly to hospital, the old man asked Paco's wife for bread in a feeble voice. He ate a little and drank a little consommé, and after a while struggled step by step to the wall, where he sat down, took a deep breath, and looked at the sea with contentment.

People like Dr Valdes believed it was a pity he hadn't died, just to teach him a lesson. Father Ignacio felt the proximity of new miracles, and caused the bells of the church to be rung as a sign of thanksgiving. The sergeant regretted that Vicente hadn't been dragged to the hospital, 'where lunatics belong'. Only Paco and the fishermen felt there was more reason than desperation in Vicente's sudden return to the wall.

'He wouldn't have gone there just because he was afraid to die,' Paco insisted, looking at the enigmatic little figure now once again where he belonged.

And, in fact, the truth was that to Vicente there was no mystery in an Albanian floating near the coast of Spain. It wouldn't have surprised him if the man had turned out to be a pygmy, or a head-hunter from Borneo. The sea is the sea, a place without frontiers and without surprises. Its rules are older and more binding than the law. A man overboard is saved whatever his race or creed, or at least his rescue is attempted, and if necessary, nothing short of heroism will satisfy tradition. A battleship may pay a friendly visit to an open city, but whoever heard of a division effecting a friendly occupation of a town?

The land is where the trouble starts. The roads that lead nowhere, the dust and the sand and the starving trees, and the people, all crushed together in a heaving marmalade, and the churches and the barracks and the rippling tides of gossip, rumour, information.

Vicente did not think all these thoughts. He didn't have to in order to know how to behave according to his lights, which were bright and clear. His intuition was infallible, and his reflections so profound they would have defied expression even by a poet.

If he now decided to return to the wall, it was because a part of him—was it the toes, or the eyes, or the inner ear, or just a mood of the heart?—told him that somewhere across the huge arena, an acquaintance had returned to his seat at the ringside, and was now perched once again on some other wall before some other disarray of pebbles, casting his senses towards some other horizon, which, while not identical, was yet very much the same.

Sources

'Initiation' by Joseph Conrad [Teodor Josef Konrad Korzeniowski] (1857–1924): first published in *Blackwood's Magazine* (January 1906); reprinted in *The Mirror of the Sea: Memories and Impressions* (1906).

'The Voyage' by Washington Irving (1783–1859): from *The Sketch Book of Geoffrey Crayon, Gent* (1820).

'A Descent into the Maelström' by Edgar Allan Poe (1809–49): first published in *Graham's Magazine* (May 1841); reprinted in *Prose Tales* (1843).

'I Have Been Drowned' by Tom [Sir Henry Thomas] Hopkinson (1905–90): from *The Transitory Venus: Nine Stories* (1948).

'Mocha Dick' by J[eremiah] N. Reynolds (?1799–1858): *The Knickerbocker Magazine* (May 1839).

'The Chase' by Herman Melville (1819–91): from *Moby-Dick* (1851).

'A Tragedy of Error' by Henry James (1843–1916): first published in the *Continental Monthly* (February 1864); reprinted for the first time in *The Complete Tales 1864–1868*, ed. Leon Edel (1962).

'High-Water Mark' by Francis Bret Harte (1836–1902): from *The Luck of the Roaring Camp* (1870).

'The Open Boat' by Stephen Crane (1871–1900): from *The Open Boat* (1898).

'Make Westing' by Jack [John Griffith] London (1876–1916): from *South Sea Tales* (1911).

'A Matter of Fact' by Rudyard Kipling (1865–1936): first published in pamphlet form (1892); reprinted in *Many Inventions* (1893).

'In the Abyss' by H[erbert] G[eorge] Wells (1866–1946): first published in *Pearson's Magazine* (August 1896); reprinted in *The Plattner Story and others* (1897).

'The Cruise of the *Willing Mind*' by A[lfred] E[dward] W[oodley] Mason (1865–1948): from *Ensign Knightley* (1901).

'The Terror of the Sea Caves' by [Sir] Charles G[eorge] D[ouglas] Roberts (1860–1943): from *The Haunters of the Silences* (1907).

'False Colours' by W. W. Jacobs (1863–1943): from *Light Freights* (1901).

'The Secret Sharer' by Joseph Conrad (1857–1924): first published in *Harper's Magazine* (August–September 1910), subtitled 'An Episode from the Sea'; reprinted in *'Twixt Land and Sea: Tales* (1912), subtitled 'An Episode from the Coast'.

'The Ghost Ship' by Richard Middleton (1882–1911): from *The Ghost Ship and other tales* (1912).

'Ambitious Jimmy Hicks' by John [Edward] Masefield (1878–1967): from *A Tarpaulin Muster* (1907).

'Poor Old Man!' by A[ylward] E[dward] Dingle (1874–1947): from *Spin a Yarn, Sailor* (1932).

'Easting Down' by 'Shalimar' (i.e. Frank Coutts Henry): reprinted in *Best Stories of the Sea*, ed. Thomas Woodrooffe (1946).

'The Story of the Siren' by E[dward] M[organ] Forster (1879–1970): first published separately by the Hogarth Press (1920); reprinted in *The Eternal Moment and other stories* (1928).

'The Rough Crossing' by F[rancis] Scott [Key] Fitzgerald (1896–1940): from *The Stories of F. Scott Fitzgerald* (1951).

'After the Storm' by Ernest [Miller] Hemingway (1899–1961): first published in *Cosmopolitan* (May 1932); reprinted in *Winner Take Nothing* (1933) and in *The Fifth Column and the First Forty-Nine Stories* (1938).

'The Bravest Boat' by [Clarence] Malcolm Lowry (1909–57): from *Hear Us O Lord from Heaven Thy Dwelling Place* (1961).

'The Boy Stood on the Burning Deck' by C[ecil] S[cott] Forester [Cecil Lewis Troughton Smith] (1899–1966): from *The Man in the Yellow Raft. Short Stories* (1969).

'Turnabout' by William Faulkner (1897–1962): first published in *Harper's Monthly Magazine* (1931); reprinted in *Dr Martino and other stories* (1934).

'The Frontiers of the Sea' by Peter Ustinov (b. 1921): from *The Frontiers of the Sea* (1967).

Source Acknowledgements

Full source information is given above.

A. E. Dingle, 'Poor Old Man!' from *Spin a Yarn, Sailor* (Harrap, 1932). Reprinted by permission of Chambers Harrap Publishers Ltd.

William Faulkner, 'Turnabout', copyright William Faulkner 1931. Reprinted by permission of Chatto & Windus on behalf of the Estate of William Faulkner, and Curtis Brown, London on behalf of Random House Inc., New York.

F. Scott Fitzgerald, 'The Rough Crossing', reprinted with the permision of Charles Scribner's Sons, an imprint of Macmillan Publishing Company, from *The Stories of F. Scott Fitzgerald*, edited Malcolm Cowley, copyright 1929 the Curtis Publishing Company; renewal copyright 1957 the Curtis Publishing Company and Frances Scott Fitzgerald Lanahan.

C. S. Forester, 'The Boy Stood on the Burning Deck'. Reprinted by permission of the Peters Fraser & Dunlop Group Ltd.

E. M. Forster, 'The Story of the Siren'. Reprinted by permission of Sidgwick & Jackson, and Harcourt Brace Jovanovich Inc.

Ernest Hemingway, 'After the Storm', reprinted with the permission of Charles Scribner's Sons, an imprint of Macmillan Publishing Company, from *Winner Take Nothing*, copyright 1932 by Ernest Hemingway; renewal copyright © 1960 by Ernest Hemingway, and the Hemingway Foreign Rights Trust.

Tom Hopkinson, 'I Have Been Drowned' from *The Transitory Venus: Nine Stories* (Horizon: Chatto, 1948). Copyright the Estate of Sir Henry Thomas Hopkinson. Reprinted by permission of Richard Scott Simon Limited, literary agents for the Estate of Sir Thomas Hopkinson.

Malcolm Lowry, 'The Bravest Boat'. Reprinted by permission of the Peters Fraster & Dunlop Group Ltd.

John Masefield, 'Ambitious Jimmy Hicks'. Reprinted by permission of the Society of Authors as the literary representative of the Estate of John Masefield.

A. E. Mason, 'The Cruise of the *Willing Mind*'. Reprinted by permission of A. P. Watt Ltd. on behalf of Trinity College, Oxford.

'Shalimar', 'Easting Down'. Copyright the Estate of Frank Coutts Henry.

Peter Ustinov, 'The Frontiers of the Sea' from *God and the State Railways* by Peter Ustinov, © Pavor S. A. 1966, reproduced by permission of Michael O'Mara Books Ltd.

H. G. Wells, 'In the Abyss'. Reprinted by permission of A. P. Watt Ltd. on behalf of the Literary Executors of the Estate of H. G. Wells.

Oxford University Press apologize for any errors or omissions in the above list and would be grateful to be notified of any corrections that should be incorporated in the next edition or reprint of this volume.